Crystallography in Molecular Biology

NATO ASI Series

Advanced Science Institutes Series

A series presenting the results of activities sponsored by the NATO Science Committee, which aims at the dissemination of advanced scientific and technological knowledge, with a view to strengthening links between scientific communities.

The series is published by an international board of publishers in conjunction with the NATO Scientific Affairs Division

A	Life Sciences	Plenum Publishing Corporation
B	Physics	New York and London
C	Mathematical and Physical Sciences	D. Reidel Publishing Company Dordrecht, Boston, and Lancaster
D	Behavioral and Social Sciences	Martinus Nijhoff Publishers
E	Engineering and Materials Sciences	The Hague, Boston, Dordrecht, and Lancaster
F	Computer and Systems Sciences	Springer-Verlag
G	Ecological Sciences	Berlin, Heidelberg, New York, London,
H	Cell Biology	Paris, and Tokyo

Recent Volumes in this Series

Volume 123—The Molecular Basis of B-Cell Differentiation and Function
edited by M. Ferrarini and B. Pernis

Volume 124—Radiation Carcinogenesis and DNA Alterations
edited by F. J. Burns, A. C. Upton, and G. Silini

Volume 125—Delivery Systems for Peptide Drugs
edited by S. S. Davis, L. Illum, and E. Tomlinson

Volume 126—Crystallography in Molecular Biology
edited by Dino Moras, Jan Drenth, Bror Strandberg, Dietrich Suck, and Keith Wilson

Volume 127—The Organization of Cell Metabolism
edited by G. Rickey Welch and James S. Clegg

Volume 128—Perspectives in Biotechnology
edited by J. M. Cardoso Duarte, L. J. Archer, A. T. Bull, and G. Holt

Volume 129—Cellular and Humoral Components of Cerebrospinal Fluid in Multiple Sclerosis
edited by A. Lowenthal and J. Raus

Series A: Life Sciences

Crystallography in Molecular Biology

Edited by

Dino Moras
Institute of Molecular and Cellular Biology
Strasbourg, France

Jan Drenth
Chemistry Laboratory of the University
Groningen, The Netherlands

Bror Strandberg
Biochemical Center
Uppsala, Sweden

Dietrich Suck
European Molecular Biology Laboratory
Heidelberg, Federal Republic of Germany

and

Keith Wilson
University of York
York, England

Plenum Press
New York and London
Published in cooperation with NATO Scientific Affairs Division

Proceedings of a NATO Advanced Study Institute,
held September 12-21, 1985,
at Bischenberg near Strasbourg, France

Library of Congress Cataloging in Publication Data

NATO Advanced Study Institute (1985: Bischenberg, France)
  Crystallography in molecular biology.

  (NATO ASI series. Series A, Life sciences; v. 126)
  "Published in cooperation with NATO Scientific Affairs Division."
  "Proceedings of a NATO Advanced Study Institute, held September 2-21,
1985, at Bischenberg near Strasbourg, France."—T.p. verso.
  Includes bibliographies and index.
  1. Molecular biology—Methodology—Congresses. 2. Crystallography—
Methodology—Congresses. I. Moras, Dino. II. North Atlantic Treaty Organization.
Scientific Affairs Division. III. Title. IV. Series.
  QH506.N378 1985   574.8'8   87-2509
  ISBN 0-306-42497-5

© 1987 Plenum Press, New York
A Division of Plenum Publishing Corporation
233 Spring Street, New York, N.Y. 10013

All rights reserved. No part of this book may be reproduced, stored in a retrieval system,
or transmitted in any form or by any means, electronic, mechanical, photocopying,
microfilming, recording, or otherwise, without written permission from the Publisher

Printed in the United States of America

PREFACE

"Crystallography in Molecular Biology", a NATO Advanced Study Institute, was held on September 12-21, 1985, at the Bischenberg near Strasbourg, France. The meeting, co-sponsored by EMBO and CNRS (Centre National de la Recherche Scientifique), was attended by participants from 18 countries.

The aim of the course was to give an overview of crystallography related to molecular biology with special emphasis on recent results and new methodological approaches. This field of research is expanding enormously, partly due to its primordial contribution to biotechnology. Protein engineering, site-directed mutagenesis, drug and vaccin design are some of the direct beneficiaries of crystallographic investigations.

The need for large quantities of structural informations and the necessity to tackle more and more complex problems are a strong incentive to develop new methods in order to overcome the various bottlenecks of a structure determination. Impressive progresses have been made in the field of data collection, phase determination and refinement technics. Despite some recent achievements crystal growth is the last serious handicap of the method. The course was a good opportunity for participants to be up-to-date with many facets of single crystal crystallography.

Due to the flood of structural results in molecular biology, selection had to be made among the numerous potential topics. Emphasis was given to relevant results in the field of nucleic acid-protein and protein-protein recognition. We will never forget a special monday evening session, when four different virus structures were presented, one of them for the first time, all of them solved since less than six months. This session was a good illustration of today's possibilities of X-ray crystallography, as was, in its own way, each conference of the meeting.

This book summarizes most of the contributions made during the ten days of the meeting and presents a significant amount of informations both in methodological developments and in structural molecular biology. Since a large part of the success of the course was due to the direct involvement of participants through poster sessions we have kept the memory of these communications by publishing relevant abstracts.

We wish to take this opportunity to thank all those who have contributed to the success of the course and to the content of this book. T. Blundell, who participated in the organizing committee and as such was a key contributor, deserves a special mention.

We also wish to express our gratitude to the members of biochemistry and crystallography laboratories of the Institut de Biologie Moléculaire et Cellulaire (IBMC du CNRS) for their cooperation during the ASI. We are greatly indebted to Mrs. Werling and Dr. J.P. Samama for their invaluable contributions to the organisation of the Institute and the publication of this book. For their excellent work and their enthusiasm we want to express our gratitude.

                                                          D. Moras
                                                          J. Drenth
                                                          B. Strandberg
                                                          D. Suck
                                                          K. Wilson

CONTENTS

1 CRYSTALLIZATION AND DATA COLLECTION

Crystallization of integral membrane proteins   3
R.M. Garavito and J.A. Jenkins

Crystallization of protein and nucleic acids: A survey of methods and importance of the purity of the macromolecules   15
R. Giegé

Synchrotron radiation and macromolecular crystallography   27
R. Fourme and R. Kahn

Data collection   45
J.R. Helliwell

2 PHASE PROBLEM

Density modification methods   63
A.D. Podjarny

Anomalous scattering in macromolecular structure analysis   81
W.A. Hendrickson

Single molecule electron crystallography   89
M. van Heel

Structure analysis of bacteriorhodopsin by electron crystallography   101
J.M. Baldwin, T.A. Ceska, R.M. Glaeser and R. Henderson

Neutron diffraction : contribution to high and low resolution crystallography   117
B. Jacrot

3 GRAPHICS AND STRUCTURE ANALYSIS

Electron density fitting with computer graphics : A review and a glimpse   125
T.A. Jones

Computer grapnics in the study of macromolecular interactions   131
A.J. Olson, J.A. Tainer, E.D. Getzoff

Prediction of protein structure from amino acid sequence   141
M.J.E. Sternberg

4 WATER AND DYNAMICS

X-ray analysis of polypeptide hormones at $\leqq$ 1Å resolution :
    Anisotropic thermal motion and secondary structure of
    pancreatic polypeptide and deamino-oxytocin   153
    A.C. Threharne, S.P. Wood, I.J. Tickle, J.E. Pitts,
    J. Husain, I.D. Glover, S. Cooper and T.L. Blundell

Water at biomolecule interfaces   167
    J.M. Goodfellow, P.L. Howell and R. Elliott

The water structure in 2Zn insulin crystals   179
    T. Baker, E. Dodson, G. Dodson, D. Hodgkin and
    R. Hubbard

The structure of water around a macrocyclic receptor : A
    Monte Carlo study of the hydration of 18-crown-6 in
    different conformations   193
    G. Ranghino, S. Romano, J.M. Lehn and G. Wipff

Simulating protein dynamics in solution : Bovine pancreatic
    trypsin inhibitor   197
    M. Levitt and R. Sharon

5 DRUG DESIGN - SITE DIRECTED MUTAGENESIS

Rational design of DNA minor groove-binding anti-tumor drugs   209
    R.E. Dickerson, P. Pjura, M.L. Kopka, D. Goodsell
    and C. Yoon

Protein crystallography and drug design   223
    W.G.J. Hol

Enzyme mechanism : What X-ray crystallography can(not)
    tell us   229
    J.N. Jansonius

The role of separate domains and of individual amino-acids
    in enzyme catalysis, studied by site-directed
    mutagenesis   241
    D.M. Blow, P. Brick, K.A. Brown, A.R. Fersht and
    G. Winter

Selected and directed mutants of T4 phage lysozyme   251
    T. Alber, T.M. Gray, L.H. Weaver, J.A. Bell,
    J.A. Wozniak, S. Daopin, K. Wilson, S.P. Cook,
    E.N. Baker and B.W. Matthews

6 VIRUSES

The structure of a human common cold virus (rhinovirus 14)
    and its functional relations to other picornaviruses   263
    M.G. Rossmann, E. Arnold, J.W. Erickson,
    E.A. Frankenberger, J.P. Griffith, H.J. Hecht,
    J.E. Johnson, G. Kamer, M. Luo, A.G. Mosser,
    R.R. Rueckert, B. Sherry and G. Vriend

The structure of poliovirus at 2.9 Å resolution :
    Crystallographic methods and biological implications   281
    J.M. Hogle, M. Chow and D.J. Filman

The structure of cowpea mosaic virus at 3.5 Å resolution 293
C.V. Stauffacher, R. Usha, M. Harrington, T. Schmidt,
M.V. Hosur and J.E. Johnson

Adenovirus architecture 309
R.M. Burnett, M.M. Roberts and J. Van Oostrum

7 PROTEIN - NUCLEIC ACIDS

A bacteriophage repressor/operator complex at 7 Å resolution 319
S.G. Harrison and J.E. Anderson

Refined structure of DNase I at 2 Å resolution and model
for the interaction with DNA 327
C. Oefner and D. Suck

Ribonuclease A and T_1 Comparable mechanisms of RNA
cleavage with different active site geometries 337
W. Saenger, R. Arn, M. Maslowska, A. Pähler and
U. Heinemann

Structural studies of ECO RV endonuclease and of its
complexes with short DNA fragments 345
F.K. Winkler, R.S. Brown, K. Leonard and J. Berriman

Errors in DNA helices : G.T. mismatches in the A, B and Z
forms of DNA 353
D. Rabinovich

8 PROTEINS

Relation between functional loop regions and intron
positions in α/ß domains 359
C.I. Bränden

The three-dimensional structure of antibodies 365
R.J. Poljak

The 3-dimensional structures of influenza virus
neurominidase and an antieurominidase Fab Fragment 373
P.M. Colman, J.N. Varghese, W.G. Laver and R.G. Webster

Periplasmic binding proteins : Structures and new
understanding of protein-ligand interactions 385
F.A. Quiocho, N.K. Vyas, J.S. Sack and M.A. Storey

Crystal structure of thaumatin I, a sweet taste receptor
binding protein 395
A. de Vos and S.H. Kim

Structures of pyruvoyl-dependent histidine decarboxylase
and mutant-3 phrohistidine decarboxylase from
lactobacillus 30A 403
M.L. Hackert, K. Clinger, S.R. Ernst, E.H. Parks and
E.E. Snell

State of X-ray structure determination of ascorbate oxidase
from green zucchini squash 413
A. Messerschmidt, M. Bolognesi, A. Finazzi-Agro,
R. Ladenstein

X-ray structure of the light-harvesting biliprotein C-phycocyanin from *M. laminus*
W. Bode, T. Schirmer, R. Huber, W. Sidler and H. Zuber ... 417

The crystal structure of the photosynthetic reaction center from *Rhodopseudomonas viridis*
J. Deisenhofer and H. Michel ... 421

Structure and function of some electron-transfer proteins and complexes Monoheme C-type and multiheme C_3-cytochromes
R. Haser, M. Frey and F. Payan ... 425

AUTHOR INDEX ... 439

SUBJECT INDEX ... 441

1 CRYSTALLIZATION AND DATA COLLECTION

CRYSTALLIZATION OF INTEGRAL MEMBRANE PROTEINS

R. Michael Garavito and John A. Jenkins

Biozentrum der Universitaet Basel
Abteilung Strukturbiologie
CH-4056 Basel Switzerland

INTRODUCTION

    Until 1980, it was considered quite unlikely that integral membrane proteins could be crystallized in three dimensions. In that year two groups (1,2) grew crystals of two different membrane proteins that yielded X-ray diffraction. With the refinements in the crystallization methods, several other groups have now reported crystals from seven membrane proteins (3,4,5,6,7,24), one of which has had its structure elucidated at high resolution (8). This rapid progress in the field of membrane protein structure analysis is not due to the use of radically new methods, but to the growing awareness and appreciation of the physical chemistry of detergent solubilized membrane proteins. In the short span of five years, much has been done but only short discussions have been published on the crystallization methods (9,10,11). We wish to review briefly the field as it stands and report on some previously unpublished observations from our group in Basel, particularly on the crystallization behavior of porin (Omp F), a pore-forming protein from the outer membrane of E. coli.

THE GENERAL SYSTEM

    Large crystals of integral membrane proteins have only been grown from monodisperse solutions of detergent solubilized protein. Once a protein has been prepared in a suitable detergent system (see Section III), the "classical" methods for protein crystallization can then be used (12,13). Both ammonium sulfate (AS) and polyethylene glycol (PEG) are effective crystallization agents in the presence of moderate concentrations of detergent. Microdialysis and large-scale vapor diffusion have yielded the largest crystals (3,11), though micro-scale methods of sitting or hanging drop vapor diffusion afford quick and economical ways to test different crystallization conditions. For the hanging drop method, it should be noted that the reduction in surface tension of the protein solution due to the presence of detergent limits the drop size to 5-8 µl. If the physical effects of the detergent on the solution are taken into account, then essentially all the published crystallization techniques can be applied to detergent solubilized proteins.

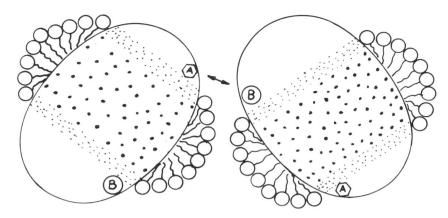

Figure 1. Schematic drawing of the interaction between two protein containing micelles. The dotted region is the hydrophobic protein surface while the lightly stippled region is the protein surface in contact with the detergent head group. Note that interactions between sites A and B can be severely hindered by the overall size of the micelle layer if the micelle is not too deformable.

Figure 2. Four crystal forms of porin grown from PEG/NaCl/detergent solutions. The tetragonal (a), monoclinic (b) and hexagonal (c) crystals were grown in the presence of β-OG. The triclinic crystals in c were grown in the presence of C_8HESO. The figure is taken from ref. 11. The bars are all 0.2 mm.

DETERGENT ASPECTS

The key factor in crystallizing membrane proteins is the judicious choice of detergent(s) for solubilizing the protein into a stable, monodisperse state. Three older reviews (14,15,16) cover, from a biochemical viewpoint, the action and behavior of detergents, while monographs by Tanford (17) and Rosen(18) as well as a review by Wennerström and Lindman (19) describe in detail the physical chemistry of detergents, solubilization and micellarization. For crystallizing membrane proteins, the detergent must not only keep the protein in a stable, solubilized state, but also not interfere in the formation of protein-protein contacts that occur during crystal nucleation and growth. This aspect can be better appreciated by looking at the molecular nature of a protein- detergent system. In an isotropic, monodisperse solution of a membrane protein in detergent, two distinct macromolecular species exist: pure detergent micelles and mixed micelles of protein and detergent. The latter species, which we wish to crystallize, is an anisotropic structure with two different surfaces (protein and detergent) exposed to the bulk solvent (Figure 1 and ref.30). The structure and behavior of the detergent layer about the protein depends on the character of the detergent(s) in it. Unfavorable detergent characteristics seem to be the primary reason that membrane proteins do not crystallize readily. The potential factors which can affect crystallization are:

1. <u>Micelle Size</u> A detergent layer of large physical size about the protein can act as a barrier to close protein-protein contact. Furthermore, detergents with large, hydrated head groups might also render regions of the hydrophilic protein surface inaccessible to intermolecular contacts.

2. <u>Monomer Fluidity and Micelle Deformability</u> The more fluid the detergent monomers are within the micellar region the more likely the micelle around the protein will be deformable. Hence, protein-protein interactions can distort the micellar surface, maximizing intermolecular contacts or exposing new contacts sites.

3. <u>Micelle Colloidal Behavior</u> Due to the hydration of the detergent head group (for nonionic detergents), the micelle surface is quite repulsive and inhibits the close approach of other micellar species. However, for many nonionic detergents, micelle aggregation occurs when the micelle surface is sufficiently dehydrated by the thermal melting of bound water (20,29) or by the addition of salts or polymers (20). This detergent phenomenon has mostly been studied in detergent/water binary systems (21,22) where aggregation and the eventual phase separation is induced by temperature. The participation of protein-detergent mixed micelles in the aggregation process has not been well studied, but it is clear that they partition into the micelle aggregates and, as shown by Bordier (23), they will also partition into the detergent phase if phase separation occurs.

How these factors affect membrane protein crystallization are discussed in the next three sections.

Table 1. Detergents Used in Membrane Protein Crystsllizations

Detergent [a]	Mw	CMC (mM) [b]	Characteristics [c]	Source [d]
β-octyl glucoside (β-OG)	292	23	UC	V
β-nonyl glucoside	306	6.5	UC	C
dodecyl maltoside	510	0.16	UC(?)	V
octyl tetraoxyethylene (C_8E_4)	306	7.0	LC	B
octyl pentaoxyethylene (C_8E_5)	350	4.3	LC	B
dodecyl octaoxyethylene ($C_{12}E_8$)	518	0.08	LC	N
decyl dimethylaminoxide (DDAO)	201	10.4	UC	O
dodecyl dimethylaminoxide (LDAO)	229	1.1	UC	F
dodecyl dimethylaminopropyl sulfoxide (zwittergent 3-12)	335	3	UC(?)	C
octyl (hydroxyethyl) sulfoxide (C_8HESO)	206	29.9	UC, X2	B
nonanoyl N-methyl glucamide (MEGA-9)	363	23	X1	O
decanoyl N-methyl glucamide (MEGA-10)	377	5	X2	O

a. Chemical name and abbreviation.

b. CMCs will vary depending on temperature and solution conditions. The CMCs quoted here are from the literature and are measured at low ionic strength.

c. LC, lower consolute boundary; UC, upper consolute boundary; X1, crystallizes under $10°C$; X2, crystallizes under $18°C$.

d. V, various suppliers; B, Bachem, Switz.; C, Calbiochem; F, Fluka, Switz.; N, Nikko Chemical, Japan; O, Oxyl, W. Germany.

DETERGENT SYSTEMS

Small, pure and chemically well-defined detergents with moderate to high critical micellar concentrations (CMC) have turned out to be the best candidates for crystallization experiments. These detergents form isotropic, monodisperse solutions of small micelles (1.6-3.0 nm radius), thus resulting in a protein-detergent micelle with a total hydrodynamic size not much larger than expected for the protein alone. Table 1 lists the detergents frequently used in successful crystallizations. Advances in detergent chemistry should increase the number of suitable detergents in the future.

Choosing a detergent for crystallization experiments remains a trial and error process and is complicated by the fact that not all membrane proteins are stable in the best detergents for crystallization. Furthermore, a protein's crystallization behavior in a detergent is not readily predictable. For example, porin from E. coli outer membranes crystallizes in the presence of many nonionic detergents, including those in Table 1, but large, single crystals suitable for X-ray analysis grow only in β-octyl glucoside (11). Michel (24) observes that bacteriorhodopsin crystallizes well in β-octyl glucoside (β-OG), but not in β-nonyl glucoside, just one methylene group longer. With the addition of small amphiphilic compounds (see section V), he notes that bacteriorhodopsin crystals now grow in the presence of more detergents.

For porin, the resulting crystal forms are influenced by the ionic environment (e.g. pH, ionic strength, ionic composition) of the solvent in the presence of many nonionic detergents. Figure 2 shows the crystals forms of porin obtainable grown from PEG/NaCl/β-OG(11). Figure 3 shows the phase diagram for porin crystallization (11). PEG seems to be the agent which induces crystal formation while the NaCl concentration determines the resulting crystal form. The crystallization boundary occurs just ahead of the PEG induced phase transition, where the solution separates into detergent-rich and detergent-poor phases. The interrelation of the two phase boundaries is discussed in Section VI. The general conclusion is that the ionic environment affects the growth and habit of porin crystals, and most of the other membrane protein crystals (3,5,6,7,24), in a manner similar to that of water soluble proteins.

The detergent environment also plays a distinct, though often subtle role in porin crystallization. For example, at pH 9 porin can crystallize in either the tetragonal or hexagonal space groups under standard conditions with β-OG (11). However, if octyl hydroxyethylsulfoxide is used instead, triclinic crystals (Figure 2d) grow. Other examples of detergent effects on the growth of certain crystal forms have been often observed in our laboratory. The most interesting involves the shift away from tetragonal crystals to the monclinic and finally hexagonal forms when crystallizations done in dodecyl dimethylaminoxide (LDAO) are increasingly doped with other 12-carbon alkyl tail detergents having different nonionic head groups (Z. Markovic and R.M. Garavito, unpublished results). These observations demonstrate that changing the detergent composition in the micellar layer directly affects protein intermolecular contacts and induces the growth of crystal forms that were otherwise specified by the ionic environment of the solvent.

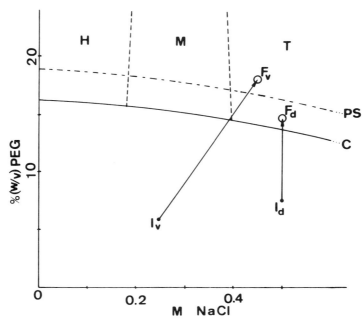

Figure 3. The phase diagram of porin crystallization in a β-OG system. The crystal forms are hexagonal (H), monoclinic (M), and tetragonal (T). PS and C are the phase separation and crystallization boundaries, respectively. Also shown are the pathways for vapor diffusion (Iv to Fv) and microdialysis (Id to Fd) experiments. The figure is taken from ref. 11.

Table 2. Some Useful Small Amphiphiles and Polar Organic Solvents

heptane-1,2,3-triol	piperidine-2-carbonic acid
triethylamine phosphate	ethyleneglycol butyl ether
glycolate butyl ester	hexane-1,6-diol
ethanol	hexanoylamide

MICELLE PERTURBATION

Michel (3,9,24) reports that membrane protein crystallization can be affected by the addition of small organic amphiphiles or co-surfactants. For the photosynthetic reaction center from R. viridis (3) and bacteriorhodopsin (24), the addition of heptan-1,2,3-triol to the crystallization medium was either essential for crystal growth or dramatically improved crystal quality. We also found that the introduction of polar organic solvents markedly affected the growth of lambda phage receptor crystals (5). In most cases, 1-5 % by weight or volume is added to the crystallization medium, resulting in a relatively high concentration of these compounds during the experiment (0.1- 1.0 M) when compared to the detergent (1-30 mM) or micelle (10-300 µM) concentration.

How these small amphiphiles act is still unclear. At the concentrations they are normally used, there will be a noticeable change in the solvent dielectric. Work on pure detergent systems (25,26,27) also reveal that polar organic solvents readily partition into the micelle phase, disrupting the micelle-bound water and increasing the apparent CMC of the detergent. These observations suggest that polar organic compounds alter the structure of the micelle, perhaps by increasing the monomer fluidity and thus micelle deformability. A less rigid micelle about the protein would enhance the formation of the weak intermolecular bonds needed for crystal growth.

Table 2 lists the commonly used small amphiphiles and organic solvents. Though they are not a panacea for membrane protein crystallization and do not work in all cases, they represent an important tool for the crystal grower.

MICELLE COLLOIDAL EFFECTS

Detergent micellar systems display phase transitions typical of colloidal systems (19,28). As the species which we want to crystallize is a protein-detergent mixed micelle, it too will be affected by the colloidal behavior of micelles. The degree to which micellar interactions influence crystal formation might be expected to depend on the amount of surface area exposed by the protein. From Figure 1, we can imagine that if the extramicellar protein surface is extensive, then the protein-protein interactions would be expected to dominate the interactions between protein containing micelles. A "buried" membrane protein, on the other hand, would not exposed much protein surface and its crystallization would be significantly affected by dispersive forces arising from the detergent colloidal effects. Given recent information about the packing of the protein molecules in crystals of photosynthetic reaction center (H. Michel, personal communication) and porin (R.M. Garavito, P. Timmins and M. Zulauf, unpublished results), the reaction center is an example of the former case and porin the latter.

When micelle interactions play a role in crystallization, an awareness of micelle colloidal interactions seems essential for developing a successful crystallization. As mentioned earlier, the phenomenon which concerns us most is micelle clustering or aggregation that under extreme conditions leads to detergent phase separation (the "cloud point" phenomenon; 20,21,22). For detergents with oligo-oxyethylene head groups

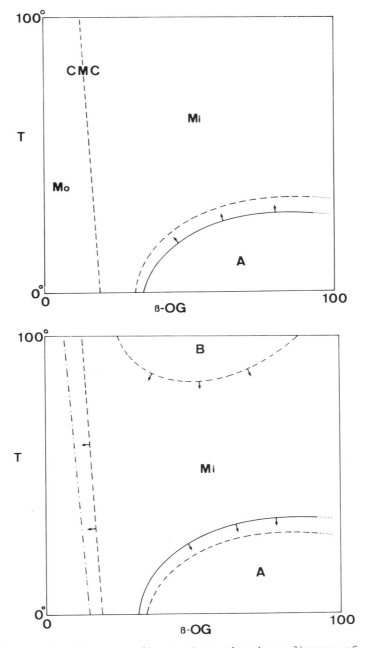

Figure 4. The upper figure shows the phase diagram of β-OG in a PEG solution. As more PEG is added the region A expands. Mo is the monomer region and CMC is the CMC boundary.

Figure 5. The lower figure shows the phase diagram of the system in Figure 4 after an oligo-oxyethylene detergent is added. Note the appearance of a new phase region B.

(e.g. Triton-X or Brij), the degree of aggregation appears to be related to the degree of dehydration of the micelle surface (20,21). The introduction of co-solutes such as salts (like AS) or polymers (like PEG) into nonionic detergent solutions can induce micelle aggregation and, at high enough concentrations, phase separation. In the presence of AS or PEG, β-OG solutions exhibit phase diagrams as shown in Figure 4. The region <u>Mi</u> is where a dispersed micellar solution exists. As the temperature decreases, the system moves towards the boundary surrounding region <u>A</u>. Micelle aggregation increases continually until the boundary is reached and phase separation begins. Though micelle aggregation can thus be induced by temperature, increasing the solute (in this case PEG) concentration will result in the same effect. As this boundary has a maximum, it is called an <u>upper consolute boundary</u> and is exhibited by some nonionic detergents like β-OG and LDAO in PEG or AS solutions. Many other nonionic detergents, particularly those with oligo-oxyethylene head groups, display a phase boundary minimum (like around region <u>B</u> in Figure 5) or <u>lower consolute boundary</u>.

The utilization of detergent phase phenomena for assisting crystallization is best developed in the case of porin (2,4,11). Figure 3 shows that at a given detergent concentration and composition, crystallization occurs when increasing PEG concentration drives the system towards boundary C ($I_d \rightarrow F_d$ in Figure 3). This crystallization boundary always just precedes the phase separation boundary of the detergent. Looking at the phase diagram differently, increasing PEG concentration raises the upper consolute boundary towards the temperature of the experiment (Figure 4). A small amount of a second detergent is usually added to porin solutions to produce the best crystals and to suppress the phase separation of the detergent solution. The detergent mixture has the phase diagram shown in Figure 5. The upper consolute boundary due to β-OG (region A) still exists but is depressed to lower temperatures by the action of the second detergent component. The second detergent also causes the appearance of a lower consolute boundary (region B), which does not interfere with crystallization, and a slight lowering of the CMC.

The participation of micelle aggregation in porin crystallization has been deduced from several experiments where temperature, PEG concentration and detergent concentration were varied. First, the temperature at which crystallization occurs is highly dependent on PEG concentration and varies in the same manner as the upper consolute boundary (region A in Figure 5) varies with PEG concentration. At 22 $^\circ$C, 13.8-14.2% PEG 2000 is required to grow crystals while only 10% is needed at 4 $^\circ$C. At temperatures above 30 $^\circ$C, crystallization requires PEG 2000 concentrations in excess of 16%. This behavior follows directly the increase of micelle aggregation as the temperature decreases. In a few experiments where the detergent C_8E_4 was used instead of β-OG, crystallization was inhibited at <u>lower</u> temperatures. This would be expected since this detergent displays only a lower consolute boundary (just region B in Figure 5) and therefore micelle aggregation <u>decreases</u> as the temperature decreases (20,21). Hence, porin crystallization, at constant ionic and PEG conditions, can be driven in part by highly temperature dependent micelle interactions.

A second line of evidence supporting this hypothesis is the effect of detergent concentration on crystal nucleation. At standard conditions for crystal growth for porin (using microdialysis, ref. 11), the rate and extent of nucleation increases markedly as the β-OG concentration rises from just above the CMC to twice the CMC (0.8% to 1.6% w/v). Crystal nucleation was estimated by the time it took for small crystals to appear and how many crystal finally appeared. The explanation for this observation is that the PEG/temperature induced micelle aggregation produces small micelle clusters which act as nucleation centers for crystallization. At low detergent concentrations just above the CMC, almost all the micelles that exist are protein containing micelles, though their molar concentration is low. Hence, the rate and extent of nucleation is small but enough to yield large single crystals. Increasing the detergent concentration increases only the concentration of pure detergent micelles since, above the CMC, detergent binding to porin is constant. This increases then the number of micelle clusters and, therefore the number of nucleation centers. This explanation predicts that at very high detergent concentrations the pure detergent micelles would begin to significantly dilute the protein containing micelles in these clusters and thus retard nucleation. Experiments have not yet been done to test this.

The role of micelle interactions in the formation of porin crystals, we feel, is very extreme and is due to porin's "buried" nature in native membrane. Proteins with more extramembranous surface should suffer less extreme detergent effects, though controling micellar interactions would definitely improve the chances for obtaining large single crystals. We had concluded earlier (4) that detergent phase separation was required for porin crystallization. This was an artefact of the way we set up the original vapor diffusion experiments. We now feel that the beneficial effects of micellar interactions have already occurred by the time phase separation has appeared. A final point on the role of detergent effects on porin crystalization concerns recent results from single crystal neutron diffraction studies on porin (R.M. Garavito, P. Timmins, M. Zulauf, unpublished results). Low resolution (∼16 Å) density maps of the tetragonal crystal form (Figure 2a) reveal that extensive contacts between detergent surfaces exist in the crystal. It is obvious from the crystal packing that any changes in the micellar behavior or detergent environment would readily disrupt the protein intermolecular contacts. This again explains this need to "fine tune" the detergent system to crystallize porin.

CONCLUSIONS AND COMMENTS

Our experiences in Basel and those of H. Michel in Martinsried suggest that crystals of many integral membrane proteins could be obtained with moderate effort. The major difficulties lie in the preparation of large single crystals for X-ray diffraction. Microcrystals of several membrane proteins have been often produced after a few weeks of effort, though improving them turned out to be a major task. One quickly realizes how many new variables are introduced when working with detergent containing solutions, particularly if stable and active protein solutions are only obtained using a multicomponent detergent system. A futher complication is that new definitions of purity and homogeneity must be used for membrane protein preparations. Not only must the sample be free of protein contaminants but also detergent or lipid contaminants. It makes little sense to spend effort optimizing a detergent system for crystallization if the amount and type of protein-bound lipid varies from preparation to preparation, thus changing the detergent environment. However, with better purification techniques and quality control coupled with more experience and patience, many membrane proteins will be amenable in the future to structural analysis by X-ray crystallography.

VIII. ACKNOWLEGDEMENTS

We would like to thank Drs. J.N. Jansonius and J.P. Rosenbusch for support and advice over the years and our many other colleagues, particularly Drs. M. Zulauf, Z. Markovic, and H. Michel, for helpful discussions and exchanges of ideas. This work has been supported by Swiss National Science Foundation grants 3.201.82 and 3.655.84 (to R.M.G.).

IX. REFERENCES

1. H. Michel and D. Oesterhelt, Proc. Natl. Acad. Sci. USA 77:1283 (1980).
2. R.M. Garavito and J.P. Rosenbusch, J. Cell Biol. 86:327 (1980).
3. H. Michel, J. Mol. Biol. 158:567 (1982).
4. R.M. Garavito, J.A. Jenkins, J.N. Jansonius, R. Karlsson and J.P. Rosenbusch, J. Mol. Biol. 164:313 (1983).
5. R.M. Garavito, U. Hinz and J.M. Neuhaus, J. Biol. Chem. 259:4254 (1984).
6. J.P. Allen and G. Feher, Proc. Natl. Acad. Sci. USA 81:4795 (1984).
7. W. Welte, T. Wacker, M. Leis, W. Kreutz, J. Shinozawa, N. Gad'on and G. Drews, FEBS Letters 182:260 (1985).
8. J. Deisenhofer, O. Epp, K. Miki, R. Huber and H. Michel, J. Mol. Biol. 180:385 (1984).
9. H. Michel, Trends Biochem. Sci. 8:56 (1983).
10. R.M. Garavito and J.A. Jenkins, In: "Structure and Function of Membrane Proteins," E. Quagliariello and F. Palmieri, eds., Elsevier Sci. Pub., Amsterdam (1983).
11. R.M. Garavito and J.P. Rosenbusch, Methods in Enzymol., in press (1986).
12. A. McPherson, "Preparation and Analysis of Protein Crystals," John Wiley and Sons, New York (1982).
13. G.L. Gilliland and D.R. Davies, Methods in Enzymol. 104:370 (1984).
14. A. Helenius and K. Simons, Biochim. Biophys. Acta 415:29 (1975).
15. C. Tanford and J.A. Reynolds, Biochim. Biophys. Acta 457:133 (1976).
16. A. Helenius, D.R. McCaslin, E. Fries and C. Tanford, Methods in Enzymol. 56:734 (1979).
17. C. Tanford, "The Hydrophobic Effect," John Wiley and Sons, New York, (1980).
18. M.J. Rosen, "Surfactants and Interfacial Phenomena," John Wiley and Sons, New York (1978).
19. H. Wennerström and B. Lindman, Phys. Rep. 52:1 (1979).
20. M. Zulauf, in: "Physics of Amphiphiles: Micelles, Vesicles and Microemulsions," V.Degiorgio and M. Corti, eds., Elsevier Sci. Pub., Amsterdam (1984).
21. M. Zulauf and J.P. Rosenbusch, J. Phys. Chem. 87:856 (1983).
22. R. Triolo, L.J. Magld, J.S. Johnson, Jr. and H.R. Child, J. Phys. Chem. 86:3689 (1982).
23. C. Bordier, J. Biol. Chem. 256:1604 (1981).
24. H. Michel, EMBO Journal 1:1267 (1982).
25. P. Stilbs, J. Colloidal Interface Sci. 94:463 (1983).
26. L. Benjamin, J. Colloidal Interface Sci. 22:386 (1966).

27. K.W. Herrmann and L. Benjamin, J. Colloidal Interface Sci. 23:478 (1967).
28. D.J. Mitchell, G.J.T. Tiddy, L. Waring, T. Bostock and M.P. McDonald, J. Chem. Soc. Faraday Trans. 79:975 (1983).
29. J.B. Hayter and M. Zulauf, Colloid Polymer Sci. 260:1023 (1982).
30. M. Le Maire, S. Kwee, J.P. Andersen and J.V. Moller, Eur. J. Biochem. 129:525 (1983).

CRYSTALLIZATION OF PROTEIN AND NUCLEIC ACIDS: A SURVEY OF METHODS AND IMPORTANCE OF THE PURITY OF THE MACROMOLECULES

Richard Giegé

Institut de Biologie Moléculaire et Cellulaire du CNRS
15, rue R. Descartes, 67084 Strasbourg Cedex, France

INTRODUCTION

Purification and crystallization of macromolecules are often challenging steps in structural projects using X-ray diffraction methods (1,2) and in some cases they can represent limiting factors. This explains the recent interest of molecular biologists and physicists to better understand crystal growth of macromolecules (3). In what follows we will discuss the role of purification in the crystallization of biopolymers and give particular emphasis to the concept of purity of preparations. Experimental set ups and properties of precipitants used to induce crystallization will be reviewed. Particular parameters affecting crystal growth of proteins (i.e. proteolysis) or of nucleic acids (i.e. nature of counter-ions, nuclease digestion, chemical fragility of RNAs) will be discussed. Examples taken from our laboratory in the field of aminoacyl-tRNA synthetases and transfer ribonucleic acids, and more generally in the field of nucleoproteins will illustrate the subject.

THE ROLE OF PURIFICATION IN THE CRYSTALLIZATION OF PROTEINS AND NUCLEIC ACIDS

The Concept of Purity

With biological macromolecules, lack of control of their purity is a common cause of unsuccessful crystallization. To be crystallized a macromolecule not only has to be pure in terms of contaminating molecules, but also in term of sequence integrity and conformational homogeneity. It appears that for crystallogenesis, the purity requirements of molecules often have to be higher than those needed in other fields of molecular biology. Therefore before trying crystallization attempts, all efforts should be directed toward the obtention of homogeneous "crystallography grade" macromolecules. It is expected that the recent breakthrough in chromatographic and genetic engineering technologies (new supports, HPLC methodologies, genetic constructions of modified and more stable proteins, and of overproducing strains) will permit to reach these requirements.

The presence of uncrontrolled contaminants in chemicals used to induce crystallization is another cause which can interfer with crystal growth and lead to unpredictable results. So for instance with trypsin-modified elongation factor from **Escherichia coli**, different polymorphic crystal forms can be obtained according to whether polyethylene glycol, the precipitant, is highly purified or is contaminated with the divalent phosphate or sulfate anions (4).

Microheterogeneities in Proteins and Nucleic Acids

Beside trivial causes leading to non homogeneous macromolecule preparations (e.g. presence of contaminant macromolecules or of precipitated material) many other subtle effects, often neglected, can interfer with crystallogenesis of biomolecules.

In proteins, incomplete post-translational modifications can be responsible for microheterogeneities (e.g. 5-8). When such modifications involve addition of charged residues, for instance as occurs with phosphorylation or aspartylation (9), the pHi of the protein will be altered, and hence its solubility properties. Microheterogeneities can appear after prolonged storage. For instance ageing of proteins can be the consequence of red-ox effects which may alter conformations in the neighborhood of cysteine residues.

Partial proteolysis is one of the most serious causes of heterogenities. It can occur during the purification process and storage of proteins, but often already occurs in vivo. Control of the physiological state of cells (10) or addition of protease inhibitors (11-13) permits to limit the degradations. Since proteases often attack exposed and flexible regions, controlled proteolysis can lead to more homogeneous protein preparations, and in some instances such degraded molecules have been successfully crystallized. This is the case for methionyl-tRNA synthetase (14) and elongation factor (15-17) from **Escherichia coli**.

Heterogeneities at the conformational level are more difficult to detect. They can be linked to different functional states of the molecules. In that case addition of ligands can freeze an enzyme in a definite conformation. Often proteins are contaminated by inactive and denatured forms generated during the purification process, for instance during dialysis or ultrafiltration, especially when the proteins are highly enriched. Aggregation occurs with proteins of hydrophobic character. High ionic strength, glycerol or non-ionic detergents, were found useful in many cases, for instance with aspartyl-tRNA synthetase (18). Thus techniques developed for membrane proteins (e.g. 19-20) might be of more general use and could permit to prevent denaturation and aggregation phenomena induced by the presence of hydrophobic domains in proteins. More generally it has been formulated that substances that lead to protein preferential hydration can also stabilize the native structure of globular proteins in aqueous solutions. This is true for many protein precipitants, salts for example used in crystallization experiments ; with polyethylene glycol or MPD, however, the effects might be more complex and for some proteins, these compounds can lead to destabilizations (21 and references therein).

Sequence and conformation heterogeneities were found in nucleic acids. These effects were mainly studied in RNAs and, in relation with crystal growth, in transfer RNAs (22). Beside effects comparable to those occuring in proteins (nuclease cleavage, variable extents of minor nucleotides (23), aggregated samples (e.g. 24)), RNAs present an intrinsic chemical fragility leading to partial hydrolysis of the alkaline type (25). The extent of degradations is usually very low (1-2%

of the molecules present nicks) and they concern mainly sequences of the pyrimidine/adenosine type or those located in flexible loop regions (26-27). To prevent chemical splitting molecules should be handled at low temperature, at pH below 7.0, and if possible in the absence of metal ions (especially lead, but also magnesium) which can catalyze under certain circumstances RNA hydrolysis (28-30)). To note is the observation that hydrolysis occurs inside of tRNA crystals. With yeast tRNA(Asp) crystal self-splitting occurs in the D-loop which is very flexible as reflected by high temperature factors ; in solution the pattern is different since the fragility is highest in the anticodon loop (27). Since crystal packing involves anticodon-anticodon interactions (GUC-GUC anticodon dimers) it can be easily understood that tRNA(Asp) samples possessing too high extents of degradations in their anticodon loop cannot be crystallized (22).

Selected Purification Procedures

In view of crystallization assays the choice of the last purification step is crucial. At this stage methods leading to dilute macromolecule solutions or to denaturations, should be avoided. This is also the case for chromatographic supports releasing adsorbed organic phases like those used for reverse phase chromatography.

Chromatography on Sepharose 4B columns (or similar supports) eluted with inverse gradients of macromolecule precipitants satisfies the requirements for crystal growth experiments : such chromatography is harmless and gives maximal recovery of active and concentrated macromolecules ; in addition it is very resolutive and permits to eliminate minor contaminants. This method is well known in the tRNA and aminoacyl-tRNA synthetase field (31,32) ; it is based on salting-out of the macromolecules on top of the inert support and elution by back-solubilization. This principle is reminiscent of the salting-out procedures used for macromolecule crystallization (1,2) and actually fractions eluted from such columns can be used for crystallization without further handling (33). Back-solubilization of aspartyl-tRNA synthetase by ammonium sulfate solutions from a Sepharose 4B matrix was used to prepare samples for the crystallization of this protein. Other precipitants (salts, organic solvents or polyethylene glycol) were used to back-elute proteins like insulin, α chymotrypsin or lysozyme from Sepharose or Sephadex matrices. Crystals were obtained from some (but not all) column fractions containing protein (33), thus illustrating the fact that minor contaminants can influence crystal growth. As a practical rule for crystallization experiments it can be recommended not to mix different batches of the same macromolecule.

CRYSTALLIZATION OF MACROMOLECULES : PRINCIPLES AND PRACTICE

The Crystal Growth Process

As far as physical principles are concerned, crystal growth of biological macromolecules is similar to that of other molecules. It is a multiparametric process which involves a nucleation phase, a growth phase, and cessation of growth (34-36). Thus to obtain crystals the first search consists to find conditions where nuclei form at the solubility minimum of the molecules. With macromolecules this search can be facilitated by taking into account recent theories of protein solubility (37). Since large crystals of high diffracting power are needed for X-ray studies, regular growth has to be maintained as long as supersaturation exists. This step is extremely delicate to control because many external causes can perturb the kinetics of crystal growth

Table 1. Most common precipitating agents used to induce crystallization of biological macromolecules

Position	Precipitating agent	Number of crystal forms	
1.	Ammonium sulfate alone.................296 mixed with others salts.......... 65		361
2.	PEG.............6000 or 8000.......... 78 4000................. 30 others................ 22		130
3.	Low ionic strength............................	79	
4.	MPD...	77	
5.	Na^+, K^+ or phosphate...........................	46	
6.	Ethanol.......................................	33	
7.	NaCl..	25	
	Other salts (lithium sulfate, sodium citrate, magnesium sulfate, ...)............	53	
	Other organic solvents (isopropanol, methanol, dioxane, ...)...........	49	

PEG : Polyethylene glycol with average Mr ; MPD : 2-methyl-2,4-pentanediol. Adapted from Gilliland and Davies (39).

and even minor quantities of impurities can have drastic effects on it (38 and references therein). Such effects, well known in materials science, likely also occur during macromolecular crystallogenesis.

What makes crystal growth of biological macromolecular compounds different from that of other materials are the consequences of their particular physical-chemical properties, i.e. their optimal stability in aqueous media and their temperature and pH stability. This restricts the search for supersaturation conditions in aqueous solutions (they can be supplemented by organic solvents) in rather narrow temperature and pH ranges. Also, because of their conformational flexibility and of their chemical versatility, macromolecules are very sensitive to external conditions and exhibit ageing phenomena. It is obvious that such behaviors interfer with crystal growth.

Precipitating Agents and Experimental Strategies

Many chemicals, or variation of physical parameters (i.e. temperature or pH), have been used to achieve the solubility minimum of macromolecules (2). But, as appears in Table 1, some precipitating agents are more commonly used than others. This is in particular the case for ammonium sulfate (2, 39), polyethylene glycol (40), and MPD (41). Thus, in a preliminary screening of conditions, those precipitants should be first assayed. The choice of a precipitant and that of additives which modulate solubility properties, should also be dictated by the biochemical properties of the macromolecules to be crystallized. Likewise the choice of heavy-atom markers should be dictated by the chemical properties of both macromolecules and precipitants (1, 42).

The good ranking of several organic solvents among the precipitants listed in Table 1 needs some explanation. As to isopropanol and dioxane this is due to their frequent use in tRNA and oligonucleotide crystallizations (22, 43). This holds true for MPD, but in addition this alcohol is also a very strong precipitant for proteins.

Because of the complexity of the behavior of macromolecules in solution, not yet well understood, several solubility minima can exist for the same protein or nucleic acid, a fact reflected a posteriori by crystalline polymorphism (2,22). As a consequence many conditions should be screened to obtain crystals suitable for diffraction studies. Since the amount of products available is often limiting, all possible combinations of parameters cannot be checked. Therefore the choice of conditions is not always rational and merely reflects the feelings of the experimentalists. A better way to approach this problem and to set up optimal conditions could be the use of statistical methods to select amongst parameters and to evaluate the results of crystallization attempts (44).

Methods and Experimental Set Ups

Because of the relative rarety of many compounds and of the great number of conditions to be screened, crystallizers of biological macromolecules had to develop micromethods. This was for instance the case of the vapor diffusion technique introduced very soon for the crystallization of tRNAs (45). At present it is one of the most widely used technique (2, 46). In the sitting drop version the method often uses the sandwich box set up (2). In the authors laboratory another type of set up is routinely used (Figure 1 and ref. 22). The hanging drop version presents advantages for screening, since the microdroplets can be as small as 5 µl (or less) instead of 20 to 40 µl for the sitting drops (47). This last method, nevertheless, was successfully applied to produce large crystals of many proteins, for instance tyrosyl-tRNA synthetase (48).

Liquid diffusion is also widely used, either in microdialyis or in free interface set ups. Microdialysis cells have been described by Zeppezauer (49) ; they are made from capillary tubes closed either by dialysis membranes or polyacrylamide gel plugs. Those cells require only 10 µl or less of macromolecule solution per assay. A modified version of the Zeppezauer cell was described by Weber and Goodkin (50). Another type of cell is known as the Cambridge buttom (2). Examination of crystals in these systems can be delicate, but solution conditions are easily varied. Free interface diffusion was introduced by Salemme (51) ; experiments can be conducted in very small capillaries for screening, but set ups have been adapted for growing large monocrystals. Interface diffusion is particularly well suited for assays under microgravity environment (52). A pulse-diffusion method combining dialysis and free diffusion has been reported (53). Varying the precipitant concentration in a pulsed fashion allows to redissolve the smallest nuclei thus favoring growth of larger crystals. Pulsed variations of parameters can be useful in many cases ; for instance with tRNA(Asp) nucleation was often induced by small positive (or negative) temperature changes using the vapor diffusion set up (22).

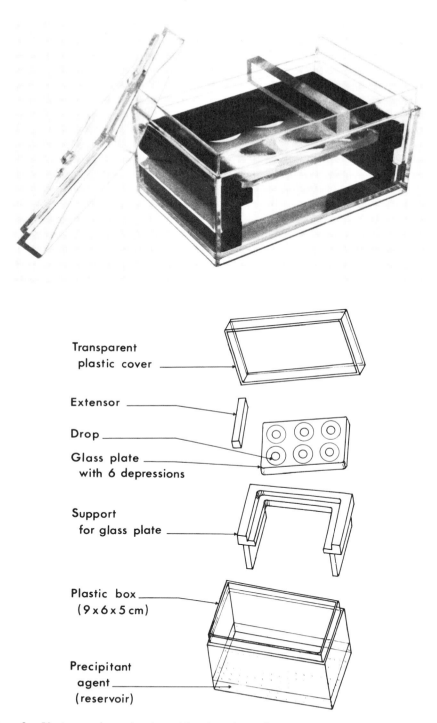

Fig. 1. Photograph and schematic drawing of a vapor diffusion apparatus.

Methods of crystallization have been developed taking advantage of the pH and temperature sensitivity of macromolecules (54). A known technique is the sequential extraction method of Jacobi (55), based on the property of proteins to be more soluble in concentrated ammonium sulfate when lowering the temperature. The method can be adapted for microassays. Usually microcrystals are grown which can be used as seeds to produce larger crystals by other means. A well known example is the crystallization of methionyl-tRNA synthetase from **Escherichia coli** (14).

Transfer of crystals from crystallization cells to X-ray capillaries might be difficult and for very fragile crystals can lead to internal damage and to mechanical cracks. To overcome such difficulties methods of crystallization in X-ray capillaries have been developed (56). A significant application of this method concerns the crystallization of the large ribosomal particle from **Bacillus stearothermophilus** (57).

Seeding is sometimes a successful technique to grow larger crystals (1, 2). Significant improvements to this approach were recently brought by seed enlargement and repeated seeding procedures (58).

SPECIFIC PARAMETERS AFFECTING CRYSTAL GROWTH : A FEW EXAMPLES

The choice of the crystal growth technique is not always preponderant in the success of a crystallization experiment. As discussed before chemical purity and conformational homogeneity of samples are important factors for crystallogenesis. In this context let us recall the use of small non-ionic detergents which permit to solubilize membrane proteins, and consequently was the clue for the crystallization of this class of proteins (19, 20). Here we will mention several other cases where specific parameters, system specific, govern crystal growth.

The choice of polyamines, mainly spermine, was the key discovery leading to the crystallization of tRNAs in forms suitable for X-ray studies (22, 59). Most probably these compounds act as specific counter-ions for neutralizing negative charges on the nucleic acid. Additionally monovalent and divalent cations, in particular magnesium ions, are needed for crystallization of tRNA. Beside their function to neutralize charges, divalent and polycations lower the concentration of precipitant needed to achieve solubility minima. Interestinly one crystal form of tRNA(Asp) was induced by spermine without addition of any other precipitant (22). Following these lines it was found that spermine and magnesium are also necessary compounds needed for the crystallization of ribo- or deoxyribonucleotides (e.g. 43) and of ribosomal 5S RNA (60, 61).

With the complex formed between aspartyl-tRNA synthetase and two molecules of tRNA(Asp) one crucial parameter which governs crystallization is the enzyme/tRNA stoechiometry (62). Measurements of accurate macromolecule concentrations is not a simple matter so that correct stoechiometries mostly can only be obtained by trial and error set ups around theoretical values. Another parameter, probably the most important for the aspartic acid complex, was the choice of ammonium sulfate as the precipitant. According to current knowledge high salt conditions should prevent complex formation between nucleic acids and proteins by decreasing the electrostatic interactions. As to ammonium sulfate this assumption turned to be wrong because high concentrations of this salt neither do prevent complex formation nor the aminoacylation reaction (63). This property is not restricted to the aspartic acid system, since ammonium sulfate also stabilizes the ternary complex EF-TU.GTP.valyl-tRNA(Val) (64) ; with this system, however, no crystals could yet be grown.

The biological origin of macromolecules is often determinant for their ability to crystallize. Beside the nature of contaminants in samples and the different intracellular levels of proteases or nucleases which can lead to fragmented macromolecules there are intrinsic physical-chemical properties which make that macromolecules from one species are more stable than those from another. The reasons for that are not clearly understood, but as a result of experimental observations it appears that macromolecules from thermophilic or halophilic species are particularly stable. Hence crystals of higher quality could be grown from material extracted from such organisms. As examples let us mention crystals of tyrosyl-tRNA synthetase (48) or of ribosomal particles (57) from **Bacillus stearothermophilus**, and of the large ribosomal subunit from the extremely halophilic bacterium **Halobacterium marismortui** (65).

FUTURE PROSPECTS

Lack of reproducibility is one of the major difficulty encountered in crystallogenesis. Therefore standardization and automatization of crystallization methods, including monitoring of experimental parameters and of crystal growth kinetics, should contribute to better results. In particular geometrical parameters, like volume and shape of crystallization cells or of macromolecule droplets, should be carefully controlled (66). Also, standardization of purification and handling methods should contribute to better control the quality of macromolecules and thus reproducibility of experiments. Furthermore new ideas should be explored, for instance crystallization in systems with phase separation (e.g. 67) or in gels (e.g. 68, 69). The opportunity which presently exists to grow crystals in space in a weightless environment should not be neglected ; beside the proper interest of microgravity research, preparation of space experiments should act as a catalyst to develop other fields of the crystal growth science, in particular theoretical and technological aspects. On the other hand a deeper understanding of the solution properties of macromolecules should allow to better dictate the choice of crystallization conditions, especially as far as additives are concerned. Likewise a better knowledge of the physical-chemical properties of precipitants should permit to rationalize experiments. It is amazing that the salting-out properties of ammonium sulfate were described one century ago (70), and that the specific properties of this salt, linked to the simultaneous presence of ammonium and sulfate ions, are not clearly understood today. Finally the physics of the different phases of crystal growth of biomacromolecules should be studied more thoroughly. All these aims can only be achieved by coordinated researches combining biochemistry, macromolecular biophysics, basic and applied crystal growth physics, and crystallography.

ACKNOWLEDGEMENTS

We thank D. Moras and J.C. Thierry for suggestions and stimulating discussions. We are also indebted to all our colleagues who contributed to the structural work, in particular A.C. Dock, D. Kern, B. Lorber, P. Romby, M. Ruff, and A. Theobald, who are presently involded in preparation, characterization, and crystal growth experiments on tRNAs and aminoacyl-tRNA synthetases. We acknowledge B. Lorber for preparing Figure 1.

REFERENCES

1. T.L. Blundell and J.L. Johnson, Protein Crystallography, Academic Press, New-York (1976).
2. A. McPherson, Preparation and Analysis of Protein Crystals, John Wiley and Sons, New-York (1982).
3. Abstracts First International Conference on Protein Crystal Growth, Stanford University, USA, August 14-16 (1985).
4. F. Jurnak, Induction of elongation factor Tu-GDP crystal polymorphism by polyethylene glycol contaminants, J. Mol. Biol. 185 : 215-217 (1985).
5. R. Uy and F. Wold, Posttranslational covalent modifications of proteins, Science (Washington) 198 : 890-896 (1977).
6. F. Wold, In vivo chemical modification of proteins, Ann. Rev. Biochem. 50 : 783-814 (1981).
7. J.H. Alix and D. Hayes, Why are macromolecules modified post-synthetically ?, Biol. Cell. 47 : 139-160 (1983).
8. F. Wold and K. Moldave (eds) Posttranslational Modifications, Methods in Enzymology 106 and 107 (1984).
9. D. Kern, B. Lorber, Y. Boulanger and R. Giegé, A peculiar property of aspartyl-tRNA synthetase from baker's yeast : Chemical modification of the protein by the enzymatically synthesized aminoacyladenylate, Biochemistry 24 : 1321-1332 (1985).
10. D. Kern, R. Giegé, S. Robbe-Saul, Y. Boulanger and J.P. Ebel, Complete purification and studies on the structural and kinetic properties of two forms of yeast valyl-tRNA synthetase, Biochimie (Paris) 57 : 1167-1176 (1975).
11. J.R. Pringle, Methods for avoiding proteolytic artefacts in studies of enzymes and other proteins from yeast, in "Methods in Cell Biology", Prescott, ed., Academic Press, New-York, 12 : 149-184 (1975).
12. T. Aoyagi and H. Umezawa, Structures and activities of protease inhibitors of microbial origin, in "Proteases and Biological Control", E. Reich et al. eds, Cold Spring Harbor Laboratory, 429-454 (1975).
13. T. Achstetter and D. Wolf, Proteinases, proteolysis and biological control in the yeast **Saccharomyces cerevisiae**, Yeast 1 : 139-157 (1985).
14. J.P. Waller, J.L. Risler, C. Monteilhet and C. Zelwer, Crystallization of trypsin-modified methionyl-tRNA synthetase from **Escherichia coli**, FEBS Lett. 16 : 186-188 (1971).
15. D. Sneden, D.L. Miller, S.H. Kim and A. Rich, Preliminary X-ray analysis of the crystalline complex between polypeptide chain elongation factor, Tu and GDP, Nature (London) 241 : 530-531 (1973).
16. W.H. Gast, R. Leberman, G.E. Schulz and A. Wittinghoffer, Crystals of partially trypsin-digested elongation factor Tu, J. Mol. Biol. 106 : 943-950 (1976).
17. F. Jurnak, A. McPherson, A.H.J. Wang and A. Rich, Biochemical and structural studies of the tetragonal crystalline modification of the E. coli EF-Tu, J. Biol. Chem. 255 : 6751-6757 (1980).
18. B. Lorber, D. Kern, A. Dietrich, J. Gangloff, J.P. Ebel and R. Giegé, Large scale purification and structural properties of yeast aspartyl-tRNA synthetase, Biochem. Biophys. Res. Commun. 117 : 259-267 (1983).
19. H. Michel, Crystallization of membrane proteins, Trends Biochem. Sci. 8 : 56-59 (1983).
20. R.M. Garavito and J.A. Jenkins, Crystallization of integral membrane proteins, this issue.

21. T. Arakawa and S.N. Timasheff, Mechanism of poly(ethylene glycol) interactions with proteins, Biochemistry 24 : 6756-6762 (1985).
22. A.C. Dock, B. Lorber, D. Moras, G. Pixa, J.C. Thierry and R. Giegé, Crystallization of transfer ribonucleic acids, Biochimie (Paris) 66 : 179-201 (1984).
23. G. Dirheimer, Chemical nature, properties, location and physiological and pathological variation of modified nucleosides in tRNAs, Recent Results in Cancer Research 84 : 15-46 (1983).
24. S.K. Yang, D. Söll and D.M. Crothers, Properties of a dimer of tRNA(Tyr1) (**Escherichia coli**), Biochemistry 11 : 2311-2320 (1974).
25. D.M. Brown, Chemical reactions of polynucleotides and nucleic acids, in "Basic Principles in Nucleic Acid Chemistry", Ts'o ed, Academic Press, New-York, vol. 2 : 259-267 (1974).
26. P. Carbon, C. Ehresmann, B. Ehresmann and J.P. Ebel, The sequence of **Escherichia coli** ribosomal 16S RNA determined by new rapid gel methods, FEBS Lett. 94 : 152-156 (1978).
27. D. Moras, A.C. Dock, P. Dumas, E. Westhof, P. Romby, J.P. Ebel and R. Giegé, Anticodon-anticodon interaction induces conformational changes in tRNA. Yeast tRNA(Asp) a model for tRNA-mRNA recognition, Proc. Natl. Acad. Sci. U.S.A. 83 : in press (1986).
28. W. Wintermeyer and H.G. Zachau, Mg^{2+} Katalyzierte, spezifische Spaltung von tRNA, Biochim. Biophys. Acta 299 : 82-90 (1973).
29. C. Werner, B. Krebs, G. Keith and G. Dirheimer, Specific cleavages of pure tRNAs by lead, Biochim. Biophys. Acta 432 : 161- 175 (1976).
30. R.S. Brown, J.C. Dewan and A. Klug, Crystallographic and biochemical investigation of the lead (II) catalyzed hydrolysis of yeast phenylalanine tRNA, Biochemistry 24 : 4785-4801 (1985).
31. W.M. Holmes, R.E. Hurd, B.R. Reid, R.A. Rimerman and G.W. Hatfield, Separation of transfer ribonucleic acid by Sepharose chromatography using reverse salt gradients, Proc. Natl. Acad. Sci. U.S.A. 72 : 1068-1071 (1975).
32. F. von der Haar, The ligand-induced solubility shift in salting-out chromatography. A new affinity technique demonstrated with phenylalanyl- and leucyl-tRNA synthetase, FEBS Lett. 94 : 371-374 (1978).
33. K. Gulewicz, D.A. Adamiak and M. Sprinzl, A new approach to the crystallization of proteins, FEBS Lett. 189 : 179-182 (1985).
34. Z. Kam, H.B. Shore and G. Feher, On the crystallization of proteins, J. Mol. Biol. 123 : 539-555 (1978).
35. R. Boistelle, Concepts de la cristallisation en solution, Actualités néphrologiques 6 : 159-202 (1985).
36. G. Feher and Z. Kam, Nucleation and growth of protein crystals : General principles and assays, Methods in Enzymology 114 : 77-112 (1985).
37. T. Arakawa and S.N. Timasheff, Theory of protein solubility, Methods in Enzymology 114 : 49-77 (1985).
38. C.H. Gilmer, Computer models of crystal growth, Science (Washington) 208 : 355-363 (1980).
39. G.L. Gilliland and D.R. Davies, Protein crystallization : The growth of large-scale single crystals, Methods in Enzymology 104 :370-381 (1984).
40. A. McPherson, Crystallization of proteins from polyethylene glycol, J. Biol. Chem. 251, 6300-6303 (1976).

41. M.V. King, B.S. Magdoff, M.B. Adelman and D. Harker, Crystalline forms of bovine pancreatic ribonuclease : Techniques of preparation, unit cells, and space groups, Acta Cryst. 9 : 460-469 (1956).
42. G.A. Petsko, Preparation of isomorphous heavy-atom derivatives, Methods in Enzymology 14 : 147-156 (1985).
43. S.R. Holbrook and S.H. Kim, Crystallization and heavy-atom derivatives of polynucleotides, Methods in Enzymology 114 : 167-176 (1985).
44. C.W. Carter Jr. and C.W. Carter, Protein crystallization using incomplete factorial experiments, J. Biol. Chem. 254 : 12219-12223 (1979).
45. A. Hampel, M. Labananskas, P.G. Conners, L. Kirkegard, U.L. RajBhandary, P.B. Sigler and R.M. Bock, Single crystals of transfer RNA from formyl-methionine and phenylalanine transfer RNA's, Science (Washington) 162 : 1384-1386 (1968).
46. D.R. Davies and D.M. Segal, Protein crystallization : Microtechniques involving vapor diffusion, Methods in Enzymology 22 : 266-269 (1971).
47. A. Wlodawer and K.O. Hodgson, Crystallization and crystal data of monellin, Proc. Nat. Acad. Sci. U.S.A. 72 : 398-399 (1975).
48. B.R. Reid, G.L.E. Koch, Y. Boulanger, B.S. Hartley and D. Blow, Crystallization and preliminary X-ray diffraction studies on tyrosyl transfer RNA synthetase from **Bacillus stearothermophilus**, J. Mol. Biol. 80 : 199-201 (1973).
49. M. Zeppezauer, Formation of large crystals, Methods in Enzymology 22 : 253-266 (1971).
50 B.H. Weber and P.E. Goodkin, A modified microdiffusion procedure for the the growth of single protein crystals by concentration-gradient equilibrium dialysis, Arch. Biochem. Biophys. 141 : 489-498 (1970).
51. F.R. Salemme, Protein crystallization by free interface diffusion, Methods in Enzymology 114 : 140-141 (1985).
52. W. Littke and C. Jones, Protein single crystal growth under microgravity, Science (Washington) 225 : 203-204 (1984).
53. R.E. Koeppe II, R.M. Stroud, U.A. Pena and D.U. Santi, A pulsed diffusion technique for the growth of protein crystals for X-ray diffraction, J. Mol. Biol. 98 : 155-160 (1975).
54. A. McPherson, Crystallization of protein by variation of pH or temperature, Methods in Enzymology 104 : 125-127 (1985).
55. W.B. Jacoby, Crystallization as a purification technique, Methods in Enzymology 22 : 248-252 (1971).
56. G.N. Phillips, Jr., Crystallization in capillary tubes, Methods in Enzymology 104 : 128-131 (1985).
57. A. Yonath, M.Z. Saper, I. Makowski, J. Mussif, J. Piefke, H.D. Bartunik, K.S. Bartels and H.G. Wittmann, Characterization of single crystals of the large ribosomal particles form **Bacillus stearothermophilus**, J. Mol. Biol. in press (1986).
58. C. Thaller, L.H. Weaver, G. Eichele, E. Wilson, R. Karlson and J.N. Jansonius, Seed enlargement and repeated seeding, Methods in Enzymology 104 : 132-135 (1985), and J. Mol. Biol. 147 : 465-469 (1981).
59. S.H. Kim and A. Rich, Single crystals of transfer RNA : An X-ray diffraction study, Science (Washington) 162 : 1381-1384 (1968).
60. K. Morikawa, M. Kawakami and S. Takemura, Crystallization and preliminary X-ray diffraction study of 5S rRNA from **Thermus thermophilus** HB8, FEBS Lett. 145 : 194-196 (1982).

61. S.S. Abdel-Meguid, P.B. Moore and T.A. Steitz, Crystallization of a ribonuclease-resistant fragment of **Escherichia coli** 5S RNA and its complex with protein L25, J. Mol. Biol. 171 : 207-215 (1983).
62. B. Lorber, R. Giegé, J.P. Ebel, C. Berthet, J.C. Thierry and D. Moras, Crystallization of a tRNA/aminoacyl-tRNA synthetase complex. Characterization and first crystallographic data, J. Biol. Chem. 258 : 8429-8435 (1983).
63. R. Giegé, B. Lorber, J.P. Ebel, D. Moras, J.C. Thierry, B. Jacrot and G. Zaccaï, Formation of a catalytically active complex between tRNA(Asp) and aspartyl-tRNA synthetase in high concentration of ammonium sulfate, Biochimie (Paris) 64 : 357-362 (1982).
64. B. Antonsson and R. Leberman, Stabilization of the complex EF-TU.GTP.valyl-tRNA(Val), Biochimie (Paris) 64 : 1035-1040 (1982).
65. A. Shevack, H.S. Gewitz, B. Hennemann, A. Yonath and H.G. Wittmann, Characterization and crystallization of ribosomal particles from **Halobacterium marismortui**, FEBS Lett. 184 : 68-71 (1985).
66. A Yonath, J. Müssig and H.G. Wittmann, Parameters for crystal growth of ribosomal subunits, J. Cell. Biochem. 19 : 145-155 (1982).
67. T. Alber, F.C. Hartman, R.M. Johnson, G. A. Petsko and D. Tsernoglou, Crystallization of yeast phosphate isomerase from polyethylene glycol. Protein crystal formation following phase separation, J. Biol. Chem. 256 : 1356-1361 (1981).
68. H.K. Henoch, Crystal growth in gels, The Pensylvania University (1968).
69. M.C. Robert, F. Lefaucheux and A. Authier, Growth and characterization of brushite and lead monetite. Simulation and results, Proc. 5th Europ. Sympos. Material Sci. under Microgravity, (ESA SP-22) 193-199 (1984).
70. F. Hofmeister, Zur Lehre von der Wirkung der Salze, Arch. Exp. Path. Pharm. 24 : 247-260 (1888).

SYNCHROTRON RADIATION AND MACROMOLECULAR CRYSTALLOGRAPHY

R. Fourme and R. Kahn

LURE (CNRS, Université Paris-Sud), Bât. 209D
91405 ORSAY Cedex, France

It is hardly necessary to underline that synchrotron radiation (SR) is a tool which is now commonly used in the field of macromolecular crystallography. During this conference, many new structures presented in oral communications or posters involved the use of a SR source. Indeed, SR and area detectors are the major progress of the last few years in the technology applied to the data collection from macromolecular crystals.

This paper first presents sources and properties of synchrotron radiation. Then, main applications of SR in the field of macromolecular crystallography are discussed on the basis of selected examples together with the relevant instrumentation. Various review articles are available[1] for further information.

I. SYNCHROTRON RADIATION SOURCES

I.1 - Synchrotron radiation from dipolar magnets of e+ e- storage rings

SR is the electromagnetic emission from relativistic charged particles submitted to an acceleration. SR sources used for macromolecular crystallography are high energy e^- or e^+ storage rings. The basic structure of these accelerators features (Fig. 1) :

- a toroidal evacuated chamber in which particles describe periodically a stable orbit.
- several dipolar magnets which deflect particles by a total of 2π radians. Particles submitted to a centripetal acceleration lose a fraction of their kinetic energy in the form of SR.
- radiofrequency (RF) cavities which produce an oscillating electric field. The energy lost by radiation is replenished by this field.

In the initial so-called injection phase, particles produced par an auxiliary linear or circular accelerator are injected randomly in the storage ring chamber. Only those particles whose motion has a suitable phase relation with respect to the RF oscillations will keep a stable orbit ; they gather into one or several bunches of particles ; others will collide chamber walls after a few revolutions. The circulation of charged particles is equivalent, on average, to a current of intensity I.

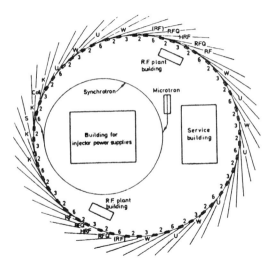
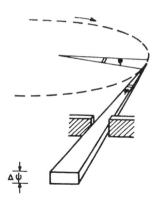

Fig. 1 : Schematic layout of a 5 GeV storage ring ; beam lines are indicated ; the injection equipment includes a microtron and a synchrotron. RF : radio-frequency cavity ; RFQ : radio-frequency quadrupole ; W wiggler U : undulator (see I.2) (after Ref. 2)

Fig. 2 : Extraction of SR from a storage ring. $\Delta\psi$ is about γ^{-1} and φ is limited by some slit system

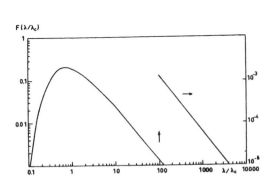
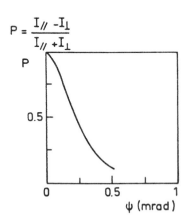

Fig. 3 : Log-log plot of F versus λ/λ_c (after Ref. 3)

Fig. 4 : Storage ring DCI at LURE (Orsay). Polarisation of SR emitted by a single 1.8 GeV electron or positron as a function of the elevation angle ψ

The current - and hence the SR emission - decays smoothly in time due to interactions between particles in each bunch and between particles and residual molecules in the chamber.

Let us introduce two important parameters attached to a given $e^+ e^-$ storage ring.

(i) $\gamma = E/E_o$ where E_o is the relativistic energy of a particle at rest (E_o = 0.511 MeV) and E is the relativistic energy of the moving particle (typically a fraction to a few GeV).

(ii) $\lambda_c = 5.59\, R/E^3$ where λ_c is called the critical wavelength (in Å unit), R is the radius of curvature (in meters) of the trajectory in the magnets and E is the relativistic energy (in GeV).

Just remember that 50% of the energy is radiated at a wavelength beyond λ_c and 50% below and that the most intense emission is close to λ_c.

The total radiated power is

$$P = 494.5\, \frac{EI}{\lambda_c}$$

(E in GeV, I in amperes λ_c in Å and P in kW)

An an example, the X-ray storage ring DCI at LURE (Orsay) has the following characteristics :

E = 1.84 GeV ; γ = 3600 ; λ_c = 3.43 Å
I = 0.250 A ; P = 66.3 kW ; one bunch, with a revolution period of 280 ns ; the intensity is divided by 2 after ~ 50 hours.

The angular distribution of the emission by a relativistic e^+ or e^- in the laboratory frame is concentrated into a forward cone with an opening half angle of ~ γ^{-1} rad. As the bunch sweeps through its closed orbit, a fan of radiation is produced which is tightly collimated in the vertical plane. In practice, SR is extracted through tangential beam lines. A fluorescent screen set on such a beam line will be illuminated at each passage of the bunch by a flash of X-rays with a maximum vertical divergence of about $\gamma^{-1}/2$ and an horizontal divergence which depends in practice of some slit system (Fig. 2). In the case of DCI, the vertical divergence is ~ 0.3 mrad ; each flash, observed every 280 ns, (frequency ~ 3.6 MHz) has a nearly gaussian shape (full width at half maximum ~ 1ns) ; the source is elliptical (FWHM H x V : ~ 5.5 x 1.2 mm²).

For a dipolar magnet, the total (i.e. integrated vertically) number of photons emitted at λ in a bandwidth of 10^{-3} Å, per second, for an horizontal fan of 1 mrad is simply given by :

$$N(\lambda) = \gamma^4\, FI/R = 0.82 \times 10^{14}\, EFI/\lambda_c$$

where F, when parameterized by λ/λ_c, is an universal function[3]. From the log-log plot of F vs $\lambda/\lambda c$, it can be seen that the optimal range of λ for a given storage ring is between ~ 0.25 λ_c and 4λ_c (Fig. 3).

Finally, polarisation properties of SR should be mentioned. For a dipolar magnet and a single particle, the emitted SR is linearly polarised in the orbit plane and elliptically polarised above and below this plane (Fig. 4). When the parameter $\gamma\psi$ is used, one gets, for a given λ/λ_c, an universal curve[3] for the horizontal and vertical components of the intensity.

In the actual case of many particles with various trajectories, the beam is not totally polarised in the mean orbit plane. For DCI, the rate of polarisation is ~ 0.9 at 1.5 Å for a sample located exactly in the mean orbit plane ; any vertical displacement will change not only the intensity but also the polarisation at the sample. One should not forget to apply a suitable polarisation correction[4] to data collected with SR.

A comparison between a SR source and a conventional X-ray source is shown in Table 1. The most important parameter for macromolecular crystallography is the spectral brilliance of the source (i.e. the number of photons emitted at a given λ, in a given bandwidth, per unit solid angle, per unit area of source, per second). The spectral brilliance of a high energy storage ring such as DCI exceeds that of any rotating anode source by roughly two orders of magnitude at the Kα wavelength and obviously much more at other wavelengths. Other properties of SR sources are comparatively less important, except in specific experiments, and may be rather troublesome for the crystallographer.

I.2 - Synchrotron radiation from insertion devices in e+ e- storage rings

It is possible to install somewhere in the straight sections of a storage ring magnetic structures which, when traversed by electrons or positrons, will produce locally SR with special characteristics : for instance a different critical wavelength and/or higher intensity, a different spectrum, a different polarisation. These devices, which do not produce a net displacement or deflection of the particles, are called insertion devices. Many conventional rings have been equipped with such devices. Machines of the latest generation include these devices in their basic design, and the magnetic lattice is accordingly optimised (Fig. 1).

Current insertion devices are called <u>wigglers</u> and <u>undulators</u>[5,6] ; the term undulator is used when there are significant interference effects between the emissions from the periodic "bumps" of the particle trajectory.

I.2.1) <u>Wigglers</u> : A structure with three magnets of alternate polarities with one central high-field magnet, such as shown in Fig. 5, will-as a first order effect-shift the SR spectrum of dipole magnets to shorter wavelengths (Fig. 7). Examples of such wigglers can be found at SRS (Daresbury) and at the Photon Factory (Tsukuba) ; both include superconducting magnets. Notice that, in the latter case, electrons oscillate vertically so that the plane of polarisation is vertical ; this feature is interesting because diffractometers with the standard horizontal equatorial plane are then not affected by polarisation losses.

Inserting N high field magnets of alternate polarities between the two low field magnets with enhance the flux by N, at least on the axis of the device and somewhat less off-axis, in addition to the wavelength shift. An obvious drawback of multipole wigglers is that the radiated power is very large ; this is a serious problem for mirrors and monochromators.

In a multipole wiggler, the angular deflection δ of electrons or positrons is large ($\delta \gg \gamma^{-1}$).

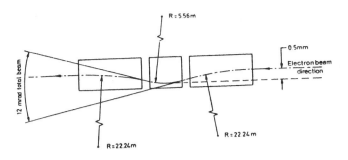

Fig. 5 : Diagram of wiggler magnet (after Ref. 6)

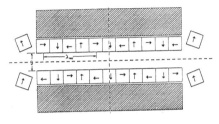

Fig. 6 : Diagram of an undulator. λ_w is the magnetic period of the undulator.

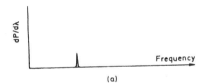

(a)

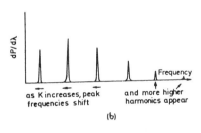

(b)

Fig. 8 : a) Single peak spectrum from an undulator. This is obtained if the undulator magnetic field is relatively low, the electron beam has a small angular spread and the radiation is viewed through a small axial pinhole.
b) The single peak becomes lower in wavelength and accompanied by harmonics as the field is increased (after Ref. 2)

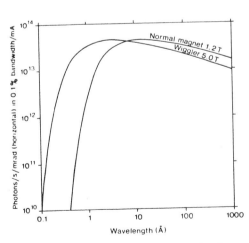

Fig. 7 : SR spectrum from dipolar magnets and from wigglers.

TABLE 1 : Comparison of X-ray tubes and SR sources

	X-RAY TUBE	SR SOURCE
emission mechanism	Transitions between electronic levels plus deceleration of electrons by matter. Kinetic energy → heat + X-Rays	Centripetal acceleration of highly relativistic e^- or e^+. Kinetic energy → X-rays
Spectrum	a few peaks over a weak background	intense, continuous and calculable spectrum
Source size	small	small
solid angle of emission	~ 2 π	ca 0.3 mrad x a few mrad
polarisation	~ no	linear or elliptical
time structure	random	pulsed

TABLE 2 : Macromolecular crystallography

FEATURES	PROBLEMS	SOLUTIONS
non-local structural method		multidetector
multiple diffraction data sets	amount of data	
large number of reflections per set	high angular resolution	
static and dynamical disorder	poor signal-to-noise ratio	parallel and monochromatic X-Ray beam + small oscillations of the crystal
diffuse scattering		
radiation damage	number of reflections per crystal to be increased	intense beam at short λ
anomalous scattering	λ to be optimised	tunable source

I.2.2) <u>Undulators</u> : An undulator is a multipole wiggler with N large and a very small angular deflection of particles ($\delta \leq \gamma^{-1}$). The discussion will be restricted to the only important case for protein crystallography, that of an experiment after a pinhole located exactly on the axis of the undulator. Interference effects concentrate the SR emission into a few peaks at wavelengths λ_i :

$$\lambda_i = \frac{\lambda_o}{i2\gamma^2} (1 + \frac{\gamma^2\delta^2}{2}) \quad \text{with} \quad \frac{\Delta\lambda_i}{\lambda_i} \sim \frac{1}{iN}$$

where λ_o is the period of the magnetic structure, $\gamma = E/m_o c^2$ and i is the harmonic number. δ can be modified at will by changing the magnitude of the undulator magnetic field. When $\delta \ll \gamma^{-1}$, (weak field case), the SR beam is monochromatic. As the field - and hence δ - increases, odd harmonics appear and the peak frequencies shift (Fig. 8).

The undulator is then a tunable source of quasi-monochromatic and extremely bright lines.

For various reasons, an undulator working properly at ~ 1.5 Å requires an optimised high energy machine. This was one of the leading goals in the design of the planned 5 GeV storage ring of the European Synchrotron Radiation Facility. At 1.5 Å, the spectral brilliance of the ESRF undulator should be 10^6-10^7 times higher than the spectral brilliance of DCI. Such quantum jumps are exceptional in physics !

II - INSTRUMENTATION

II.1 - Requirements for an effective data collection

Table 2 shows (i) a few important specific features of protein crystallography (ii) main technical problems to be solved for an effective collection of diffraction data and (iii) goals to be achieved in terms of X-ray source, optics and detector. Without further comments, it is clear that the ideal X-ray source must have a high spectral brilliance over a wide range of wavelengths. The optical system of the experimental set-up should deliver with a good yield a monochromatic and quasi-parallel X-ray beam of an easily and accurately tunable wavelength. The detector should be an area detector of high efficiency during data collection; the signal-to-noise ratio has to be maximised and, as discussed later, this is a major reason to use electronic detectors rather than films. Getting data of high accuracy is a requirement which is more and more important, especially with respect to the phase problem (use of anomalous scattering, direct methods) and for closely specifying and refining molecular models ; this involves the X-ray source, the optics and the detector as well as data collection strategy and data reduction.

II.2 - Optical systems

Most photographic cameras which are currently available at various facilities to collect high resolution data are equipped with a curved crystal-curved mirror system. Focusing in the horizontal direction is achieved by a curved perfect crystal of germanium or silicon which intercepts 1 ~ 3 mrads of the SR fan at 10 ~ 20 meters from the source. The crystal is asymmetrically cut so as to provide short crystal-to-focus distance and a demagnification of the horizontal source extension. The bending system is based on a principle first introduced in the D12 instrument at LURE[4] (Fig. 9). Focusing in the vertical direction and elimination of the higher order harmonics is achieved by the total

Fig. 9 : Curved crystal bender on the monochromator of the D24 experiment at LURE (Orsay). The triangular-shaped germanium crystal is 19 cm long (courtesy of P. VACHETTE)

Fig. 10 : Two-crystal fixed exit slit monochromator on the D21 experiment at LURE (Orsay). The second crystal is sagittally bent for line focusing (courtesy of J. GOULON)

Fig. 11 : Rotation camera D12 at LURE (Orsay)

Fig. 12 : Area detector diffractometer D23 at LURE (Orsay). The detector is a spherical drift multiwire proportional chamber with a fast digital position encoder, built by G. CHARPAK at CERN.

reflection of the beam on a curved mirror. This monochromator-mirror system produces a small focal spot even for a rather large source ; but it is not easily tunable.

Another possibility is the combination of a (+, -) double-crystal (for instance X31 at the EMBL outstation) with a double curvature mirror, giving a 1 : 1 imaging. Since there is no demagnification, it requires a source of a small size (as available in the latest generation of dedicated storage rings) if a very bright focal spot is looked for. As the monochromatic beam has a fixed direction, the instrument is rapidly tunable and is well adapted to area detector diffractometers which are preferably in a fixed setting. An interesting variant is the saggital bending of the second crystal of the monochromator[7], giving a vertical focal line (for instance the D21 instrument at LURE, Fig. 10) plus a single curvature mirror for harmonic rejection and double focusing.

To fix ideas, with similar collimation and at 1.54 Å, the intensity at the sample with the curved crystal-curved mirror system of the D42 camera on the wiggler line at LURE (in preparation) will be in excess of 10^{12} photons s^{-1} mm^{-2} at λ = 1.54 Å, instead of ~ 5 x 10^{10} photons s^{-1} mm^{-2} for the old D12 camera and ~ 10^9 photons s^{-1} mm^{-2} for an Elliott GX6 rotating anode of 1.6 kW with a 0.2 x 2 mm^2 electron focus.

II.3 - Detectors

The usual data collection method with SR source has been up to now the screenless oscillation technique[8] using photographic films (Fig. 11). Film is a detector with contrasted characteristics. It has a very good spatial resolution, limited in practice by the scanning step of the densitometer and a virtually unlimited count rate. But the information acquired per scattered photon is poor[9], so that most of the high intensity of focusing instruments simply counter-balances the inefficiencies of films. To keep the number of film packs within reasonable limits, each pack corresponds to an oscillation of the crystal over 0.5-2° ; the signal-to-noise ratio of each reflection is then not optimal because the background is integrated during a period of time which is much larger than the one during which each reflection is activated. This default is worse with SR sources than with conventional sources ; in effect, the extreme collimation and monochromaticity of SR reveal the true rocking curve of each reflection. We have frequently observed reflection profiles with a FWHM of 0.01-0.02° for fresh macromolecular crystals and a normal Lorentz factor.

Electronic area detectors have fundamental advantages over films because they derive more information per photon and because frames corresponding to very small oscillations (~ 0.01°) can be recorded so as to optimise the signal-to-noise ratio. Making a good area detector for protein crystallography with a SR source is a real challenge. There are only two area detector diffractometers which are now operational on a SR beam line, at LURE (Orsay)[10] and at SSRL (Stanford)[11]. Both use multiwire proportional chambers (MWPC) and count single photons (Fig. 12). These detectors are limited for both the total and the local maximum count rate and could not cope with SR from wigglers or undulators. TV image intensifier systems have a higher count rate capability in the integration mode ; a commercial instrument is being evaluated at SRS (Daresbury).

III - HIGH RESOLUTION DATA COLLECTION

It is in general possible to collect, using SR, diffraction data to a resolution and with an accuracy higher than otherwise possible. This is due to the synergy of several reasons (i) due to the high brilliance of SR sources, monochromatic, quasi-parallel and nevertheless intense beams are available, which maximises the signal-to-noise ratio on the detector (ii) the sample radiation damage is reduced, for a given total dose, when the dose rate is increased[12] (iii) the brilliance is still sufficient at relatively short wavelengths (say 0.5 - 1 Å), at least for SR sources with a critical wavelength of around 1 Å, to allow the use of the harder part of the spectrum ; this improves the ratio of the diffracted intensity per photon absorbed by the sample ; it also minimises absorption and obliquity corrections (iiii) dramatically shorter exposure times reduce experimental constraints such as low temperature and permits an easy selection on the site of best crystals from a given batch.

A few selected results will now be discussed :

III.1 - Crystal structure of DNA fragments

The structure of the self-complementary octanucleotide d(G-G-^{Br}U-A-^{Br}U-A-C-C), crystallized in the A-DNA conformation, has been refined to a R-factor of 0.136 using 1.7 Å resolution data collected by the oscillation method at LURE. The deoxy ribose conformations and thermal motions of individual atoms were refined. About 70 ordered water molecules were located ; the hydratation shells around the duplex display distorted pentagonal arrangements in the major groove[13].

III.2 - Crystal structure of proteins

The crystal structure of human deoxyhemoglobin at 1.74 Å was refined, starting from a 2.5 Å model, using very accurate photographic data collected at LURE. 56,287 independent reflections (99.6 % of all possible data in the range 20-1.74 Å) were measured, i.e. about three times more than the 2.5 Å diffractometer data previously available. The R-Sym factor on the intensities of 54,034 Friedel pairs was 3.6 %. The final R-factor was 16.2 %, with 221 water molecules, 2 phosphate ions and a constant solvent density outside of the molecular volume. The estimated average error in atomic positions is ~ 0.2 Å, and 0.14 Å for the main chain atoms of internal segments[14].

III.3 - Crystals with long unit-cell edges

A few modifications to the D12 camera at LURE and a careful adjustment of its optical system allowed high resolution data collection on crystals of Δ^5-3-ketosteroid isomerase (hexagonal cell with a = b = 60 Å and c = 504 Å) and the solution of the structure at 2.6 Å resolution[15]. The monochromator was set at an accurate achromatic setting at λ = 1.4 Å to minimise the size of diffraction spots. The divergence in the horizontal was reduced to 1mrad by closing the slits of the monochromator ; the divergence in the vertical direction was the natural divergence of the SR beam (0.3 mrad). A collimator with two adjustable sets of slits was essential for the accurate shaping of the beam. Data were collected on cylindrical cassettes and low background films (Fig. 13). Scattering by air was minimised by a cone filled with helium placed between the sample and the film. Useful data were subsequently obtained by the same technique for a challenging problem, cowpea mosaic virus crystals (hexagonal cell with a = b = 450 Å and c = 1044 Å)[16]. Several of the new virus crystal structures presented at this conference were solved using SR : rhinovirus, from data collected at CHESS (Cornell) after preliminary work at DORIS (Hamburg) and SRS (Daresbury), black beetle virus (Fig. 14) and the cubic form of cowpea mosaic virus, from data collected at Orsay by J. JOHNSON. Another example is the complex of t-RNA^{Asp} t-RNA^{Asp} synthetase (cubic cell with a = 354 Å) collected by D. MORAS and J.C. THIERRY at Orsay.

III.4 - Small crystals

Native and derivative data sets to 2 Å resolution have been been obtained at Hamburg on HLA crystals of 20-40 μm thickness. The structure is in progress, as discussed during this conference by D.C. WILEY.

Diffraction patterns of bovine growth hormone crystal with dimensions 120 x 120 x 15 μm^3 have been obtained at CHESS ; it was claimed that these are the most weakly diffracting single crystals of any compound to have been successfully examined by X-ray diffraction techniques[17].

III.5 - Use of short wavelengths

On the wiggler line of SRS (Daresbury), a full high resolution data set has been collected from crystals of nitrogenase at $\lambda = 0.60$ Å[18,10].

IV - TIME DEPENDENT X-RAY DIFFRACTION STUDIES

The high brilliance of SR sources allows one to get a significant fraction of the diffraction spectrum in short times. Either monochromatic beams or polychromatic beams can be used. In fact, the venerable X-ray Laue technique has been revisited, opening new prospects for time-dependent studies.

IV.1 - Monochromatic techniques

With a wiggler source, a full data set at medium resolution can be collected from strongly diffracting protein crystals in a few minutes by the standard oscillation method. As an example, 3 Å data could be collected in ~ 20-30 minutes from crystals of phosphorylase b (PPb), a protein with a molecular weight of ~ 100.000 daltons, using the SRS wiggler at Daresbury. This allowed to collect several sets from one crystal of PPb at several points in a pathway of PPb-substrate intermediates and to derive the corresponding structural information[19].

Along the same lines, the effect of temperature on a given structure can be more readily observed. The first application of SR in this field concerned the variation with temperature of refined temperature factors of trypsinogen[20].

A much better time resolution is possible when using apparatus and techniques which were initially developped for kinetic studies of muscle contraction. Test data have been collected on a millisecond time scale at DORIS (Hamburg) on crystals of carbon monoxymyoglobin. The time-course intensities of a few reflections were measured with a linear detector before and after laser photolysis of the CO ligand[21] ; to follow changes on a larger part of the diffraction pattern, a special hardware and software data handling for a 2D detector has been developed[22].

IV.2 - Polychromatic techniques

Polychromatic radiation spanning a wide wavelength range is derived from the continuous spectrum of an SR X-ray source by the application of suitable optical elements such as filters, X-ray mirrors and layered

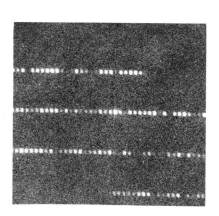

Fig. 13 : Enlarged portion of a rotation photograph of Δ^5-3-ketosteroid isomerase obtained with the D12 camera at LURE (Orsay). The shortest spacing between spots corresponds to 504 Å in the direct space (courtesy of P.B. SIGLER)

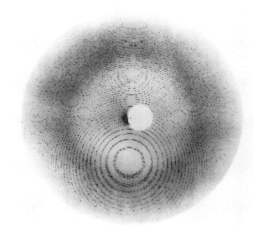

Fig. 14 : 2.8 Å rotation photograph of BBV (Black Beetle Virus) obtained with the D12 camera at LURE (Orsay) (Courtesy of J. JOHNSON)

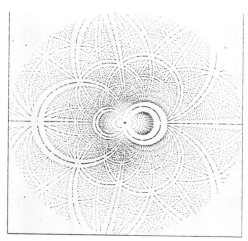

Fig. 15 : Laue pattern prediction for the protein pea lectin (a = 50.7 Å, b = 61.2 Å, c = 136.7 Å), wavelength between ~ 0.3 and 2.4 Å. Sample resolution limit of 2.6 Å is assumed. Total number of spots : 10494 ; measurable spots (omitting overlap) : 7112 (from Ref. 25)

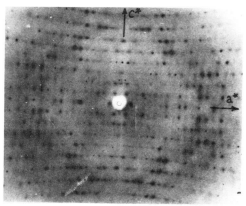

Fig. 16 : "Monochromatic Laue" pattern of hen-egg lysozyme, form B (orthorhombic) obtained with the camera D16 at LURE (Orsay) ; λ = 1.61 Å ; still sample ; film normal to the incident beam at 75 mm from the sample ; a^* and c^* perpendicular to the incident beam ; exposure time 1800 sec.

synthetic micro-structures (or multilayers)[23]; the full white beam can also be used[24]. There are several advantages : an optimal use of the natural polychromatic SR spectrum ; a dramatic reduction in exposure time ; the simultaneous recording of many reflections and the direct production of integrated intensities with a stationary crystal (Fig. 15). The technique has also disadvantages : there are complicated wavelength correction factors to be measured or calculated to enable structure amplitudes to be derived from the measured Laue intensities ; multiple orders of a reflection may be in diffracting position simultaneously and give rise to overlapped reflections.

It seems possible to overcome most of these problems and at Daresbury, an essentially complete chain of programmes as been developed from prediction of the Laue pattern (to match observation) to scaling and merging of film packs[25].

As an example, at Daresbury, the above mentioned study of the catalytic reaction of PPb in the crystal has been also tackled by the Laue method. A typical exposure time for one Laue photograph which contains a large fraction of the full data set is 200-500 ms[25]. In that experiment, the full white beam of the wiggler is used. The incident spectrum extends from 0.5 to 3 Å with an intensity of 10^{14} photons s^{-1} mm^{-2}. At CHESS, using a bending magnet and no focusing optics, the intensity is comparable ; for lysozyme crystals, exposures as short as 4 ms were obtained with Polaroid films and 64 ms on radiographic films[17].

Using focusing optics and more powerful SR sources, exposure times in the millisecond range will be at hand. This opens fascinating possibilities.

V - MOLECULAR DYNAMICS

V.1 - Temperature dependence of individual atomic temperature factors

The analysis of temperature factors, possibly coupled with molecular dynamics simulations, is a source of information on the motions of macromolecules in the crystal. The high brilliance of SR sources permits fast data collection either at low temperature[20] or at high temperature (study of the denaturation process). Monochromatic or polychromatic methods may be used.

V.2 - Diffuse scattering

The exploitation of diffuse scattering, out of Bragg reflections, with SR, is a virtually virgin field. Diffuse scattering is observed commonly during high resolution data collection ; with respect to conventional sources, it appears to be enhanced, as a result of improved signal-to-noise ratio. In one case, a dependence of the intensity of diffuse scattering with the time-structure of SR has been suggested[12], but more quantitative studies are awaited.

Diffuse scattering experiments have been undertaken at LURE with tests on various crystals such as hen-egg lysozyme. A monochromatic technique and a stationary crystal were used, with a special care to avoid parasitic scattering[26]. Various sections of the reciprocal space of lysozyme have been recorded and the analysis of the data in terms of correlated motions of molecules is in progress (Fig. 16).

VI - THE PHASE PROBLEM

VI.1 - Introduction

Phase determination is a crucial step of structural analysis, for which any improvement in the quality of data is beneficial. As an illustration of this point, many structures have now been solved in surprisingly short times from precise and accurate data measured with an area detector on a conventional X-ray source and using, in several cases, only one heavy atom derivative. SR sources, as shown previously, allow one to collect data of unprecedented quality, both quantitatively (resolution) and qualitatively (accuracy and precision). This is certainly useful for conventional methods of phase determination such as isomorphous replacement or symmetry averaging and may be crucial for the successful application of direct methods which are currently developed.

But there is a property of SR, tunability, which is of particular interest for the phase problem ; and indeed, the advent of SR sources has stimulated research on anomalous dispersion methods. The general idea is the solution of the phase problem from diffraction data at several optimal wavelengths. The idea is simple and elegant, but its practical realisation is rather challenging. It requires a sophisticated equipment - from the monochromator to the electronic area detector - which is operated in the rather hectic and constrained context of a SR facility. At the moment, the real output of SR sources is still limited, although it gives ground to optimistic prospects. To be fair, it is also quite clear that a lot of work can still be done with careful measurements at a single wavelength with a rotating anode source. In effect, the determination of the crambin structure from the anomalous scattering of a few sulfur atoms demonstrated that the two-fold ambiguity of phase determination from Bijvoët pairs at a single wavelength can be largely overcome by using the prior localisation of the anomalous scatters[27] ; further, previously unused informations are available to solve the ambiguity without recourse to a prior localisation of anomalous scatters[28] (at least when anomalous scattering is due mostly to a single atomic species).

VI.2 - Anomalous scattering factors

The atomic factor of an atom is given by

$$f = f_o + f' + if''$$

The wavelength-dependent components of f, f' and f", are much smaller than the "normal" scattering f_o, except in the vicinity of an absorption edge where the amplitude and phase of the elastic scattering change substantially. f_o depends on $s = \sin\theta/\lambda$, whereas f' and f" are assumed to be independent of s, especially in the case of macromolecular crystallography where the range of accessible s is limited.

For a given energy E, f"(E) is proportional to $\mu(E).E$ where μ is the absorption coefficient of the element. f' and f" are related by a Kramers-Kronig relation[29].

Since the largest variations of f" and f' are observed close to an absorption edge and because the edge structure depends on the chemical environment of the anomalous scatter, it is necessary to measure experimental values of f" and f' in the very crystal used for diffraction measurements and with the same collimation. The usual procedure begins with the determination of f"(E) in the region of interest. This has been done in two ways : (i) the measurement of the X-ray fluorescence of the sample, which is proportional to µ(E), using possibly the area detector set at 2θ = 90° (ii) the measurement of the difference between the intensities of the mates of a selected Bijvoët pair ; in effect, this difference is proportional to f" (an absorption correction must be done).

Then f' is calculated by the Kramers-Kronig relation. f" and f' are merged with the calculated factors at points remote from the vicinity of the edge.

VI.3 - Multiwavelength diffraction data collection and reduction

Complete sets of data are collected with an electronic detector at each of the selected wavelengths. In order to reduce errors due to sample degradation and long-term drift of the detector, small portions of the reciprocal space are scanned at each wavelength. Then, scaling procedures are used to reduce systematic errors in differences between Bijvoët mates and equivalent reflections at different wavelengths.

VI.4 - Phase determination

Two approaches have been used for phase calculation :

(i) an algebraic analysis[30] is the basis of a least squares approach to solves simultaneously for the "normal" structure factor modulus from the anomalous scatterers $|^0F_A|$ and the total structure factor $|^0F|$ and for the phase difference between the total and anomalous scattering parts ($^0\phi - ^0\phi_A$). From the $|^0F_A|$, a Patterson map of the anomalous scatters is determined. Phases calculated from the resulting anomalous scatterer structure, $^0\phi_A$, then yield directly the phases for the total normal scattering of the crystal.

(ii) Two extensions of the Multiple Isomorphous Refinement (MIR) method have been developed at LURE. In these methods, prior to phase calculations, the anomalous scatters must be localized for instance using a Bijvoët or a dispersive difference Patterson. The first method[31] is derived from a published work[32]. In the second method, both the moduli and phases of the structure factors as treated as unknown parameters to be optimally determined[33]. In both cases, the "best" Fourier map is calculated.

VI.5 - Results

(i) The algebraic analysis has been applied to the structure of lamprey haemoglobin at 5.5 Å. Data sets at 4λ close to the Fe K edge were measured at SSRL (Stanford) with on area detector. An electron density map has been obtained, which compares favorably with the MIR map from an earlier study[34]. The extension to higher resolution is in progress.

The same approach was applied to avian pancreatic polypeptide, a small protein containing one zinc atom. Data were collected on films at DORIS (Hamburg). The zinc atom was localised, but the electron density map could not be interpreted[35].

(ii) The statistical approach has been applied to solve the structure of Opsanus Tau Parvalbumin at 2.3 Å[31] in which the two native Ca^{2+} ions were replaced by Tb^{3+} ions (one site is poorly substituted). 3 data sets were measured around the TbL_{III} absorption edge which exhibits a sharp resonance so-called "white line".

These results are encouraging, especially in view of the data collection which was in all cases not optimal, in terms of systematic errors and counting statistics. Simulations on molecules of increasing molecular weight suggest that the multi-λ methods will hopefully yield, directly, after reasonably precise and accurate measurements, phases of unprecedented accuracy.

REFERENCES

1 - a) H.D. BARTUNIK, R. FOURME and M.H.J. KOCH, Macromolecular Crystallography Using Synchrotron Radiation in : "Uses of Synchrotron Radiation in Biology" H.B. STUHRMANN, ed., Academic Press, London (1982)

1 - b) J.R HELLIWELL, Synchrotron x-radiation protein crystallography, Rep. Prog. Phys 47 : 1403 (1984)

2 - D.J. THOMPSON and M.W. POOLE eds., European Synchrotron Radiation Facility, Suppl. II, ESF, Strasbourg (1979)

3 - G.K. GREEN, Report BNL 50522, BNL Laboratory, Upton, NY 11973, USA, (1976)

4 - R. KAHN, R. FOURME, A. GADET, J. JANIN and D. ANDRE, Macromolecular crystallography with synchrotron radiation : Photographic data collection and polarization correction, J. Appl. Cryst. 15 : 330 (1982)

5 - G. BROWN, K. HALBACH, J. HARRIS and H. WINICK, Wiggler and undulator magnets - a review, Nucl. Instrum. Meth. 208 : 65 (1983)

6 - G.N. GREAVES, R. BENNET, P.J. DUKE, R. HOLT and V.P. SULLER, X-Ray optics and spectral brightness of the superconducting SRS wiggler Nucl. Instrum. Meth 208 : 139 (1983)

7 - B.W. BATTERMAN and L. BERMAN, Saggital focusing of synchrotron radiation, Nucl. Instrum. Meth. 208 : 327 (1983)

8 - U.W. ARNDT and A.J. WONACOTT eds., The rotation method in crystallography, North Holland, Amsterdam (1977)

9 - G.E. SCHULTZ and G. ROSENBAUM, The MWPC as an area detector for protein crystallography in comparison with photographic films : guidelines for future development of area detectors, Nucl. Instrum. Meth. 152 : 205 (1978)

10 - R. KAHN, R. FOURME, R. BOSSHARD, B. CAUDRON, J.C. SANTIARD and G. CHARPAK, an X-ray diffractometer for macromolecular crystallography based on a spherical drift chamber. Hardware, software and multiwavelength data acquisition with synchrotron radiation, Nucl. Instrum. Meth 201 : 203 (1982)

11 - R.P. PHIZACKERLEY, C.W. CORK, R.C. HAMLIN, C.P. NIELSEN, W. VERNON, N.H. XUONG and V. PEREZ MENDEZ, Progress report on the development of an area detector data acquisition system for X-ray crystallography and other X-ray diffraction experiments, Nucl. Instrum. Meth. 172:393 (1980)

12 - K.S. WILSON, E.A. STURA, D.L. WILD, R.J. TODD, D.I. STUART, Y.S. BABU, J.A. JENKINS, T.S. STANDING, L.N. JOHNSON, R. FOURME, R. KHAN, A. GADET, K.S. BARTELS and H.D. BARTUNIK, Macromolecular crystallography with synchrotron radiation II, results, J. Appl. Crystallogr. 16 : 28 (1983)

13 - O. KENNARD, W. CRUSE, J. NACHMAN, T. PRANGE, Z. SHAKKED, D. RABINOVITCH, Ordered water structure in an A-DNA octamer at 1.7 Å resolution, Nucl. Aci. Res. (in press)

14 - G. FERMI, M.F. PERUTZ, B. SHAANAN and R. FOURME, The crystal structure of human deoxyhaemoglobin, J. Mol. Biol. 175 : 159 (1984)

15 - E. WESTBROOK and P.B. SIGLER (private communication)

16 - R. USHA, J.E. JOHNSON, D. MORAS, J.C. THIERRY, R. FOURME and R. KAHN J. Appl. Crystallogr. 17 : 147 (1984)

17 - K. MOFFAT, D. BILDERBACK, W. SCHILDKAMP and K. VOLZ, Laue diffraction from biological molecules, Nucl. Instrum. Meth. (in press)

18 - M.G. ROSSMANN and J.R. HELLIWELL (private communication)

19 - J. HADJU, K.R. ACHARYA, D.I. STUART, T.J. McLAUGHLIN, D. BARFORD, H. KLEIN, L.N. JOHNSON, Time-resolved studies on catalysis in the crystal with glycogen phosphorylase b, (submitted)

20 - J. WALTER, W. STEIGEMANN, T.P. SINGH, H.D. BARTUNIK, W. BODE and R. HUBER, On the disordered activation domain in trypsinogen : chemical labelling and low temperature crystallography, Acta Crystallogr. B38 : 1462 (1982)

21 - H.D. BARTUNIK, Low temperature and time-resolved protein crystallography using synchrotron radiation, Nucl. Instrum. Meth. 208 : 523 (1983)

22 - H.D. BARTUNIK and C. BOULIN, DACOM : hardware data handling for stationary and time-resolved data acquisition with area detectors in "Structural Biological Applications of X-ray Absorption, Scattering and Diffraction", H.D. BARTUNIK and B. CHANCE eds., Academic Press, New York (1985)

23 - K. MOFFAT, D. SZEBENYI and D. BILDERBACK, Science 223 : 1423 (1984)

24 - B. HEDMAN, K.O. HODGSON, J.R. HELLIWELL, R. LIDDINGTON and M.Z. PAPIZ, Proc. Nat. Acad. Sci. USA (1985) (in press)

25 - P. MACHIN ed., Information Quarterly for Protein Crystallography, Daresbury Laboratory, Daresbury, Warrington WA4 4AD, England, 15 (1985)

26 - J. DOUCET and J.P. BENOIT, Rapport d'activité LURE 1984-1985, LURE, Bât. 209D, 91405 ORSAY Cedex, France (1985)

27 - W.A. HENDRICKSON and M.M. TEETER, Structure of the hydrophobic protein crambin determined directly form the anomalous scattering of sulphur, Nature 290 : 107 (1981)

28 - J. KARLE, Unique or essentially unique results from one-wavelength anomalous dispersion data, Acta crystallogr. A41 : 387 (1985)

29 - R.W. JAMES, The Optical Principles of the Diffraction of X-rays, Cornell University Press (1948)

30 - J. KARLE, Some developments in anomalous dispersion for the structural investigation of macromolecular systems in biology. Intl. J. Quant. Chem. 7 : 357 (1980)

31 - R. KAHN, R. FOURME, R. BOSSHARD, M. CHIADMI, J.L. RISLER, O. DIDEBERG and J.P. WERY, Crystal structure of Opsanus tau parvalbumin by multiwavelength anomalous diffraction, FEBS Lett. 179(1) : 133 (1985)

32 - J.C. PHILLIPS and K.O. HODGSON, The use of anomalous scattering effects to phase diffraction patterns from macromolecules, Acta Crystallogr. A36 : 856 (1980)

33 - R. FOURME, R. CHIADMI and R. KAHN, Rapport d'activité LURE 1984-85, LURE, Bât. 209D, 91405 ORSAY Cedex, France (1985)

34 - W.A. HENDRICKSON, J.L. SMITH, R.P. PHIZACKERLEY, E.A. MERRITT and W.E. LOVE, Anomalous dispersion crystallography of lamprey hemoglobin, SSRL Progress Report, SSRL, Stanford University, USA
(1984)

35 - T.L. BLUNDELL, I.D. GLOVER, J.E. PITTS, G.E. TAYLOR, I.J. TICKLE, W.G. TURNELL, S.P. WOOD, K. BARTELS and H.D. BARTUNIK, The determination of the position of the zinc ion in a small protein and the calculation of protein phases exploiting anomalous dispersion by using multiwavelength measurements with synchrotron radiation (in press)

DATA COLLECTION

J.R. Helliwell[+]

SERC, Daresbury Laboratory
Daresbury
Warrington WA4 4AD, U.K.

ABSTRACT

A survey is given of the data collection requirements for protein crystal structure analysis. The available hardware is reviewed including x-ray source and beam conditioning but specifically detector technology is discussed as the principal theme.

INTRODUCTION

The foundations of protein crystal structure analysis involve the collection of data. Initially, the search for derivatives and measurement of isomorphous series complemented by anomalous scattering data is needed for a medium resolution image of the structure. Subsequent interpretation is followed by a need for high resolution data in model refinement. Finally, it may be possible to collect very high resolution data to perform anisotropic temperature factor refinement. By this time binding studies of substrates or inhibitors are often of consuming interest. A desire may form to image <u>in real time</u> an active complex structure of a particular enzyme molecule.

The feasibility of these studies rests with technology and the sophistication of computer hardware and software. The collection of vast quantities of data has led to a need for automation and maximum efficiency of data collection. The armoury of equipment available includes automatic single crystal single counter diffractometers and screenless film camera techniques. Upon us now is the revolution brought by "electronic film" in area detector diffractometry and that brought by intense, tunable synchrotron radiation sources. The combination of electronic area detector systems with conventional x-ray sources is finding increasing use in the medium resolution stage of structure analysis leading to the first molecular model. Speed and reliability of such a system are paramount in such an application. However, weakly scattering material requires more intensity at the specimen and intense synchrotron radiation (SR) is a crucial need. The weak scattering is associated with large

---

[+]Present address: Department of Physics, University of York, Heslington, York YO1 5DD, U.K. and SERC, Daresbury Laboratory.

unit cells, large solvent content, small sample volume, disorder, severe radiation damage or combinations of all these. Additionally, high intensity is needed for the weak high resolution data. These are now traditional applications of intense SR. In addition, however, new fields are opening up where data sets can be collected so rapidly that it is feasible to probe a kinetic state. Both monochromatic and white beam techniques are being developed in this area. The tunability of white x-ray sources is used to aid crystal structure analyses where isomorphous derivatives cannot be made easily. Variable wavelength anomalous dispersion is a growing area. Fine collimation is essential to yield the necessary geometric conditions for large unit cell studies. However, this is important also in the largely undeveloped area of the measurement and use of thermal diffuse scattering (TDS) data. Unit cell correlated motions result in TDS peaks at the reciprocal lattice positions. These broad diffraction peaks introduce errors into the intensity measurements of the elastic Bragg peaks. The latter can be made so fine in angular width with SR that the Bragg peak is not contaminated by the TDS peak. Avoiding such errors will improve model refinements. More speculative is the possibility to use inter-Bragg peak TDS. In principle this data contains information on vibrations of the molecular motif via the molecular Fourier transform.

Data collection is a broad area in so much as the diffraction data required is varied and the technology available is increasingly diverse and sophisticated.

THE TASK AND SOME FACTS OF LIFE AS A DIFFRACTIONIST

The lack of a suitable lens for x-ray wavelengths means that the production of the image of a biological molecule at atomic or near atomic resolution requires measurement of the diffraction pattern. In so doing, the reciprocal space or Fourier transform of the molecular crystal is sampled. The intensity distribution can be surveyed but there is no knowledge of the phases of individual reflections which are scrambled both by x-ray source incoherence and the randomly (mis-)aligned mosaic block nature of a protein crystal. The phase problem is overcome indirectly by stimulating <u>intensity changes</u> for each reflection from which the phase can be derived either by multiple isomorphous heavy atom replacement (MIR) or anomalous dispersion (AD-wavelength dependent changes in phase and amplitude of these specific heavy atoms). Ultimately the static image of the molecule is computed via the Fourier summation of amplitudes and phases. We are all very familiar with this but let me illustrate in a graphical way the important constraints that protein crystal samples impose in terms of their scattering efficiency and also that the solution of the phase problem imposes in terms of accuracy of intensity measurements.

Scattering and Measurement Efficiency

A protein crystal is composed of atoms of low atomic number (C, N, O, H) and some sulphur atoms and in some classes of proteins, metal cofactors. In addition there is a high solvent content; extreme values are 10%-80%. On average, as a rule of thumb, for N photons incident onto the sample approximately $10^{-5} \times N$ are scattered to a resolution limit of ≈ 2.5 Å. We can call this factor of 10^{-5} a scattering efficiency factor. The larger the solvent content fewer photons appear in the spots and more in the background. Obviously, the intensity of the incident beam and the signal-to-noise characteristics of the detector are important quantities. Additionally, if the photon energy is varied the scattering efficiency changes ($\sim \lambda^2$), sample(s) absorption changes as $\sim \exp(-k_s \lambda^3 t_s)$, where

k_s is a constant of the crystal sample with $k_s\lambda^3$ the linear absorption coefficient (~ 1 mm^{-1} at 1.5 Å for a typical protein crystal), and the detector (D) quantum efficiency changes (also as ~ $\exp(-k_D\lambda^3 t_D)$. The quantities k_s and k_D show discontinuities at elemental absorption edges present in the material. The quantity t is the thickness of sample or detector.

The sample scattering efficiency as a function of all the relevant parameters is embodied in the Darwin formula for the integrated reflection intensity of a reflection from a mosaic crystal:-

$$\text{Intensity}_{hk\ell} \sim \frac{V_x}{V^2} |F_{hk\ell}|^2 L_\lambda P_\lambda \lambda^3$$

where V_x is the sample volume, V the unit cell volume, $|F_{hk\ell}|$ the structure amplitude for a given hkℓ, $L_\lambda P_\lambda$, Lorentz Polarization factor (vary with λ). Taking account of the extreme values of V_x, V and λ and solvent content means that the scattering efficiency can vary over <u>seven orders of magnitude</u>. It is not surprising therefore that there is a great interest in intense x-ray sources.

<u>Accuracy</u>

To obtain the image of a protein via crystallography I have already mentioned that indirect methods such as MIR and AD are required. I illustrate the scale of the problem using a model of the enzyme 6-phosphogluconate dehydrogenase (6-PGDH) whose structure has been determined (Adams et al., 1983) with three heavy atom derivatives [$K_2 Pt(CN)_4$, $K Au(CN)_2$ and $Pt(NH_3)_2Cl_2$]. There are, for example, two gold binding sites on the monomer. Hence, we use the changes in reflection intensities of these two gold sites in 50,000 Daltons of protein. It is simple to calculate that the expected average intensity change due to the addition of these atoms (assuming full occupancy) is ≈ 20%. One can clearly see that errors (random and systematic) must not exceed ≈ 5% on intensity (this implies an error in the differences in intensity of 10%) if we are to stand a chance of deriving accurate phases. The minimization of the random errors requires a "decent" number of counts (i.e. for 3% Poisson statistics, 1000 counts and for 1%, 10,000 counts) in the case of a single- or multi-channel counter detector. In the case of an analogue detector, the signal has to be digitised with an ADC (analogue-to-digital convertor) of sufficient bits (usually 8) and additional averaging (via multiple film exposures or frame averaging for the TV detector).

The wavelength dispersion effects of heavy atoms are smaller by a factor of ≈ 5-10 than the isomorphous intensity differences. Hence, more accurate data are needed, especially if attempts are being made to use these effects alone (and as a function of wavelength) for phase determination if, for instance, no isomorphous derivative can be prepared. The use of photographic film in these instances is limited somewhat compared with an electronic area detector. The exceptions to this rule of thumb are where Friedel (f" based) intensity differences can be measured on the same film from a well-aligned crystal in a favourable space group. This is especially true if the f" has been wavelength optimised and where the absorption edge is at a short wavelength (e.g. $\lambda \lesssim 1$ Å) so that the protein crystal absorption surface is flat (i.e. good signal and low systematic error). Fortunately, the common heavy atom derivatives in protein crystallography (e.g. Au, Pt, U) satisfy these conditions simultaneously. Interestingly, in the Laue method (section 2.3) the short wavelengths (< 1 Å) also generate the highest percentage of singlet reflections (for a given θ limit).

There are also systematic errors, which affect individual reflection measurements and may appear as part of a measured difference. Common systematic errors include sample absorption errors and thermal diffuse scattering contributions to a reflection intensity. These can be eliminated by careful measurement. The sample absorption surface needs to be measured especially for asymmetric crystal shapes (Kopfmann and Huber, 1968) and/or use of long wavelengths (Helliwell et al., 1984); this measurement is relatively easy to make and can be done either after the crystal has been totally damaged by irradiation in data collection by monitoring the transmission of a collimated direct beam with $CuK\alpha$ (Bartels, 1977) or SR (Helliwell et al., 1985) or while the crystal still diffracts (North, Phillips and Matthews, 1968). Thermal diffuse scattering corrections are very rarely if ever applied to the Bragg intensities. This could become a very regular feature of data reduction programs for electronic area detectors. Even with film there is a considerable amount of information available for a TDS correction to be applied.

Measurement of sample absorption is the exception rather than the rule and TDS is almost never measured. Instead it is the norm to rely on measuring, for instance, isomorphous derivative series under as identical conditions as possible so that systematic errors subtract out or for local scaling of data to smooth out slowly varying error trends. Nevertheless, the lack of absorption correction can cause an error in the magnitude of an isomorphous difference (even if the sign is correct) of up to a factor of 10! Experimentally, the clear solution to this is the use of short wavelengths (e.g. < 1.0 Å) which flattens an absorption surface to within a few per cent and at $\lambda = 0.6$ Å completely (Helliwell et al., 1984); this is now routinely feasible with the advent of SR wiggler-based data collection instruments (Helliwell et al., 1984, 1985; Phizackerley et al., 1985; Sato et al., 1985) which can offer a high enough intensity to compensate for the worse sample scattering efficiency and detector quantum efficiency at short wavelengths.

Finally, another rule of thumb. Though it is common in the field to measure only a few (say two to four, or even only one) equivalent reflections, similar counting statistics can be obtained from the composite measurement of many equivalents (say 10) as can be had by counting for a long time for one equivalent but with the added advantage of averaging out systematic errors.

Surveying Reciprocal Space in Quantitative Crystallography: Monochromatic and White Beam Approaches

Traditionally quantitative x-ray crystal structure analysis is based on the monochromatic beam rotating crystal method. Integration of reflection intensities is done over an angle known as the "rocking width" of each reflection. The value of the rocking width is determined by the intrinsic mosaic of the crystal, beam crossfire and spectral spread ($\delta\lambda/\lambda$) (Greenhough and Helliwell, 1982a, b). As the sample is rotated the leading and trailing edges of the Ewald sphere (of essentially single radius) sweep through reciprocal space. The optimum signal-to-noise ratio for a given reflection obviously occurs if the detector collects only counts during the angular rotation in the particular reflection rocking width.

Polychromatic techniques are finding increasing application at SR sources. In the Laue method a wide range of wavelengths fall on the sample. The sample and detector are held stationary and integration of reflection intensities is done over wavelength. The advantages of this method over the monochromatic method lie (a) in the very high intensity available in the white beam from a storage ring accelerator and the concomitant short exposure times and (b) in the case of a broad bandpass (0.2 Å ≤ λ ≤ 2.5 Å) a large sweep of reciprocal space is achieved in a single sample orientation; it can be shown that the "equivalent rotation in the monochromatic case" is about 15° of sample rotation; in favourable cases of high space group symmetry and a careful choice of beam direction with respect to the crystal axes a single "snapshot" can yield the bulk of the independent data. The principal disadvantage of the method is that a single reflection with an intrinsic rocking width of 0.2° accumulates ≈ 15° of background hence the signal- to-noise ratio is not as good as for the monochromatic method. However, the Laue patterns of protein crystals recorded on the SRS do show very good signal-to-noise characteristics. Why should this be? The large rotation background mentioned is offset by the fact that the beam size is extremely well-collimated down to ≈ 0.2 mm. This is necessary to avoid spot spatial overlap and as a result very little extraneous matter (mother liquor and glass capillary) scatter the beam. It is possible to preserve very short exposure times (i.e. 0.25 s, 20,000 reflections on the SRS wiggler protein crystallography work station) even with such a small beam spot (and associated sample volume) because of the very high intensity ($\approx 10^{14}$ polychromatic photons s^{-1} mm^{-2}) in the beam compared with say the monochromatic focused wiggler beam of $\approx 10^{12}$ photons s^{-1} mm^{-2} or a rotating anode of $\approx 10^{9}$ CuKα photons s^{-1} mm^{-2}.

Why have Laue methods of surveying reciprocal space not been developed sooner? The power in the beam is high as one can illustrate vividly if one places a highly absorbing sample such as lead in the SRS wiggler beam; it melts. Unfortunately, early experiments with the white beam on the NINA synchrotron with crystals of the enzyme 6-phosphogluconate dehydrogenase were totally unsuccessful (Bordas and Helliwell, 1976 (unpublished); Greenhough and Helliwell, 1982). Recently, Moffat et al. (1984) used the so-called modified Laue ($\delta\lambda/\lambda \lesssim 0.2$) with an SR beam from CHESS with successful results from myoglobin and haemoglobin crytals as a preliminary to laser photolysis studies of carbon monoxide debinding. Building on this work especially, Helliwell (1984) exposed a pea lectin crystal in the full (0.2 Å ≤ λ ≤ 2.5 Å) SRS wiggler white beam with beautiful results (see also Clifton et al.,1985). It seems to be the case that the more robust protein crystals survive several (3-10) exposures (define robust as a property of relatively good radiation lifetime in the monochromatic beam) whereas less robust crystals withstand a white beam of more limited wavelength range (modified Laue).

The interpretation of Laue patterns has required the development of new software for prediction, spot integration, film-to-film scale factor determination, harmonic spot unscrambling and wavelength normalization. The results on the pea lectin Laue data used as a test are very promising and compare favourably with monochromatic oscillation film data. Hence, data of useful statistical quality can be collected in very short times. This is of potential use for the study of protein crystal transient states. The technique is being actively developed at Cornell on CHESS and at Daresbury on the SRS.

THE COMPONENT PARTS OF OUR EXPERIMENT

These can be briefly (!) stated as follows:

(a) <u>The x-ray source</u>. (Sealed tube, rotating anode, synchrotron bending magnet, wiggler, undulator).
(b) Beam conditioning materials (<u>beam optics</u>) to define <u>wavelength spread</u> onto sample i.e. monochromator (Bragg reflection from perfect crystal (Ge,Si) $\delta\lambda/\lambda \approx 10^{-4}$ or mosaic crystal such as graphite

$\delta\lambda/\lambda \approx 10^{-2}$), define <u>wavelength cut off</u> (critical reflection from a mirror gives a sharp λ_{min}, transmission through a thin mirror such as a soap film gives a sharp λ_{max}, an absorbing foil gives a change in a "soft" way to λ_{max}) and finally, define the <u>crossfire in the beam</u> (with slits). The <u>polarisation state</u> of the beam can also be controlled via the monochromator and choice of source.

For a discussion of (a) and (b) see Helliwell (1984).

(c) <u>The sample environment</u> to define the temperature state of the sample with a <u>cooling cell</u>, (Bartunik and Schubert, 1982; Hajdu et al., 1985), the state or constituent of the mother liquor with a flow cell (Wycoff et al., 1967). <u>Pressure cells</u> have not found use in protein crystallography owing to the fragility of most protein crystals.

(d) <u>The sample holder</u>: the need for the protein crystal to be bathed in mother liquor is long established (Crowfoot and Bernal, 1934). Traditionally a <u>glass capillary</u> sealed with wax was used. Recent work by Mahendrasingam et al. (1985) at longer wavelengths (2.6 Å at present) has led to the use of <u>thin film capillaries</u> (e.g. Hostaphan, Hoechst Ltd.) since the thinnest glass capillaries available absorb a $\lambda = 2.6$ Å beam by a factor of 20. These capillaries will give important benefits in microcrystal data collection.

(e) <u>The detector</u>: single channel detector (scintillation counter on a diffractometer); film; imaging plate; MWPC; TV detector; CCD.

I will now discuss in more detail the subject of detectors.

DETECTORS

As an area-sensitive detector, film has been used extensively for quantitative x-ray crystallography. The oscillation camera has been, until recently, the standard for larger unit cell protein crystal (> 100 Å) data collection on conventional and SR sources. The trend is now for electronic area detectors to replace film to provide real time monitoring and data acquisition as well as greater accuracy. With the increasing availability of electronic area detectors on conventional sources the speed of film data collection at a SR source will not be so important and special, but important, problems will occupy the SR facilities. These special problems were listed in the introduction, but briefly, will involve studies of weakly scattering crystals; measurement of weak high resolution data; measurement of data where time is a crucial factor, e.g. if radiation damage is dose rate dependent (such as virus crystallography) or for enzyme kinetic or laser photolysis experiments; for wavelength-optimised anomalous dispersion where isomorphous derivatives cannot be made easily. Put another way, the technical basis for access to SR sources is changing. Film will not be made completely obsolete by electronic area detectors even if it is only a case of, at present, always being able to think of experimental situations which are beyond the technical specification of currently available technology. One instance of this is in the SR Laue diffraction. The ideal electronic area detector for SR Laue work would have the properties listed in Table 1, a specification which is beyond present technology.

Of course, the main disadvantage of film in SR Laue crystallography is the lack of real time monitoring via the diffraction pattern (e.g. of substrate binding to the enzyme in the flow cell mother liquor), yet many of the characteristics listed in Table 1 are present with film.

Table 1.

Active area	100×100 mm^2
Pixel size	25 μm $\times 25$ μm
(Number of pixels	4000×4000)
Whole area count rate	10^9-10^{10} cps
Local area peak count rate	10^5-10^6 cps (in $(200$ μm$)^2$)
Wavelength range of sensitivity	0.5 Å – 2.5 Å
Detective quantum efficiency in this wavelength range	100%
Dynamic range	12 bits (10 bits minimum)
Readout time for the data:	
for enzyme crystal kinetics	slow e.g. seconds to minutes
for laser photolysis	msec

Another major limitation of film is the dynamic range. The use of new photo-stimulable imaging plates offers to overcome this restriction (Amemiya, 1985).

Electronic Area Detectors (EAD's)

There are several very important advantages of electronic area detectors over photographic film. Firstly, the diffraction pattern is brought on line. This is essential to achieve optimum signal-to-noise of the reflection intensity. Since the rocking width, w, of a given hkl reflection from a protein crystal is quite narrow ($\sim 0.3°$) on a conventional source and very narrow on a synchrotron radiation source with a well-collimated beam ($\sim 0.005°$-$0.05°$!) it is possible with an EAD to predict and collect the peak with very little background (x-ray and any detector) noise superimposed. This should be contrasted with the case of the oscillation film method where a relatively large angular range $\Delta\psi$ is chosen (say 1°) to reduce the total number of film packs needed to survey the region of reciprocal space required. On a conventional source $\Delta\psi$ is $\gtrsim 3$ w in the oscillation method whereas on an SR source $\Delta\psi$ is $\gtrsim 20$ w. Hence, it is particularly true with an SR source that crystal sample rotations per 'exposure' of 0.05° and less will give major enhancements in data quality. Furthermore, an analogue detector like the TV detector is also best used, in terms of signal-to-noise (S/N), by reading out the reflection intensity profile with a fine angular sampling to avoid the build up of detector noise that would occur over a wide angular scan. Additionally, one can see why collimation of the beam for a TV detector on a conventional source is important to reduce the reflection rocking width, w, to reduce not only x-ray background but also detector background noise. Such a concern is not so critical with a counting detector such as an MWPC (multiwire proportional chamber) with a very low intrinsic noise.

The second advantage of EAD's over film is that they are more sensitive than film. This comes about in several ways. Firstly, the absorption efficiency of film (e.g. CEA reflex 25) at 1.5 Å wavelength is 60%, i.e. the film factor is 2.5, whereas the equivalent factor for 1 atmosphere of xenon with a gap of 12 mm is 80% dropping to ≈ 3% at 0.5 Å and a solid state detector with a 10 μm thick layer of Gd_2O_2S, gadolinium oxysulphide, of 10 mg cm^{-2} has an efficiency of ~ 100% dropping to 25% at a wavelength of 0.5 Å (the latter increases to ≈ 80% if 50 mg cm^{-2} is used). At a wavelength of 0.5 Å, CEA Reflex 25 has an efficiency of 10%.

The third advantage of a photon counting EAD (or an analogue EAD which can be shown to behave as a Poissonian counter (Arndt and Gilmore, 1979) is that it requires fewer photons to achieve a given set of statistics. For example, for counting statistics

Peak to background = $N/\sqrt{N}$

thus, for a 3% error, $0.03 = 1/\sqrt{N}$ gives $N \approx 900$.

With film 1 O.D. blackness (which is considered typical for say 5% statistics overall) requires 10^6 photons per mm^2 (Morimoto and Uyeda, 1962) at $CuK\alpha$. With photon counting, 10^6 photons would give 0.1% statistics. However, this is an unfair comparison because in practice on film the spots are normally smaller than 1 mm; e.g. 0.3 mm for a typical protein, hence 1 O.D. is produced by $9 \cdot 10^4$ photons. For a virus the spots are deliberately reduced to avoid spatial overlap to ≈ 0.15 mm, i.e. 1 O.D. $\equiv 2.25 \times 10^4$ photons. Hence, in such cases as virus crystallography, film is beginning to approach the performance of an ideal counter. Another area where this occurs is microcrystal (e.g. 20 μm) diffraction; for (20 $μm^2$) spots 1 O.D. is produced by 400 photons. Unfortunately microcrystal data collection with film is beset with several problems, (a) a diffraction spot of $\approx$ 20 μm is approaching the grain size so that this shows up as a non-uniformity of the emulsion where the "film factor" would vary quite rapidly from spot to spot, (b) even if this were ignored a very fine scanning raster would be needed to avoid Wooster effects and (c) a 10 μm raster for a single 100 × 100 mm piece of film would generate an array of 100 Mbytes. Microcrystal diffraction which is very feasible with strong beams at an SR source is best tackled with an electronic detector instead of film.

Finally, there are additional benefits of an EAD, since the pattern is on line, sample assessment is very quick and convenient and so is sample alignment which minimises the dose of the specimen prior to data collection. Of course, the additional benefit of no densitometry represents a great saving of time and effort for workers in the field.

Having reitereated all these benefits, well discussed before, as you know, there is then the question of the choice of detector system. This question has to be considered in terms of a whole system which, in addition to the detector, includes electronics, computers, storage media, prediction and processing algorithms and software. What are the system performances now, how might they improve over the next five years and what are our needs?

Needs
‾‾‾‾‾

We have discussed earlier that accuracy is important in the range of reproducibility of 1-5% (on intensity); this sets limits on, for instance, the stability of the detector over long periods of time.

Additionally, the reflection measuring rate in some experiments can be essential. Firstly, some enzyme crystal systems and many virus crystals suffer badly from radiation damage even to the extent of preventing high resolution data collection unless a very high reflection measuring rate is obtained by using a very high incident intensity. For instance, in the study of purine nucleoside phosphorylase reflection measuring rates at the SRS to allow measurement of the 3 Å data were up to 100 reflections per second (Ealick et al., 1985). In the study of rhino virus at CHESS by Rossmann and coworkers, reflection measuring rates of 250 reflections per second are typical. This dose rate dependence of radiation damage does not apply to all virus crystal studies. Obviously, some have relatively stable diffraction characteristics and so data can be measured over a long period of time with a weak source, provided of course the diffraction spots are resolvable spatially one from the next.

In the same way that I have outlined a specification for a future detector for Laue work the requirements for an EAD for SR virus crystallography are essentially the properties of film with a high reflection read out rate. The benefits of such an EAD would be higher accuracy of the data than that obtained with film. Additionally, the extra sensitivity of an EAD over film (see earlier section) means that for equivalent statistics 1/20 of the incident intensity is required to produce the necessary 250 reflections per second to by-pass time-dependent radiation damage. This output rate of data is far beyond the capability of current systems (Xuong et al., 1985). It is a good question, however, as to how a rhinovirus crystal, for example, would behave in a beam of 1/200th of the full incident intensities used so far, giving a measuring rate of 25 reflections per second. A rate which is within sight of current EAD systems.

A second area where measuring rate is all important is that of kinetic crystallography. In this case a transient state of the molecule is of interest. (In the above the time-dependent properties of the crystal were the concern.) For example, in a structural study of the phosphorylase b mechanism by Johnson et al. at the Daresbury SRS, using a flow cell and some cooling, data sets are collected as rapidly as possible to study the stages of binding and release of substrate and product. On the SRS wiggler protein crystallography station with the monochromatic beam a data set of $\approx 10^5$ reflections to 2.8 Å resolution can be collected in ≈ 30 min of which 10 min is the actual exposure time of the 32 film packs and the other 20 min is the time needed to reload the oscillation camera carousel after every eight exposures and to translate the sample to a fresh piece of crystal. Each entry into the instrument hutch has an overhead of approximately 5 min. Obviously a carousel designed to hold 32 cassettes and a motorised sample translation would allow the 10^5 reflections involved to be collected in a total time only slightly greater than the 10 min exposure (each cassette would need ≈ 5 s to move into position). In any case, the measuring rate during exposure is ≈ 170 reflections per second. With this arrangement it has been possible to collect 10 different data sets with different soak times via the flow cell to trap the transient states of the enzyme active complex. Effectively a 10-frame cartoon of the structural reaction has been produced.

However, in order to probe the time sequence further and to examine faster substrates the use of the focussed white beam in Laue diffraction data collection is being investigated. It is feasible to record in a single snapshot with an exposure time of 0.25-0.5 s on the SRS wiggler approximately 30,000 reflection intensities (i.e. $\sim 10^5$ per second). High quality diffraction patterns have been obtained. This data is being processed using programs developed and used in the pea lectin Laue statistical tests.

It is worth pointing out that the study of enzymatic states by crystallography had previously been possible in only a few cases. The use of low intensity conventional source x-ray beams meant that, even with electronic area detectors 10^5 reflections would require ≈ 10 hours to collect (Xuong et al., 1985). To reduce the enzyme turnover rate sufficiently to prolong the lifetime of the transition state required very low temperatures (e.g. $-100°C$) and the change of mother liquor to a non-freezing liquid. Unfortunately, promising though the area of "cryo-enzymology" was, its application was limited to very few cases since it is usually difficult to replace the mother liquor without disrupting the crystal. With these very intense monochromatic beams only modest (> 0°C) cooling is required for many enzymes to be studied kinetically in this way. The use of SR Laue techniques offers the prospect in this area of studying enzymes with faster turnover but still for temperatures not below 0°C.

In the area of kinetic crystallography, Moffat (1985) has discussed possible time resolved experiments including enzyme-substrate kinetics, photo-activation of an inert substrate and thermally induced reversible unfolding and refolding of proteins. The study of enzyme-substrate states was dismissed by Moffat (1985) stating that diffusion is slower than the chemical turnover. If the phosphorylase b experiment is considered, however, a suitable choice of temperature (> 0°C) and substrate varied the reaction half-life from tens of minutes to hours. The use of Laue diffraction or fast monochromatic data collection in the study of enzyme-substrate structural states seems very appropriate.

Moffat (1985), on CHESS, has succeeded in recording strong diffraction patterns of lysozyme in 4 msec in a temperature jump experiment. Bartunuk (1981) has studied the time course of approximately 30 reflections of myoglobin with a msec time resolution on a linear detector. The extension of this to study the time course of all reflections via a 2D chamber is under development (Bartunik, 1985). Such a system based on parallel processing architecture could give reflection measuring rates 1-2 order of magnitude more than is currently feasible ($\sim$ 1 to ten reflections per second).

Clearly, reflection measuring rates in the range 10^2-10^5 with monochromatic or white beams are a definite need in a whole variety of studies either to overcome radiation damage or to capture short lived structural states.

Electronic Area Detector Facilities at Daresbury

I have now established, as a background, the needs, possibilities and limitations of electronic area detectors for crystallography and indeed of film. The accompanying papers in this session detail the specific technology and configurations, based primarily around MWPC and TV detector technology. I will concentrate on detailing the background to the choices I have made for the work with synchrotron radiation from the SRS in protein crystallography and where necessary giving details of the Daresbury systems.

The Daresbury MWPC system: system choice and comparison of some options

The physical basis of photon detection and event position encoding of MWPC and the electronic configurations for converting the time of an event and/or amplifying signals is given in Helliwell et al. (1981) for the Daresbury MWPC. A major feature of MWPC's is the need for a reasonably good efficiency; 1 atm of xenon with a 12 mm depth has an 80% probability of absorbing a CuKα photon. With an oblique angle of incidence (2θ say for normal beam geometry) of a diffracted beam striking the chamber a considerable lengthening of a diffraction spot occurs $\sim$ g cot2θ where g is the gap between cathode planes). With a flat chamber the simple trick of inclining the detector towards the sample by an angle θ and positioning it above the direct beam reduces the parallax broadening from g cot2θ to g cotθ with normal incidence (and no streaking) at the centre of the detector and for the median resolution.

There are two technical solutions of the parallax broadening of the 1 atm flat chamber. Firstly, the gas can be pressurised thus reducing the gap appropriately; the effect of this is to reduce the active area for the same Be window thickness or support it with a grid or to increase the thickness; this does not matter too much for CuKα work but is a serious problem for work at longer wavelengths. Secondly, the flat chamber can be fronted by a drift chamber with radial field lines which are designed to bring the charge induced by the photon absorption normally incident onto the rear flat chamber; such a drift space has to be long enough to guarantee 100% conversion of the photons before striking the MWPC otherwise severe parallax broadening will occur with a gap g which is the sum of the drift space and the MWPC gap (the effect of this is to limit the minimum operating wavelength of the LURE system to ≈ 1.07 Å (Kahn, personal communication).

At Daresbury, the bending magnet station of the SRS for x-ray diffraction serves protein crystallography, small angle scattering and fibre diffraction. A MWPC design specified in table 2 was decided on for this station (Helliwell et al., 1981). This was based on the flat chamber configuration with 1 atm gas and no drift space. It was considered a conservative design except in two respects. Firstly, an anode wire to wire spacing of 1 mm was ambitious and was decided on to benefit the low angle scattering where parallax is not a problem due to the near-normal incidence of the pattern striking the chamber and hence where the major resolution component was the wire to wire spacing (in one direction). Secondly, a sealed chamber was used for convenience and was based on the advice that a similar sealed chamber had been used for many years without gas change for medical imaging of 40 keV photons with count rates of 300 kHz.

Table 2. Design Specification of the Daresbury MWPC System (from Helliwell et al., 1981)

(a) Detector (Rutherford Laboratory)

Active area	200×200 mm^2
Window material	Be; 1 mm thick
Cathode-anode-cathode gaps	6 mm : 6 mm
Anode wire pitch	1 mm
Cathode wire pitch	1 mm
Active gas	70% Xe, 30% isobutane at 1 atm pressure
Expected operational wavelength range	1.28 Å $< \lambda < 2$ Å
Sealed chamber[a]	

(b) Readout electronics
Pre-amplifiers
Constant fraction discriminators
Lecroy time to digital convertors

Daresbury histogramming memory 16 bits deep[b], 1 width of CAMAC[c] 64 K, expandable to 4 Mbytes, Mean rate 2 MHz, peak rate 10 MHz with FIFO buffer.

[a] Now gas flow, [b] Now 24 bits, [c] Now 1 width is 256 K.

The wavelength sensitivity of the chamber between 1.3 and 2 Å is very good, the aperture of the chamber of 20×20 cm and the pixel size $\sim 1 \times 0.5$ mm is well matched to the requirements of protein crystal diffraction in this wavelength range. The count rate capacity of the chamber of a maximum of 300 KHz though modest by comparison with what is achievable by amplifier per wire devices ($\sim$ 1 MHz) is actually well matched to the use of a well-collimated, fine spectral resolution

($\delta\lambda \geqslant 3 \times 10^{-4}$) mode on the SRS protein crystallography station on the bending magnet. The development of a 16 bit large capacity CAMAC memory (4 MWds max) at Daresbury was done to allow use of time slicing techniques essentially for the non-crystalline biological diffraction and for the protein crystallography to allow rapid dumping of whole images for a reasonable rotation range of the sample. Hence 32 frames of 256 × 256 × 16 bits rapid local storage was feasible. This memory, developed in Daresbury, has been used in EMBL Hamburg and a new 24 bit deep version has been used at LURE. Both are now marketed commercially by Nuclear Enterprises.

What about the performance of this system? The system has been used primarily so far for the 2D recording in real time of fibre diffraction data, for example various states of natural and synthetic DNA fibres as a function of various parameters (relative humidity and salt). For protein crystallography, test data were collected from the detector on the protein crystal pea lectin and zinc:avian pancreatic polypeptide (APP). The pea lectin crystals are used at Daresbury for detector and methods evaluation because (a) they are very stable in the SR beam, (b) they contain two natural anomalous scatterers, Mn and Ca at two sites in the asymmetric unit, which contains 50,000 Daltons of protein and (c) the unit cell parameters are markedly asymmetric (~ 50 Å × 60 Å × 130 Å, $P2_12_12_1$) so that simply by rotating the crystal about the a or b axis very different spot reflection density patterns are produced. The Zinc:APP is used to test the MWPC because (a) the zinc edge (1.28 Å) is the shortest wavelength accessible absorption edge for which the bending magnet station provides a reasonable flux of photons and (b) the zinc concentration in this case is 1 zinc in 6000 Daltons, unlike the pea lectin case of low concentration Mn and Ca.

The whole area count rate can be monitored as a function of wavelength at the zinc absorption edge from a single crystal of zinc:APP which revealed that the chamber could be used effectively as a fluorescence detector for setting the wavelength in anomalous dispersion experiments. Several problems were encountered with the chamber on the beamline. Firstly, the chamber repeatedly 'tripped' off the anode high voltage due to a high leakage current on the anode wire and electrical breakdown; the effect of this was to lose completely the chamber sensitivity. Secondly, one area of the chamber was particularly prone to tripping and showed a definite lack of sensitivity in that region (indicated by the absence of recorded background photons). The combined effect of these problems was to prevent the recording of a non-uniformity of response over the pixels of the chamber within one continuous period without tripping i.e. it was not possible to show that the uniformity was constant with time.

The stability of the chamber has now been investigated off beam line. Several changes have been made which have improved the reliability enormously, on a radioactive source. The changes made so far are (a) the end few wires on the edge of the 20 cm × 20 cm area have been shielded from irradiation - the chamber area has been reduced to ≈ 195 × 195 mm² and (b) to allow a gas flow a hole has been drilled in the side of the chamber and a valve fitted.

There is a major difference in the character of patterns produced with SR compared with a conventional source in that the diffraction spots are exceedingly well collimated and hence the local count rate can be much higher. This has probably exacerbated the problems due to the close spacing of the anode wires. It may yet be necessary to relax the anode wire to wire spacing from 1 mm to 2 mm; this will allow a lower anode voltage to be used, reduce the leakage current and the tripping problem and also carbon contamination of the wires.

The Enraf-Nonius FAST system on the SRS Wiggler Protein Crystallography Station

This station is fed by x-radiation from a 5 T single bump (3 pole) wiggler magnet with a critical wavelength of 1 Å which gives a strong flux of photons down to a wavelength of ≈ 0.1 Å for the SRS at 2 GeV. The optical design and the various modes of operation have been described by Helliwell et al. (1984) and Helliwell et al. (1985). Briefly, these include (a) point focused monochromatic tunable beams with wavelengths selected in the range 0.5 Å $< \lambda <$ 2 Å with maximum intensity 10^{12} photons s^{-1} mm^{-2}, (b) monochromatic rapidly tunable beams over the same wavelength range with maximum intensity 10^{10}-10^{11} and (c) line focused white beam with 10^{14} polychromatic photons s^{-1} mm^{-2}. The experiments and data collection options that become feasible on this line include very rapid (sub-second) enzyme crystal kinetics, use of microcrystal and very radiation labile virus crystallography. Additionally, and very importantly, the wavelength range available now encompasses the absorption edges of the usual heavy atom derivatives of protein crystallography such as U, Hg, Pt, Au as well as the biologically important molybdenum K-edge. Hence, this makes feasible the use of multi-wavelength anomalous dispersion on derivatives and renders usable those that are non-isomorphous.

The utilisation of these absorption edges for anomalous dispersion phasing requires accurate data. The use of microcrystals with photographic film has enormous problems in the densitometry stages and even if overcome the data quality is not very good. Both these applications produce modest count rates at the detector, the former due to narrow collimation and $\delta\lambda/\lambda$ and the latter due to sample volume.

An area detector is therefore required at short wavelengths which has (a) a high absorption efficiency (for the anomalous dispersion work), and (b) the correct geometric match of the Bragg cone solid angle with the detector aperture and pixel size. The need for accurate statistics implies a need for good stability. These criteria can be met by a solid state detector system with an appropriate phosphor.

We are customers of the Enraf-Nonius FAST system, modified in some ways, which include mounting of the diffractometer on its side to reduce the effect of the linear polarisation of the beam. The diffractometer is mounted on the beam line but with the important option of an x-ray tube Cu$k\alpha$ source being available. To date the following has been achieved. Firstly, statistical tests have been made on pea lectin crystals using the intense beam in so-called "camera mode" which is the case of dumping a whole image without predictions. It is clearly necessary to use a fine angular range (0.1°) to avoid the effect of detector noise piling up especially at the weak reflections. Unfortunately without on-line prediction this mode of data collection is very slow and virtually impractical. Secondly, the resolution characteristics of the detector have been tested; the point spread factor is $\approx$ 8 pixels for a contour drawn at 2% of the peak value. At any one crystal setting the maximum resolvable number of orders is therefore $\approx$ 60 × 60 maximum. The detector can be placed at distances from the sample of 30 to 1000 mm. In the case of ribosome crystals (unit cell maximum, 900 Å) the resolution limit is $\approx$ 15 Å. Hence for a detector distance setting of 500 mm (with helium tube) the full diffraction resolution range of ∞ to 15 Å can be collected at one setting and the 900 Å axis resolved. The beam is well collimated (< 1 mrad) for this experiment and a fine collimator was used. Finally, microcrystal tests have been successfully carried out in that the small number of overall counts in a diffraction spot produces a measurable diffraction peak on the diffractometer.

An important parameter to give is the current performance of the system hardware (detector and computer) and software. On the basis of the concept of processing an image and recording a new image and taking account of the overheads of transferring a 256 kbyte image in the system approximately 0.25 reflection/sec can be measured with acceptable signal-to-noise, the major limitation being a slow transfer rate of $\approx$ 20 seconds from the acquisition memory to the computer CPU. A faster data bus or on line prediction and processing of reflection intensities would be much quicker and limited by the speed of prediction. The latter procedure is much more amenable to hi-tech computer hardware such as a vector or array processor to yield large blocks of reflections with computed x,y and sample rotation angle coordinates. In such a case in excess of 100 reflections per second may be realised. The fast rotation rate or large unit cell of the sample that this implies, would be feasible only if sufficient statistics could be accumulated in a short time - a high intensity source is also necessary to support high reflection measuring rate.

ACKNOWLEDGEMENTS

The SERC and the University of Keele are thanked for support. The Director and staff of the SERC Daresbury Laboratory are thanked for the provision of synchrotron radiation and facilities at the SRS. I am very grateful to many colleagues and collaborators at Daresbury, in the SRS User Community and from other SR facilities around the world for useful discussions.

REFERENCES

ADAMS, M.J., ARCHIBALD, I.G., BUGG, C.E., CARNE, A., GOVER, S., HELLIWELL, J.R., PICKERSGILL, R.W. & WHITE, S.W., 1983, EMBO, J. 2 : pp1009-1014

AMEMIYA, Y., 1985, paper presented at the Stanford Synchrotron Radiation Instrumentation meeting.

ARNDT, U.W. & GILMORE, D.J., 1979, J. Appl. Cryst. 12: 1.

BARTELS, K.S., 1977, in "The rotation method" edited by U.W. ARNDT & A.J. WONACOTT, North Holland.

BARTUNIK, H.D. & SCHUBERT, P., 1982, J. Appl. Cryst. 15: 227-231.

CROWFOOT, D. & BERNAL, J.D. 1934 Nature.

GREENHOUGH, T.J. & HELLIWELL, J.R., 1982a, J. Appl. Cryst. 15: 338-351.

GREENHOUGH, T.J. & HELLIWELL, J.R., 1982b, J. Appl. Cryst. 15: 493-508.

GREENHOUGH, T.J. & HELLIWELL, J.R., 1983, Prog. in Biophys. and Molec. Biology 41: 67-123.

HAJDU, J., McCLAUGHLIN, P., HELLIWELL, J.R., THOMPSON, A.W. & SHELDON, J., 1985, J. Appl. Cryst. in press.

HELLIWELL, J.R., HUGHES, G., PRZYBYLSKI, M.M., RIDLEY, P.A., SUMNER, I., BATEMAN, J.E., CONNOLLY J.F. & STEPHENSON, R., 1981, Nucl. Instrum. and Meth. 201: 175-180.

HELLIWELL, J.R., 1984, Reports on Progress in Physics 47: 1403-1497.

HELLIWELL, J.R., MOORE, P.R., PAPIZ, M.Z. & SMITH, J.M.A. 1984a J. Appl. Cryst. 17: 417-419.

HELLIWELL, J.R., CRUICKSHANK, D.W.J., ELLIS, G.H., HABASH, J., PAPIZ, M.Z., RULE, S. & SMITH, J.M.A., 1984b, Daresbury Study Weekend Proceedings DL/SCI/R22, p.41-59.

HELLIWELL, J.R., PAPIZ, M.Z., GLOVER, I.D., HABASH, J., THOMPSON, A.W., MOORE, P.R. CROFT, D. & PANTOS, E., 1985, Nucl. Instrum. and Methods, in press.
JOHNSON, L.N., HAJDU, J., STUART, D. & coworkers, 1985, to be published.
KOPFMANN, G. & HUBER, R., 1968, Acta Cryst. A24: 348.
MAHENDRASINGAM, A., SOWERBY, A., & HELLIWELL, J.R., 1985, unpublished results.
MOFFAT, K., SZEBENYI, D. & BILDERBACK, D., 1984, Science 223: 1423.
MOFFAT, K., 1985, Nucl. Instrum. and Methods, in press.
MORIMOTO, H. and UYEDA, R., 1963, Acta Cryst. 16: 1107.
NORTH, A.C.T., PHILLIPS, D.C. & MATTHEWS, F.S., 1968, Acta Cryst. A24: 351.
PHIZACKERLEY, R.P., MERITT, E., EICHORN, K. et al, to be published.
SATO, Y. et al, to be published.
WYCOFF, H. et al, 1967, J. Mol. Biol. 27: 563.
XUONG, N. et al, 1985, Acta Cryst. B41: 267-269.

2 PHASE PROBLEM

DENSITY MODIFICATION METHODS

A. D. Podjarny

Dept. of Biochemistry and Molecular Biology, U. of Chicago
920 East 58th. St., Chicago, IL, 60637

Macromolecular crystallography is growing rapidly as a tool to image the structure of proteins, nucleic acids and their interactions. In many of these projects, however, MIR (Multiple Isomorphous Replacement) phasing methods have not been successful in imaging macromolecules within their crystals. These problems have motivated efforts to supplement or replace MIR as a phasing technique. Various approaches have been tried, both in real and reciprocal space, and among them DM (density modification) methods have proven to be quite succesful in improving phases using real-space constraints.

DEFINITION OF THE PROBLEM

DM methods aim to improve the agreement of a density map with a set of physically meaningful constraints. While the density map represents the Fourier transformation of the diffraction pattern of its crystal, the constraints include both experimental observations, like the native amplitudes, and general physical criteria, like positivity of the electron density. We can thus classify electron density constraints into two types: experimental constraints and physical constraints. Experimental constraints are those coming from the diffraction experiment, such as observed amplitudes and MIR phase distribution. Physical constraints are those based on our a-priori knowledge of the characteristics of any density function, and include positivity; high-resolution atomicity; boundedness; uniformity of solvent regions; continuity of the bio-polymer chain; and known non-crystallographic symmetry of the macromolecule.

A starting map to be processed by DM, such as an MIR map, is usually obtained from our experimental data, and therefore it will fully agree with our set of experimental constraints. If the map also agrees with the set of physical constraints, then there is no room for DM techniques, and map interpretation should be attempted. It is more often the case that the map does not fully agree with all physical constraints. In that case there is opportunity to improve this agreement, which should diminish the errors present in the map density.

ALGORITHMS

There are several possible ways to design DM algorithms, which can roughly be divided into two categories:

a) Minimisation algorithms, in which agreement with all constraints is measured and optimised. This approach can be strictly defined mathematically, and together with an extra constraint that tends to minimize the amount of spurious information introduced by the algorithm itself, is globally known as the "Maximum Entropy Method". Since this method is discussed in detail elsewhere in this volume, we will not develop it here.

b) Iterative algorithms. In these algorithms, constraints are imposed alternatively in real and reciprocal space, so that one searches for a map that has an optimal agreement with them. The major features of this procedure were originally proposed by Hoppe and Gassmann (1968) and are outlined in the following diagram (ρ= electron density map, F=diffraction amplitude; $\emptyset$= diffraction phase) :

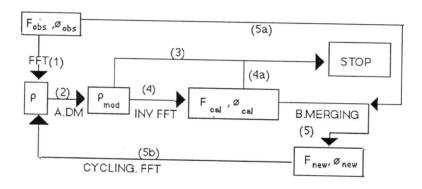

The steps are:

1) An original set of SFs (structure factors), F_{obs}, $\emptyset_{obs}$, is obtained from observed amplitudes and initial trial phases. The electron density map ρ is calculated by Fourier transform of F_{obs}, $\emptyset_{obs}$.

2) A modified electron density is obtained from the current map by one of many possible methods discussed below. For example, Hoppe and Gassman used the following formula:

$$\rho_{mod} = a \cdot \rho^2 + b \cdot \rho^3 + c \cdot \rho$$

3) Map ρ_{mod} is checked for the presence of correct structural features. If it can be interpreted, the procedure was succesful and the the method stops at this point.

4) The map ρ_{mod} is inverted by Fourier transform to calculate SFs,

F_{cal}, $\varnothing_{cal}$. If the mean difference between consecutive sets of calculated phases is less than a preset threshold, the method has converged and stops at this point (4a).

5) The information from the calculated SFs is merged with the experimental SFs (5a) to produce a new set of SFs, F_{new}, $\varnothing_{new}$, as discussed below. For example, Hoppe and Gassman created a new structure factor by using the calculated phases and the experimental amplitudes. A density map is calculated from these merged SFs (5b) and the method cycles back to step (2).

There are two fundamental steps in this procedure that distinguish the different algorithms:

a) DM step. The original map, which agrees with the experimental constraints in reciprocal space is modified to agree with the physical constraints in real space.(step2)

b) Merging step. The calculated SFs and phases from the modified map are merged back with the with the experimental amplitudes and phases, thus reinstating the experimental constraints (step 5). An additional step can also be imposed: that of building a composite map after the DM step (Bhat, 1984,1985).

DM algorithms can be judged to have converged when a given set of parameters, normally the phases, do not change more than a given small amount per cycle. Convergence is not analytically assured, since oscillation or divergence of phases or other parameters is possible.

A. DENSITY MODIFICATION IMPLEMENTATION

Iterative DM algorithms start with a set of experimentally obtained SFs and phases, as Fourier transform coefficients which are used to calculate a trial density map. This map is then modified by applying one of a number of physical constraints. Whereas the original Hoppe and Gassman modification was an ad-hoc procedure designed to be applied fully in reciprocal space and therefore did not explicitly use physical properties, the generalized use of Fast Fourier transforms (Barrett and Zwick, 1971), togeher whith more powerful computers and computational algorithms, has now enabled us to impose explicit constraints in real space. In general, the phasing power of a given constraint is related to the number of density points it affects. These constraints include many of the following concepts :

a1) Positivity

This is a necessary physical property of any electron density. It will be a powerful phasing constraint when the original map has deep negative holes, like heavy atom "ghosts", and extended negative regions due to phase errors. To impose positivity, negative density is deleted or attenuated (See fig. 1 a). In order to avoid sharp boundaries, attenuation is generally chosen. Proper implementation requires the knowledge of the value of F(000). In cases where this value is difficult to obtain, a rough measure can be inferred from the observed solvent level and a crude overall scale. A problem which might arise from the attenuation of negative regions is the appearance of excessively high positive peaks. This is solved by positive density truncation (see a3). It must be pointed out that, since many neutron scattering amplitudes (notably that of deuterium) are negative, this constraint does not apply to neutron diffraction studies.

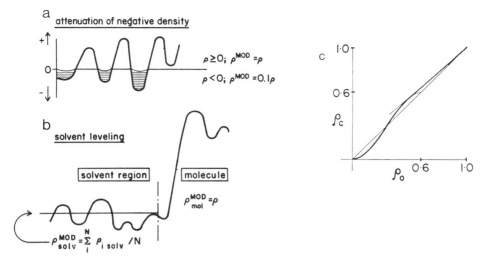

Fig. 1. Density modification constraints.
a) Density regions affected by attenuation of negative density.
b) Density regions affected by solvent flattening.
c) Relation between modified density, ρ_{mod}, and observed density, ρ_{obs}, when the 3-2 rule is applied to the light atoms of the structure and a heavy atom is present (from Collins et al., 1976).

a2) Atomicity

This constraint can be applied when very high resolution data are available, and the corresponding map truly resolves atoms. Sayre (1952) observed that an atomic density has the property of being roughly proportional to its square, $\rho_{mod} = K.(\rho^2)$. This consequence of atomicity is readily incorporated into a density modification algorithm, but it has the problem of excessively increasing large densities. In order to solve this problem, the "(3-2)" rule was proposed (Hoppe and Gassman, 1968; Collins at al., 1975) where a normalized density with values between 0 and 1 is modified according to:

$$\rho_{mod} = 3.\rho^2 - 2.\rho^3$$

This rule works well for quasi-equal atom structures. In cases where there is a heavy atom, like iron in rubredoxin, it must be modified so that the heavy atom density does not grow excessively (See fig. 1 c). (Collins at al., 1976; Raghavan and Tulinsky, 1979). As it will be discussed in the results, this modification has been succesful for improving maps with resolution better than 2.5 A. The atomicity constraint can also be adapted to medium or low resolution data by defining groups of atoms which diffract as a single unit.

a3) Maximum Value

The electron density has a true maximum value related to the chemical character of the structure. This maximum value can be calculated once the density is expressed on an an absolute scale. In high resolution structures that have an intrinsic heavy atom, maximum value regions are not modified to avoid the development of excessive density values (Collins , Brice, La Cour and Legg, 1976; Raghavan and Tulinsky, 1979).

In medium and low resolution structures, the application of positivity and solvent flatness constraints often leads to an increase of the high density values. To avoid this problem, Cannillo, at al. (1983) have devised a density modification technique where the density can have only two values; a maximum value, corresponding to the protein region, and a minimum value, corresponding to the solvent. This has been succesfully used for phase extension from 4.0 A to 1.8 A in Sperm Whale myoglobin. After some trial and error, a boundary level of .6 e/A between the two values was found to work adequately.

A similar problem was found while improving a 25 A neutron diffraction map of the nucleosome histone core (Bentley at al., 1984). In this case, the original map phased by the solvent contrast variation technique had extensive negative regions. Attenuation of these regions by a multiplicative factor of 0.1 improved the map but accentuated peaks of positive density beyond reasonable limits. Therefore, another constraint was added, by which density values above a given value $\rho_{threshold}$ were modified according to

$$\rho_{mod} = \rho_{threshold} + 0.1 \cdot (\rho - \rho_{threshold}).$$

The optimal value of $\rho_{threshold}$ was found by trial and error to be 0.7 of the maximum density of the original map. Though this modification affected only 1.5 % of all points in the unit cell, it prevented the excessive growth of positive peaks during density modification.

a4) Solvent flatness

As observed by Bricogne (1974), the existence of flat solvent regions in a crystal structure places strong constraints on the structure factor phases. The constraints are more powerful if more of the crystal volume is occupied by solvent. Hendrickson (1981) showed that the existence of a solvent region implies the following relation between SFs (where $\mathbb{F}$ represents the complex structure factor $F.e^{i\emptyset}$):

$$\mathbb{F}_h = (1/V) \cdot \sum_k \mathbb{F}_k \cdot \int_U e^{(2\pi i (h-k) r)} d^3 r$$

h,k= reciprocal vectors, r = real space vector, U= Molecular volume, V= Unit cell volume.

The smaller the molecular volume, the larger the number of SFs related by this equation.

In macromolecular crystals, between 30% and 85% of the crystal volume might be occupied by solvent. For the cases of highest solvent content, this constraint becomes quite powerful. In order to implement it the molecular boundaries are identified and the density in the solvent region is replaced by its mean value (see fig. 1 b). The following procedures have been used for boundary definition (roughly in chronological order):

a) Hand digitalization of a minimap with the aid of a graphic tablet. (Hendrickson at al., 1975; Schevitz at al., 1981). This procedure showed the potential of the method and encouraged further developments. However, it is labor intensive, prone to subjective judgemental errors and is not easy to repeat.

b) Definition of the molecular region as regions of linked high density. The solvent region is its complementary volume (Bhat and Blow, 1982). This has been applied succesfully and will be discussed in more detail in connection with the constraint of connectivity (see a5).

c) The map itself can be used to create a mask function. The mask is obtained by replacing every point of the map by a weighted average over all neighbours within a sphere of given radius (B.C. Wang, 1983). This is equivalent to a convolution between the map and a radially-weighted sphere. Density above a specific value is considered within the molecular envelope. This procedure is strongly related (but not exactly equivalent) to lowering the resolution of the electron density map in order to clearly see the molecular region. It is automatic but quite computer intensive, and it can be repeated several times during the course of the DM. It has been observed (Podjarny at al., 1985) that SFs of a resolution 3 to 5 times less than the measured limit are needed in order to properly define the molecular mask. Though this is not a problem for high resolution structures, it can become a limiting factor for medium and low resolution ones.

d) The convolution defined in (c) can be performed by reciprocal space multiplication. This procedure has all the properties of (c), but can be executed in a fraction of the time (E. Westbrook, 1984; A. Podjarny, 1984). Though it is esentially an identical procedure to (c), the large difference in necessary computer time allows for more repetitions or variations. Fig. 3c shows an example of a mask obtained by this method.

e) Definition of the molecular volume by identifying large diffracting elements at low resolution (Rees, 1984; Podjarny at al., 1985). This procedure is effective when solving the phase problem without a starting set of phases. In this case, DM is combined with low-resolution modelling of the density. The low-resolution model is obtained by positioning and refinement of large diffracting elements. This defines the molecular volume which is then used for DM.

f) The molecular boundary can be obtained from a consistency map. This is part of a procedure called "consistent electron density" (Bhat,1984, 1985) where the correct density is identified by the property of invariance under the replacement of the map with a composite of fragment maps .

It is interesting to note that the evolution of techniques dealing with the solvent uniformity constraint has practically marked the evolution of density modification algorithms. An interesting property of the different ways of defining envelopes is that techniques have gone from fixed envelopes obtained from a given map in (a) to an envelope which is redefined in every cycle in (f).

a5) Map continuity corresponding to macromolecular stereochemistry

An important property of any density map corresponding to a biological macromolecule is that,at medium resolution, the density displays single-chain connectivity, corresponding to the single-chain stereochemistry of proteins and nucleic acids. This feature is essential for the interpretation of a map that does not display atomic resolution. Stereochemical constraints are also very useful in refinement. It is however quite difficult to implement these constraints without a molecular model. Bhat and Blow (1982) have successfully applied this constraint for in a crystal where a partial model was known and could be used to define connectivity. They also proposed a method to define connectivity without a model, by examining the densities of neighbouring density. This procedure is now being tested.

a6) Non-crystallographic symmetry

When it is present, non-crystallographic symmetry is a very powerful constraint that relates different density points within a map. Because of the intrinsic accuracy of this method, it practically assures the solution of the phase problem when the redundancy in the data is high enough, e.g., for virus crystals. It can be formulated either in reciprocal space (Rossman and Blow,1962) or in real space (Buehner et al, 1974; Bricogne, 1974; Argos et al, 1975) where it follows the overall scheme of density modification.

After modification is completed, the map can be searched for meaningful features. Though this was routine in early applications, in current ones maps are analyzed after the iterations have converged.

Inverse Fourier Transform

An inverse Fourier transform is now applied to the modified map. Calculated structure factor amplitudes and phases are obtained at this step. The important criteria in this step are the speed and accuracy of the calculation. To exploit the speed of the FFTs, algorithms had to be developed that could us crystallographic space group symmetry (Ten Eyck, 1977; Bantz and Zwick,1974). These Fast Fourier transforms are now used routinely with remarkable sucess. However, in some cases of very high symmetry and very large unit cells Super Fast Fourier transforms (Auslander at al., 1981) might be needed and are currently being developed.

We test for convergence before proceeding with the generation of new coefficients for the next iteration. If the difference between phase sets in consecutive iteration steps is smaller than some preset threshold the process is stopped.

B. MERGING OF CALCULATED AND OBSERVED SFS

In this step, the structure factor information obtained from the modified map is combined with the experimental data. To perform this recombination correctly, we must merge the structure factor information calculated from the modified map with the experimental data and obtain new SFs with which to synthesize a new improved map. Since the error in phase is normally much larger than the error in amplitude, the observed structure factor can be considered in the first approximation as having a fixed value. The first merging procedure consisted simply in taking the observed amplitude and the calculated phase, as shown in fig. 2a (Hoppe at al.,1968; Barrett and Zwick, 1971). This procedure ignored completely the experimental and calculated phase probability distributions, as well as the calculated amplitude, and produced maps which were strongly biased towards the effect of the modification in the calculated phase. To improve the situation, three alternatives have been proposed:

b1) Merge the amplitudes

To diminish the bias introduced by the calculated phases, Collins at al. (1975) used $F_{new}=(2F_{cal}-F_{obs})$, $\emptyset_{new} = \emptyset_{cal}$. Main (1979) analyzed the characteristics of the resulting map in detail. This procedure was also used by Zwick at al.(1976) and Raghavan and Tulinsky (1979). This approach suffered from an important drawback. The original phase information is still not fully used, since only its centroid and figure of merit go into the original map. For the cases where the probability distribution cannot be reconstructed from the centroid and figure of merit (e.g., SIR bimodal phase probability distributions) this represents a loss of important experimental information which is not biased towards any preconception of the structure.

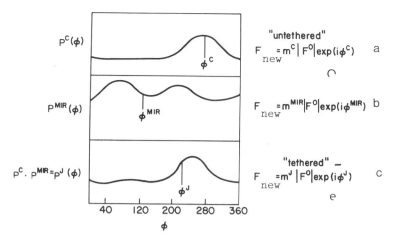

Fig. 2. Merging phase information.
a) "Untethered" case. The new phase is ϕ_{cal}, with a figure of merit obtained from the Sim probability curve.
b) MIR phase distribution curve and corresponding SF.
c) "Tethered" case. The new phase probability curve is obtained by multiplication of the two previous ones. The new phase is its centroid.

b2) Merge the phases

To overcome the problem of multimodal phase probability distributions and excessive bias towards the calculated phase, Hendrickson (1975) multiplied the experimental and calculated phase probability distributions, and obtained a merged phase, ϕ_{mer}, from the centroid of the product distribution (see fig. 2c). A similar idea had been proposed by Rossman and Blow (1961) to combine the phases from MIR with that of a partial model. For the partial model case, Sim (1959) showed that the proper phase probability distribution for the calculated phase, ϕ_{cal}, is:

$$\text{Prob}(\phi_{cal}) = K \cdot \exp((F_{obs} \cdot F_{cal} \cdot \cos(\phi_{cal} - \phi_{cal}^{mean})/E)$$

where K is a normalization constant an $E = \sum_j f_j^2$ where the sum goes over the form factors f_j of the unknown atoms.

Assuming a gaussian distribution of errors, Hendrickson (1975,1981) showed that a similar formula is applicable to the case where the calculated structure factor is obtained from Fourier inversion. In this case, the value of E is obtained empirically from root mean square lack of closure errors at best phase angles. A similar result had been obtained by Bricogne (1976) for applications of non-crystallographic symmetry real space refinement.

In order to improve calculation efficiency and store probability distributions compactly, the phase probability curves are expressed as:

$$\text{Prob}(\phi) = K \cdot \exp(A \cdot \cos(\phi) + B \cdot \sin(\phi) + C \cdot \cos(2\phi) + D \cdot \sin(2\phi))$$

(Hendrickson and Lattman, 1970); thus one need only keep record these coefficients.

Multiplying two curves is the carried out simply by adding the coefficients A,B, C and D. Clearly, the Sim probability distributions can be expressed by the coefficients A and B only. SIR phase distributions can also be expressed quite accurately with this approximation.

The merged phase, ϕ_{mer}, is the centroid of the combined phase probability distribution. The associated figure of merit, fm_{mer}, is also obtained from this distribution. Both are used in conjunction with the observed amplitude to produce a new SFs as:

$$F_{new} = fm_{mer} F_{obs} \; ; \; \phi_{new} = \phi_{mer}$$

This approach has been used succesfully in various density modification phase improvement and extension schemes by Schevitz at al. (1981), Keith at al. (1981), Bhat and Blow (1982), Westbrook at al. (1984) and B.C Wang (1983). It has the advantage of using all the available information, including the original phase probability distribution and the calculated amplitude. However, it does not produce the theoretically optimal map at every step, as shown in the next alternative.

b3) Merge the amplitudes and the phases

While applying DM methods as part of a procedure for obtaining ab-initio phases for the complex of Asp-tRNA and Asp-RS from yeast, Podjarny at al. (1985) found that it was useful to use the following coefficients : $F_{new} = 2F_{obs} - F_{cal}$, $\phi_{new} = \phi_{mer}$. This procedure improved the final map interpretability, as compared with the map obtained using $F_{new} = F_{obs}$.

Use of this type of coefficient has been prevously reported in the context of MIR and model phase calculation. While solving the structure of Phosphoglycerate Kinase, Rice (1981) did an exhaustive analysis of different ways to combine experimental and model SFs. While the $F_{obs} \phi_{mer}$ SF proposed in (2) came close to the optimal, he found that the best map was synthesized with a coefficient $2F_{obs} - F_{cal}$, ϕ_{mer} Following a similar line of thought, Stuart and Artymiuk (1985) have proposed a coefficient of the form

$$F_{new} = fm_{mer} (F_{obs} + Q_{mer} (F_{obs} - F_{cal})), \quad \phi_{new} = \phi_{mer}$$

where Q_{mer} is a function of the MIR figure of merit, calculated figure of merit, and number of correctly placed atoms. In this case, F_{new} tends to $2F_{obs} - F_{cal}$ when most of the atoms are known and the calculated figure of merit is close to 1. For other cases, it tends to produce a more accurate map. Otwinowski (1985) arrived to a similar result by analyzing bias terms. He analyzed the variational dependance of the mixed structure factor F_{obs} , ϕ_{mer} on the complex variable $F_{cal} \exp(i\phi_{cal})$. He found that this dependance has a linear term in $F_{cal} \exp(i\phi_{cal})$ .He assumed that this term is due to bias, since it has the property of reflecting directly the modifications, and substracted it, ending up with the following result:

$$F_{new} \exp(i\phi_{new}) = F_{obs} \exp(i.\phi_{mer}) - A \cdot F_{cal} \exp(i. \phi_{cal})$$

where $A = <F^2_{obs} (1-fsim^2)/E>$ over the given resolution zone, E being the variance parameter in the Sim distribution and fsim being the figure of merit associated with this distribution. From this last analysis, he concluded that the bias in the coefficient F_{new} exp($i\phi_{new}$) arising directly from F_{cal} exp($i\phi_{cal}$) can be cancelled independently of the source of calculated SFs. This provides a theoretical background for the observation of Podjarny at al.(1985), and also shows ways of improving it.

Fourier transform. Cycling

After obtaining a new set of merged Fourier coefficients is obtained, a new map is calculated. Optimally, this map carries the improvements due to density modification with a minimum bias. This new map is then modified again, according to step B, and the procedure cycles until the phase change between two consecutive steps falls below a certain threshold. At this point, the final map is calculated and examined for improvements.

C. BIAS MINIMIZATION. COMBINED OMIT MAP TECHNIQUE

The evolution, just reviewed, of the original idea of Hoppe and Gassman underscores the delicate way in which relative weights for the different constraints must be defined. Care must be taken to maintain the agreement with the experimental data, and avoid producing a map in perfect agreement with all physical constraints but with poor connection to reality. On the other hand, to give the experimental data an excessive weight might result in a procedure that produces no improvement. Ideally, the information obtained from the modified map should be merged with the experimental data without introducing any bias Bhat (1984,1985) has designed a technique to specifically approach this condition, and apply it to ab-initio phase determination.

The modified map obtained from step (A) is normally an improvement over the original map from experimental phase data. However, to the degree that the original phases are in error, noise will persist in the modified map. In the limit of the ab-initio phase determination, where experimental phase data is effectively random, this problem becomes crucial.

Bhat's algorithm adds an extra step after density modification in order to stringently filter the noise coming from the original phase data. He suggested that the unit cell be divided in contiguous boxes, called "phasing volumes". The density in a phasing volume is replaced by a new one with minimum bias from the original phase set by the following technique:

1) The density in the phasing volume is set to a constant value and the map is back-transformed. Calculated "omit" phases are obtained.

2) The calculated "omit" phases are combined with the original amplitudes and a new map is calculated. The density in the phasing volume is obtained from this map.

By repeating this procedure for the phasing volumes in the asymmetric unit, a new map is obtained wherein no volume element has been created by phases representing the structure (and its modification) within that element. All of the structure information for a particular element comes from the observed diffraction amplitudes and the phases derived from all the other elements. By introducing this procedure in the overall density modification scheme, Bhat has developed a new algorithm, called "consistent electron density". This algorithm produces a map where wrong peaks are not echoed, and correct peaks are echoed with weights of 0.5. In doing so, any spurious bias introduced by imposing the constraints and any errors in the original phase set, are minimized without seriously diminishing the overall phasing power.

RESULTS

The results of several applications of DM that have been published or which have been reported at scientific meetings are shown below. This list does not pretend to be exhaustive, but is compiled to illustrate some of the most important trends in the application of this method.

a) High resolution. Maximum resolution <2.5 A

1968. Hoppe at al. Solved many small molecules .Constraint: Atomicity. Merging:
$F_{new} = F_{obs}$, $\emptyset_{new} = \emptyset_{cal}$

1971. Barrett and Zwick. Extended phases in Myoglobin from 3 to 2 A. DM constraint: Atomicity (squaring), positivity. Merging: $F_{new} = F_{obs}$, $\emptyset_{new} = \emptyset_{cal}$. Mean error: 78 Deg.

1976. Collins at al. Extended phases in Rubredoxin from 2 A to 1.5 A. DM constraint: Atomicity (3-2 rule), positivity, maximum value Merging:
$F_{new} = 2F_{obs} - F_{cal}$, $\emptyset_{new} = \emptyset_{cal}$ Mean error:39 Deg.

1976. Zwick at al. Decreased error in 32 atom structure, from 85 to 32 Deg. DM constraint: Atomicity (3-2 rule), positivity. Merging:
$F_{new} = Max(2F_{obs} - F_{cal}, KF_{obs})$, $\emptyset_{new} = \emptyset_{cal}$

1976. Nixon and North. Human lysozyme phase refinement.DM constraint: Atomicity (3-2 rule), positivity. Starting set :HEW lysozyme phases.No merging done. No improvement.

1979. Raghavan and Tulinsky. Extended 2.8 A MIR phases in chymotripsin to 1.8 A. DM constraint: Atomicity (3-2 rule), positivity. Merging:
$F_{new} = 2F_{obs} - F_{cal}$, $\emptyset_{new} = \emptyset_{cal}$. Map improvement.

1981. Keith at al. Diminished error in 2.5 A MIR PLPA2 phases from 55 to 52 degrees. DM constraint: Solvent flatness, positivity, bounds. Merging:
$F_{new} = F_{obs}$, $\emptyset_{new} = \emptyset_{mer}$.

1983. Cannillo at al. Extended 4 A model phases (with errors) in myoglobin to 1.8 A with an error of 22 deg. in strongest F's. DM constraint: Solvent and protein levelling. Merging: $F_{new} = F_{obs}$, $\emptyset_{new} = \emptyset_{cal}$.

1984. Podjarny,Sussman at al. Refinement of 2.2 A SIR phases from Dead Sea Ferredoxin. DM constraints: Solvent flatness, positivity, boundedness. Merging: $F_{new} = F_{obs}$, $\emptyset_{new} = \emptyset_{mer}$. Increased density map detail.

b) Medium resolution. 2.5 A < Max. resolution < 8 A.

1975. Hendrickson at al. Refinement of 5.5 A myohemerythrin phases. DM constraints: Solvent flatness, positivity. Merging: $F_{new} = F_{obs}$, $\emptyset_{new} = \emptyset_{mer}$. Mean fig. of merit improved from .89 to .92.

1977. Podjarny and Yonath. Refinement of 5 A tRNA Phe phases.DM constraint: Positivity. Merging: $F_{new} = F_{obs}$, $\emptyset_{new} = \emptyset_{cal}$. Provided a correct starting phase set for matricial phase extension.

1981. Schevitz at al. Refinement of 4 A tRNA fMet MIR phases. DM constraint: Positivity, solvent flatness. Merging: $F_{new}=F_{obs}$, $\emptyset_{new}=\emptyset_{mer}$ first, $\emptyset_{new}=\emptyset_{cal}$ later. Mean phase error diminished from 68 to 43 deg.

1982. Bhat and Blow. Refinement of MIR 2.7 A Tyr synthetase data. DM constraint: Positivity, solvent flatness, continuity. Merging : $F_{new}=F_{obs}$, $\emptyset_{new}=\emptyset_{mer}$. Overall map improvement led to correct chain tracing

1983. Wang at al. Refinement of SIR 3 A Bence-Jones protein Rhe phases. DM constraint: Positivity, solvent flatness (automatic masking, real space) Merging. $F_{new}=F_{obs}$, $\emptyset_{new}=\emptyset_{mer}$. Overall phase error improved from 51 to 32 deg.

1984. Westbrook at al. 6 A refinement of ketosteroid isom. MIR phases.
DM constraint: Positivity, solvent flatness (automatic masking, rec. space). Merging. $F_{new}=F_{obs}$, $\emptyset_{new}=\emptyset_{mer}$ Improved protomer boundaries and map interpretability.

1985. Bhat. 4 A Ab-initio determination of Creatine Kinase phases. DM constraint: Consistent density. Positivity. Solvent flatness. Merging: $F_{new}=F_{obs}$, $\emptyset_{new}=\emptyset_{mer}$. Obtained a phase set accurate enough to properly phase a heavy atom difference map.

c) Low resolution. 8 A < Max. resolution

1984. Bentley at al. 16 A refinement of neutron diffraction phases of the histone core of the nucleosome particle, originating from the solvent contrast variation technique and the DNA model. DM constraint: positivity, solvent flatness. Merging: $F_{new}=F_{obs}$, $\emptyset_{new}=\emptyset_{mer}$. Increased definition of histone dimer boundaries.

1985. Podjarny at al. 25 A refinement of neutron diffraction phases of the Asp tRNA synthetase complex, originated from a 4 gaussian sphere model.
DM constraints: Positivity, solvent flatenning, boundedness. Merging:
$F_{new}=F_{obs}$, $\emptyset_{new}=\emptyset_{cal}$. Increased interpretability and allowed recognition of tRNA shape.

1985. Podjarny at al. 10 A refinement of X-Ray diffraction phases of the Asp tRNA synthetase complex, originated from a low resolution model including the tRNA phosphate backbone. DM constraints: Positivity, solvent flatness, boundedness, non-crystallographic symmetry. Merging:
$F_{new}=F_{obs}$, $\emptyset_{new}=\emptyset_{mer}$. Increased continuity in synthetase region and defined the tRNA positions more clearly.

Examination of the list shows that the emphasis of DM applications has shifted from high resolution structures, where atomicity was the most important constraint, to medium resolution structures were solvent flatness is the driving force. An example of a DM application in this resolution range is shown in fig. 3, taken from Schevitz at al. (1981).

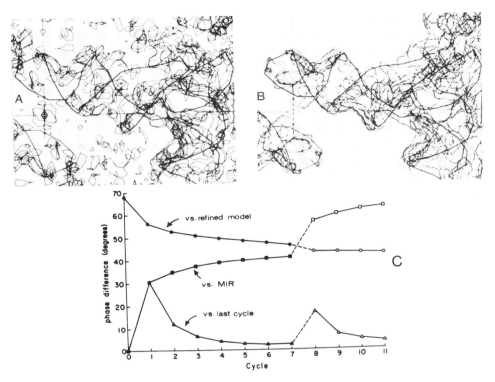

Fig. 3. Application of density modification at medium resolution.
A) 4.5 Å MIR map of tRNA$^f_{Met}$ from yeast.
B) Same as (A), after density modification using negative density truncation, solvent levelling and phase merging.
C) Course of density modification as measured by the phase changes.
7 cycles of DM and phase merging were performed, followed by 4 cycles of "untethered" refinement using $F_{new} = F_{obs}$, $\emptyset_{new} = \emptyset_{cal}$. Note the improvement of phase error.

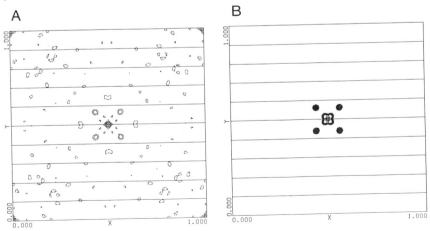

Fig. 4. Comparison of heavy atom difference Patterson synthesis for Bovine Heart Creatine Kinase X-ray diffraction data.
A) Difference Patterson synthesis using the observed amplitudes.
B) Calculated difference Patterson from the heavy atom position located using the CED phases.

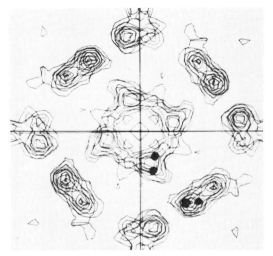

A) Sections of neutron scattering density map projected down 4-fold axis. Positive contours only. Space group I432, a=354 A. Positions of model spheres are marked by heavy dots.

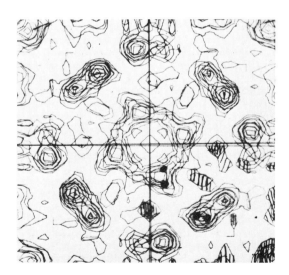

B) Same as (A), including contours below solvent level. These contours are shaded in the bottom-right quadrant

Fig. 5. Continued

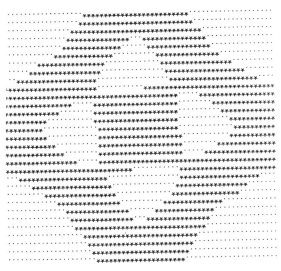

C) One section of density modification mask, obtained from map (A) by the method outlined in a4.c. Heavy dots (*) correspond to the molecule, and light dots (.) correspond to the solvent.

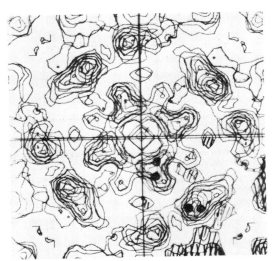

D) Same map sections after density modification. Note how the L-shape of density superimposed with the two dots nearest to the cente has been accentuated. This density was correctly interpreted as corresponding to a tRNA molecule, and led to a proper interpretation and refinement up to 10 A resolution using X-ray data.

Fig. 5. Application of density mofication to a neutron scattering map at 25 A resolution of the complex of Asp-tRNA Asp-tRNA synthetase from yeast. Phases were obtained from a model consisting of four gaussian spheres.

The consistent electron density method (CEDM), developed by Bhat (1984,1985) shows a clear possibility of ab-initio phasing. Currently, a set of phases is being obtained from the X-ray diffraction amplitudes of Bovine Heart Creatine Kinase. These phases were used to locate a heavy atom position in a difference Fourier map. Fig. 4 shows that these positions seem to be correct. This preliminary result points towards the applicability of this method in macromolecular crystallography. The work of Podjarny at al. (1985),see Fig. 5, also shows how density modification can be a tool in ab-initio phasing. In this case, DM is used to complement low resolution modelling and refinement in order to provide new density information and increase model complexity.

CONCLUSIONS

It is quite clear that DM has the power to improve a MIR map over a wide range of resolutions when a suitable protein boundary can be found. It appears to be the method of choice when an MIR map cannot be interpreted in terms of a molecular model, but shows some recognizable features. For cases where a map can be correctly interpreted, experience shows that a combination of graphical interpretation and constrained-restrained least squares refinement is the best approach. However, for those cases where there is some map interpretation but it is not sufficient to start a model refinement, DM is still very helpful, if not essential, in defining the correct molecular structure. A most interesting recent development is the possibility of solving the phase problem ab-initio.

The work of Bhat (1984,1985) suggests that step (C), in which the density map is replaced by a composite omit map , is powerful enough to force the procedure to converge to the correct solution even if one is starting from a random set of phases. The work of Podjarny at al. (1985) shows that DM can be used to improve a very approximate map obtained from a model of minimum complexity. This has lead to a reasonable map for the Asp-tRNA AspRS complex at 10 A resolution. In this case, the previous knowledge of the overall structure of the tRNA component was very helpful for map interpretation at low resolution. These developments point towards the possibility of eventually replacing MIR as the main phasing method for difficult cases where it is not easily applicable.

AKNOWLEDGMENTS

The author thanks the Director and the Organizing Committee of the "Crystallography in Molecular Biology" lecture course for the invitation to present this review and for providing an outstanding opportunity for scientific discussion.
Funds for the author's participation and for the author's research concerning the tRNA-synthetase complex were provided by the NIH, Grant R01 GM 34942-01. The author also wishes to thank P. Alzari, T.N.Bhat, D. M. Collins, Dino Moras, J. Navaza, P. B. Sigler, Z. Otwinowsky , B. Rees, R. Schevitz, D. Sayre, Jean Claude Thierry, E. Westbrook and M. Zwick for the many hours of scientific discussion that made possible the comprehension of this subject. Last but not least, the author wishes to thank Christianne Rether and the staff of Plenum publishing company for their endless patience.

REFERENCES

Argos, P., Ford, G.C. and Rossmann, M.G., (1974) Acta Cryst., A31, 499-506.
Auslander, L., Feig, E. and Winograd, S. (1981) In Computational Crystallography ,edited by D.Sayre, pp. 451-469. Ottawa: Clarendon Press.

Bantz, D. and Zwick, M. (1974). Acta Cryst. A30, 257-260.
Barrett, A. N., and Zwick, M. (1971). Acta Cryst. A27, 6-11.
Bentley, G. A., Lewit-Bentley, A., Finch, J.T., Podjarny, A.D and Roth, M. (1984). J. Mol. Biol, 176, 55-75.
Bhat, T.N. and Blow, D.M. (1982). Acta Cryst., A38, 21-29.
Bhat, T.N. (1984). Annual Meeting of the ACA, Lexington, Kentucky, Section Q.
Bhat, T.N. (1985). Annual Meeting of the ACA, Stanford, California. Section H.
Bricogne, G. (1974). Acta Cryst., A 30, 395-405.
Bricogne, G. (1976). Acta Cryst., A 32, 832-847.
Buehner, M., Ford, G.C., Olsen, K.W., Moras, D., Rossmann, M.G. (1974). J. Mol. Biol., 90(1), 25-49.
Cannillo, E., Oberti, R and Ungaretti, L. (1983). Acta Cryst., A39, 68-74.
Collins, D. M., Cotton, F. A., Hazen, E.E. Jr., Meyer, E.F., and Morimoto, C.N. (1975) Science, Vol 190, 1047-1053.
Collins, D.M., Brice, M.D., La Cour, T.F. M. and Legg, M. J. (1976) In Crystallographic Computing Techniques, edited by F. R. Ahmed, K. Huml and B. Sedlacek, pp. 330-335. Copenhagen:Munskgaard.
Hendrickson, W. A. (1981). In Structural Aspects of Biomolecules, edited by R. Srinivasan and V. Pattabhi, pp. 31-80. New Delhi:Macmillan.
Hendrickson, W. A., Klippenstein, G. L., and Ward, K. B. (1975) PNAS, USA, 72, 2160-2164.
Hendrickson, W. A. and Lattman, E.E. (1970) Acta Cryst., B26, 136-143.
Hoppe, W., and Gassman, J. (1968). Acta Cryst. B24, 97-107.
Keith, C., Feldman, D., Deganello, S., Glick, J., Ward, K., Oliver Jones,E. and Sigler, P.B. (1981). JBC, 256, 8602-8607.
Main, P. (1979). Acta Cryst., A35, 779-785.
Nixon, P. E., and North, A.C.T. (1976) Acta Cryst., A32, 325.
Otwinowski, Z. (1985). Personal communication.
Podjarny, A. D., and Yonath, A., Acta Cryst. (1977) A33,655-661.
Podjarny, A.D., Thierry, J.C., Cavarelli, J., Rees, B., Lewit-Bentley, A., Roth, M., Lorver, B., Romby, P., Ebel, J.P, Giege, R. and Moras, D. (1985). PNAS, sent for publication.
Podjarny, A. D., Sussman, J. L., Bhat, T. N., Westbrook, E. M., Harel, M.,Yonath, A., and Shoham, M. (1984). Acta Cryst. A35s, Abstract 01.4-2.
Raghavan, N. V and Tulinsky, A. (1979) Acta Cryst., B35, 1776-1785.
Rees, B. (1984). Personnal Communication.
Rice, D.W. (1981). Acta Cryst., A37, 491-500.
Rossman, M. G. and Blow, D. M. (1961). Acta Cryst., 14, 641-647.
Sayre, D. (1952). Acta Cryst., 5, 60-65.
Schevitz, R. W., Podjarny, A. D., Zwick, M., Hughes, J.J. and Sigler, P. B. (1981). Acta Cryst., A37, 669-677.
Sim, G. A. (1959). Acta Cryst., 12, 813-815.
Stuart, D. and Artymiuk, P. (1985). Acta Cryst., A40, 713-716.
Ten Eyck, L.F. (1977) Acta Cryst., A33, 486-492.
Wang, B.C. In Notes of the School on Direct Methods and Macromolecular Crystallography, 1983. Edited by H. Hauptman. Buffalo.
Westbrook, E. (1984).Personal Communication.
Westbrook, E., Piro, O. and Sigler, P. B. (1984). JBC, 259, 9096-9103.
Zwick, M., Bantz, D. and Hughes, J. (1976). Ultramicroscopy, 275-277.

ANOMALOUS SCATTERING

IN MACROMOLECULAR STRUCTURE ANALYSIS

Wayne A. Hendrickson

Department of Biochemistry and Molecular Biophysics
Columbia University
New York, NY 10032

INTRODUCTION

Anomalous scattering is a phenomenon in fundamental physics that proves to have a special role in macromolecular crystallography. When x-rays impact matter they are scattered from the electrons of atoms. Conventional diffraction analysis takes the scattering as being directly proportional to electron density. Thus this "normal" scattering is insensitive to atom type. However, such a description is only adequate for the lightest of elements (e.g. carbon and nitrogen). In general, resonance between electronic vibrations induced by the incident x-ray wave and the characteristic frequencies of bound atomic electrons modifies the scattering. This change, the "anomalous" scattering, has two especially pertinent features: the magnitude of effect varies with x-ray wavelength and each type of atom gives a distinctive response. Anomalous scattering is particularly pronounced at absorption edges; that is, when the x-ray energy approaches the characteristic energy levels of atomic orbitals.

The utility of anomalous scattering in structure analysis was first recognized by Bijvoet[1] and since the early days it has played an important, albeit secondary, role in protein crystallography. Three major kinds of applications are conventional: it is used for phase evaluation in conjunction with isomorphous replacement by heavy atoms, it is used to determine absolute configuration, and it is used to locate the positions of native metal centers. More recently it has become apparent that anomalous scattering can also offer the more significant possibility of direct determination of phase angles from the diffraction data of a single crystal or crystalline species. It is these latter developments that are the focus of this article. The discussion here will be essentially qualitative, but more detail can be found in other recent contributions.[2-5]

THEORETICAL BACKGROUND

The distinctiveness of an anomalous scattering center can be expressed in its atomic scattering factor. The normal scattering factor, $f°$, that would pertain in the absence of resonance effects is modified by a wavelength dependent correction -- the anomalous dispersion -- that involves a phase shift δ with respect to the normal scattering factor.

Thus,
$$f = f° + f^{\Delta}e^{i\delta} = f° + \Delta f' + i\Delta f''. \tag{1}$$

The most direct impact of the imaginary component, $\Delta f''$, is seen in a breakdown of Friedel's law of symmetric diffraction. Effects of the real component, $\Delta f'$, are evident in intensity differences at different wavelengths. The manner in which these effects are exploited in phase determinations depends on the application.

In the event of experiments at a single wavelength, it is convenient to consider the separate real and imaginary components of a structure factor. Thus,
$$F(h) = F'_T(h) + F''_A(h) \tag{2}$$
where $F'_T = |F'_T| \exp(i\phi)$ is the structure factor based on the real parts ($f' = f° + \Delta f'$) of the scattering from all atoms in the total structure and $F''_A = |F''_A| \exp[i(\psi+\omega)]$ is that from the imaginary components ($\Delta f''$) of the set of anomalous scattering centers. It follows that, when the anomalous scatterers are of the same atomic type and contribute relatively weakly to the total scattering then to good approximation the Bijvoet difference is given by
$$\Delta F = |F(h)| - |F(-h)| = -2|F''_A| \sin(\psi-\phi). \tag{3}$$

This relationship can be used to deduce and refine the atomic positions of anomalous scatterers. Then, given calculated values of $|F''_A|$ and ψ together with the observation of ΔF, two possible values of ϕ obtain. Probabilistic methods have been devised to take errors into account in this analysis.[6,7]

In the case of multiple wavelength analyses it is advantageous to separate scattering contributions somewhat differently. This is motivated by the need to refer all phasing estimates to a common reference point. Karle[8] has shown that by choosing the normal scattering terms as this point of reference, the analysis is simplified. Now it is important to distinguish the wavelength of measurement as well as the Friedel mate. Thus we have
$$^{\lambda}F(h) = °F_T + \sum_k (\Delta f'(\lambda)/f°)\, °F_{A_k} + i\sum_k (\Delta f''(\lambda)/f°)\, °F_{A_k} \tag{4}$$
where $^{\lambda}F(h)$ is a structure factor for reflection h at a particular wavelength λ, $°F$ corresponds to the normal scattering terms ($f°$) of the total structure, and $°F_{A_k}$ corresponds to the normal scattering contribution from the kth kind of anomalous scatterer in the structure.

This formulation permits an elegant separation of wavelength-dependent and wavelength-independent variables. Again the case of a single kind of anomalous scatterer illustrates the point. If we denote the phase of $F(h)$ by $°\phi_T$ and that of $°F_A(h)$ by $°\phi_A$, then
$$\begin{aligned}|^{\lambda}F(\pm h)|^2 = & |°F_T(h)|^2 + a(\lambda)|°F_A(h)|^2 \\ & + b(\lambda)|°F_T(h)||°F_A(h)|\cos(°\phi_T - °\phi_A) \\ & \pm c(\lambda)|°F_T(h)||°F_A(h)|\sin(°\phi_T - °\phi_A)\end{aligned} \tag{5}$$

All wavelength dependence in Eq. (5) is embodied in coefficients that are known from scattering factors that can be calculated on measured experimentally. These are

$$a(\lambda) = (f^{\Delta}/f^{\circ})^2 \qquad (6a)$$

$$b(\lambda) = 2(\Delta f'/f^{\circ}) \qquad (6b)$$

and

$$c(\lambda) = 2(\Delta f''/f^{\circ}). \qquad (6c)$$

Given the values of a, b and c, then from suitable measurements at ± h and at multiple wavelengths, a constrained system of equations can be constructed to solve for $|^{\circ}F_T|$, $|^{\circ}F_A|$ and $(^{\circ}\phi_T - ^{\circ}\phi_A)$. The structure of the relatively few anomalous scatterers can be determined from the $|^{\circ}F_A|$ data. This structure in turn gives $^{\circ}\phi_A$ and thereby yields $^{\circ}\phi_T$. The values $^{\circ}F_T$ are then the proper coefficients to produce a true electron density distribution for the complete structure.

SINGLE WAVELENGTH EXPERIMENTS

The procedures for practical application of anomalous scattering in the direct determination of phases for proteins were worked out in the structure analysis of crambin.[7] The first step is the critical one of making accurate diffraction measurements. This was essential in the case of crambin since the anomalous scattering from sulfur atoms is rather slight at the Cu Kα wavelength. Experimental conditions designed to eliminate common systematic errors from Bijvoet differences and local scaling to reduce residual, smoothly varying errors are important considerations in this step. The second step concerns the location and refinement of the anomalous scatterer structure. The Bijvoet difference Patterson function[9] is an important tool for arriving at the initial model, and a least-squares procedure based on the strongest differences is used for refinement. The phase determination step involves the combination of information from anomalous scattering with that from another source to resolve the ambiguity indicated by Eq. (3). In the case of crambin, the partial structure given by the real part of scattering from the anomalous centers was used to make this resolution. Other sources of information have been used in other cases. The final step is, of course, that of interpreting density maps as atomic structures. Except in one respect, this is standard. Here, unless the array of anomalous centers is itself centrosymmetric, the two enantiomorphous structures must be considered. The chemical reasonableness of features in one of the alternative maps is usually an adequate guide.

The weakest aspect of the resolved anomalous phasing procedure used for crambin is the reliance on the partial structure information for ambiguity resolution. Oftentimes the array of anomalous centers is not a dominating phasing influence; although the six sulfurs in crambin present a weaker situation than pertains in many heavy-atom complexes with proteins. Other sources of complementary phasing have also been used or proposed. These include molecular averaging[10], a partially known protein model[11], direct methods[7,12] and solvent flattening[13].

Quite separate analyses of anomalous scattering have also been initiated in the context of traditional direct methods. Hauptman[14] has derived conditional probability distributions for three-phase structure invariants that take anomalous scattering into account. Karle[15] has given an alternative analysis by which simple rules are deduced to evaluate triplet phase invariants. In both cases, this incorporation of anomalous scattering into direct methods leads to unambiguous phase indications, which should not be surprising as distinctive kinds of information are combined. What is unexpected is that these results do not hinge on first evaluating the structure of anomalous scattering centers.

Very recently, Karle[16] has suggested that single wavelength experiments might be analyzed by means of the separation of variables adopted for the multiwavelength case. Although one has only two observational equations (at $\pm h$) and three independent variables ($|°F_T|$, $|°F_A|$ and ($°\phi_T - °\phi_A$)), by including a third relationship based on the approximation of Eq. (3), Karle finds a "essentially unique" least-square solution from ($°\phi_T - °\phi_A$). This dubious generation of additional equations from existing ones seems unnecessary when the anomalous scatterer structure can be solved from Bijvoet difference Pattersons. In that case, only $|°F_T|$ and $°\phi_A$ remain as unknowns. In principle, then, a direct solution should obtain. The effect of errors in practice must be tested, but this appears to be a promising new direction.

MULTIPLE WAVELENGTH EXPERIMENTS

The multiple wavelength experiment offers exceptional promise for the direct analysis of macromolecular structures. By working near absorption edges the magnitude of effects can be greatly enhanced and from the differences betwen wavelengths information complementary to that from Bijvoet differences is obtained. Experience in the analysis of measurements at multiple wavelengths is not as advanced as for the conventional single wavelength experiments. Kahn et al[17], have recently published a three wavelength study of a terbium parvalbumin crystal and other work is in progress at various places. We have recently carried out a successful analysis of synchrotron data measured at four wavelengths near the iron edge of lamprey hemoglobin and have also solved the structure of d-selenolanthionine from rotating anode data at four wavelengths from the Bremstrahlung continuum. The lamprey hemoglobin work (done in collaboration with Janet Smith, Paul Phizackerley and Ethan Merritt) illustrates our normal plan of attack and the selenolanthionine experiment (done in collaboration with Jan Troup and Göran Zdansky) suggests an interesting alternative approach.

The effectiveness of the anomalous dispersion experiments depends critically on the choice of wavelength. The imaginary component is proportional to absorption and thus is maximal at the peak of the x-ray absorption spectrum. The real component of anomalous scattering reaches an extreme value at the inflection point of the rise at the absorption edge and is only useful in contrast to data taken at remote points away from the edge. The exact position of the edge and the magnitude of the effects depend on the chemical state of the metal center. Thus experimental evaluation of scattering factors from absorption data is essential. We measure these data by fluorescence detection from the same crystal used for the diffraction measurements. By fitting to theoretical values away from the edge, values of $\Delta f"$ can be extracted directly from the absorption spectrum. The relationship to produce $\Delta f'$ values involves the Kramers-Kronig transformation, but can be done quite straightforwardly, again with reference to the theoretical values away from the edge.

Diffraction data were measured from D2 crystals of lamprey hemoglobin on the multi-wire proportional chamber constructed at SSRL by Phizackerley. We chose four wavelengths: a remote pre-edge point (1.8000 A), the inflection point (1.7402 A), the absorption peak (1.7381 A) and a remote high-energy point (1.5000 A or 1.6500 A). The crystal and detector were aligned so that Bijvoet pairs related by mirror symmetry were detected simultaneously. Complete data to 3.0A spacings have been measured.

The data analysis was carried out using a newly devised least squares analysis based on Eqs. (5) and (6). The first step in this multi-wavelength anomalous dispersion (MAD) phasing process is a linear least-squares solution for the four wavelength invariant factors given the experimental values of a, b and c. Results from this step are used to initiate a non-linear refinement that imposes the trigonometric identity restricting sine and cosine values through a Lagrange constraint. The resulting $|°F_A|$ values give extraordinarily clean Patterson maps and phases computed from the resulting refined iron structure then produce the protein phases from the $\Delta\phi$ values. Analysis of the newer 3A data is incomplete, but maps from an earlier 5.5A data set measured with the detector in a noisy state are in good agreement with the MIR map. MAD and MIR phases have a 56.4° average phase discrepancy in this experiment.

The selenolanthionine problem was chosen as an unknown structure to test the possibility of using conventional sources for MAD phasing. It also serves as a chemical analog of selenomethionine which is of interest as a general anomalous center. This experiment used graphite monochromatized radiation from a molybdenum anode. Four data sets were measured to 0.98A spacings or beyond. One was at 0.71A (MoKα), a second was at 1.21A and the other two were at points very near the absorption edge at about 0.98A. In this case the Se anomalous scattering is a dominant influence and approximations such as Eq. (3) are not valid. However, Eq. (5) is exact and served well in the analysis. Experimental values of scattering factors could not be obtained with the present instrument, but a survey of possible values led to tentative values. Relative scaling to take account of scattering factor variation was crucial in this case. The first map gave an extraordinarily clean results with atomic peak densities ranging from 3.3 to 6.1 times the next highest non-ripple peak. Atomic density variation perfectly reflects the atomic numbers of C, N and O). The average diffracted intensity in this continuum experiment is such that experiments with protein will probably be possible with area detectors.

PROSPECTS FOR A GENERAL PHASING VEHICLE

The theoretical basis for MAD phasing is solid and capabilities for making measurements at multiple wavelengths are coming into place. Absorption edges for elements of atomic numbers in the range from 20 (Ca) to 47 (Ag) and from 50 (Sn) to 92 (U) are accessible to x-ray wavelengths in the range of 0.5 to 3A. This includes many atoms that can be attached to proteins or nucleic acids, but except for in metalloproteins few are naturally occurring. Thus, although the range of possibilities is extended, it might seem that heavy-atom searches will remain an essential feature of macromolecular crystallography. However, there is a prospect for a more general introduction of appropriate anomalous scattering centers.

Selenium is in the same group of elements as oxygen, sulfur and tellurium and it has properties that are particulary close to those of sulfur. Selenium is a biologically essential trace element yet it can also be toxic, probably because of the enhanced reactivity of organoselenides over their sulfur analogs. However, whereas terminal selenium groups (as in selenocysteine) are readily oxidized, organoselenides (e.g. selenomethionine) are more stable. Indeed, the selenolanthione (also an R-Se-R selenide) used in our structure analysis had been prepared 17 years earlier. Selenium compounds also tend to be structurally very similar to their sulfur conterparts; biotin and selenobiotin crystal structures are isomorphous as are those of d,l-methionine and d,l-selenomethionine. The C-Se bonds are typically about 0.14A longer than comparable C-S bonds.

The striking similarity of selenium to sulfur and the stability of selenomethionine would suggest that selenomethionyl proteins might be quite viable. Many experiments indicate that selenomethionine can be incorporated in place of methionine during biosynthesis and the first of these may be the most convincing. Cowie and Cohen[18] showed in 1957 that a methionine auxotroph of E. coli grows exponentially for at least 100 generations in selenomethionine. Apparently the only adverse consequences of total replacement of methionine by selenomethionine is a 33% increase in division time. Recently, essentially complete replacement of selenomethionine into azurin purified from a Pseudomonas auxotroph has been demonstrated[19]. The obvious extension of the earlier studies is to use a methionine auxotroph as the host for a recombinant plasmid designed to express a particular protein of interest in a selenomethionine medium.

The phasing potential of selenomethionyl proteins is high. Methionine abundance on average is 1 in 58 residues and at that level both Bijvoet differences and dispersive differences (changes between wavelengths) are expected to be at least 5% of the total diffraction. The MAD refinement procedure is expected to produce very accurate $|°F_A|$ estimates and, since the selenium sites must necessarily be separated by at least 4A, excellent atomic resolution is present even with 3A resolution data. Thus even if a large protein has 20-50 methionines, direct methods programs such as MULTAN stand a good chance to solve these structures.

Much remains to be tested, both on the side of molecular biology and on the side of diffraction analysis, but the prospect is bright that selenomethionyl proteins produced by recombinant DNA techniques might serve as a rather general phasing vehicle.

ACKNOWLEDGEMENTS

I thank the various participants in the anomalous scattering projects described in this report, and I am particularly grateful to Janet Smith for her collaboration in the lamprey hemoglobin project, to Jan Troup for his role in the selenolanthionine project, and to John Horton for his help in developing the selenomethionine idea. This work was supported in part by a grant, GM-34102, from the National Institutes of Health.

REFERENCES

1. J. M. Bijvoet, Phase determination in direct Fourier-synthesis of crystal structures. Proc. Acad. Sci. Amst. B52: 313-314 (1949).
2. W. A. Hendrickson, J.L. Smith, and S. Sheriff, Direct phase determination based on anomalous scattering, Methods in Enzymology 115:41-55 (1985).
3. W. A. Hendrickson, Anomalous scattering in protein crystallography, in "Proceedings of a symposium on new methods in x-ray absorption scattering and diffraction, H.D. Bartunik and B. Chance, eds., Academic Press, in press.
4. W. A. Hendrickson, Anomalous dispersion in phase determination for macromolecules, in "Computational Crystallography 3", G. Sheldrick, ed. Oxford University Press, in press.
5. W. A. Hendrickson, Analysis of protein structure from diffraction measurements at multiple wavelengths, Trans. Amer. Cryst. Assn. 21:in press.
6. W. A. Hendrickson, Phase information from anomalous-scattering measurements, Acta Cryst. A 35:245-247 (1979).
7. W. A. Hendrickson and M.M. Teeter, Structure of the hydrophobic protein crambin determined directly from the anomalous scattering of sulphur, Nature 290:107-113 (1981).
8. J. Karle, Some developments in anomalous dispersion for the structural investigation of macromolecular systems in biology. Int. J. Quant. Chem. 7:356-367 (1980).
9. M. G. Rossmann, The position of anomalous scatterers in protein crystals, Acta Crypt. 14:383-388 (1961).
10. J. L. Smith, W.A. Hendrickson and A.W. Addison, Structure of trimeric hemerythrin, Nature 303:86-88 (1983).
11. J. L. Smith and W.A. Hendrickson, Resolved anomalous phase determination in macromolecular cyrstallography, in "Computational Crystallography," D. Sayre, ed. Oxford University Press, pp. 209-222 (1982).
12. H.-f. Fan, F-s. Han, J-z. Quian and J.-x Yas, Combining direct methods with isomorphous replacement and anomalous scattering data. I., Acta Cryst. A40:489-495 (1984).
13. B. C. Wang, Methods in Enzymology, in press.
14. H. Hauptman, On integrating the techniques of direct methods with anomaolous dispersion I. the theoretical basis, Acta Cryst. A38:632-641 (1982).
15. J. Karle, Rules for evaluating triplet phase invariants by use of anomalous dispersion data, Acta Cryst. A40:4-11 (1984).
16. J. Karle, Unique or essentially unique results from one-wavelength anomalous dispersion data, Acta Cryst. A41:387-394 (1985).
17. R. Kahn, R. Fourme, R. Bosshard, M. Chiadini, J.L. Risler, O. Dideberg and J.P. Wery, Crystal structure study of Opsanus tau parvalbumin by multi-wavelength anomalous diffraction, FEBS Lett. 79:133-137 (1985).
18. D. B. Cowie and G.N. Cohen, Biosynthesis by Escherichia coli of active altered proteins containing selenium instead of sulfur, Biochim. Biophys. Acta. 26:252-261 (1957).
19. P. Frank, A. Licht, T.D. Tullius, K.O. Hodgson and I. Pecht, A selenium-containing azurin from an auxotroph of Pseudomonas aeruginosa, J. Biol. Chem. 260:5518-5525 (1985).

SINGLE MOLECULE ELECTRON CRYSTALLOGRAPHY

Marin van Heel

Fritz Haber Institute of the Max Planck Society

Faradayweg 4-6, D-1000 Berlin-Dahlem, West Germany

INTRODUCTION

Electron microscopy provides us directly with images of biological macromolecules rather than with diffraction patterns. This implies that we not only obtain the amplitudes of the objects' Fourier components but also their phases. Moreover, the instrumental resolution of modern electron microscopes is better than 2 Angstrom which is sufficient for the recognition of the amino-acid residues of a polypeptide chain. With the electron microscope being such an ideal instrument for the elucidation of protein structure, one would wonder why X-ray crystallography still exists.

A major problem in electron microscopy of biomacromolecules is the radiation sensitivity of the specimens. The biological molecules rapidly disintegrate during the electron radiation exposure needed to register a good image. Good images are those with a signal-to-noise ratio (SNR) that is sufficiently high for direct high-resolution visual interpretation. We can either register a good SNR image of the "ashes" of a protein molecule, or alternatively, register an image of the intact molecule that is so noisy that all detail is totally obscured.

Staining of the molecules in the electron microscopical preparations with heavy metal salts like uranyl acetate, enhances the contrast in the electron micrographs. The higher contrast implies a larger signal with more or less the same amount of noise. The SNR values are thus better in negative stain preparations rendering smaller image details visible. However, even in negatively stained images one seldomly achieves a reliable, directly interpretable resolution of better than 60 Angstroms. One of the reasons for this is that the negative stain crystallites, which surround the protein molecules, are themselves distributed in a random manner therewith introducing new statistical uncertainties.

Since noise is one of the major problems in electron microscopy of stained or unstained biomacromolecules, much of the research effort in the last 20 years has been focussed on enhancing the SNR values of the images. Averaging of large numbers of noisy images from individual molecules was the main basic idea for obtaining good SNR images. By recording a very large number of individual molecular images using a low total electron exposure per molecule, we can obtain images of intact molecules and yet have a good SNR result in the form of the total average. In X-ray crystallography, the total radiation load is shared by all individual molecules in the crystal during the registration of the diffractograms. Post-registration averaging of molecular images is the electron microscopical equivalent of this same principle of "shared suffering".

The use of periodic structures such as helices /1/ or two dimensional crystals - a natural extension of the X-ray crystallographic tradition - has led to the elucidation of significant structural information down to the 6 Angstrom level /2/, and recently even below 4 Angstrom /3/. This crystallographic approach, however, shares some of the fundamental limits associated with X-ray crystallography /4/: suitable 2D crystals may be difficult or impossible to obtain; it may be difficult to investigate an enzyme in relation to a substrate; etc. The purpose of this contribution is to review an alternative and complementary EM technique in which the individual molecules are not crystallized at all, and in which we fully rely on computer image processing techniques to extract the information present in large populations of "randomly oriented" individual molecular images.

ALIGNMENT PROCEDURE

A prerequisite for any averaging procedure is that the images of the individual molecules be aligned relative to each other. In a 2D crystal, this alignment is given a priori (the unit vectors of the crystal lattice can readily be determined) and one only needs to extract the individual unit cells and add them together to obtain the wanted average. In contrast, single molecules are not kept in a fixed position by their neighbours and they hence exhibit a large orientational freedom in the electron microscopical preparation. Let us first consider the relatively simple situation in which the molecule to be analysed exhibits a clear preferential orientation relative to the plane of the supporting carbon foil. An example of such a molecule is the rather flat glutamine synthetase molecule /5/. To bring the various copies of the molecule into register, prior to averaging, we have to apply translations ($\Delta x, \Delta y$) and rotations ($\Delta \gamma$) to the molecular images in the plane of the support film /5/.

Translational alignment between a molecular image and a reference image is achieved by searching for the position of the maximum in the cross-correlation function (CCF) /5/. The image to be aligned is then shifted by ($-\Delta x, -\Delta y$) to the position of maximum translational overlap. The CCF is calculated by multiplying the Fourier transform (FT) of the image with the complex conjugate of the FT of the reference; the CCF is obtained from this result by an inverse FT operation (Figure 1).

Rotational alignment is performed similarly by searching for a maximum in the rotational correlation function (RCF). The image and reference are both converted to cylindrical coordinates, and the cylindrical images are then Fourier transformed in the tangential direction. The resulting one-dimensional FT's are (conjugate) multiplied with each other; the RCF is the inverse FT of this product, integrated over the radial coordinate. The sequence of rotational and translational alignment (Figure 1) is normally repeated iteratively until an optimum is obtained /6,5,7/. This method of aligning images is very sensitive and its theoretical noise limits are known /8/.

In general molecules can lie in more than one position relative to the the support film, giving rise to entirely different projections through the molecule. The molecular orientation is then characterised by the shift vector ($\Delta x, \Delta y$) and three Eulerian angles ($\Delta \alpha, \Delta \beta, \Delta \gamma$). Such orientational freedom has to be accounted for before averaging the aligned molecules; it is only permitted to average molecular views that are very similar. Eigenvector eigenvalue analysis methods in the form of correspondence analysis /9,10/ have been introduced as an instrument to help sort mixed populations of images /11-17/.

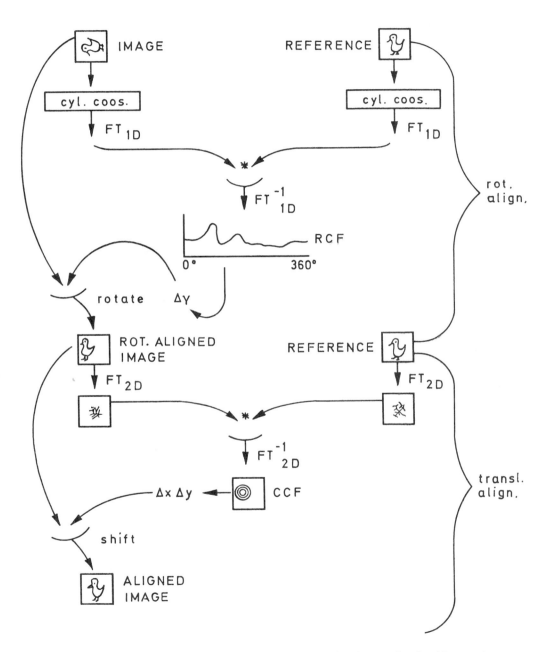

Figure 1. Iterative aligment procedure to obtain optimal alignment between an image and a reference image. For details see text.

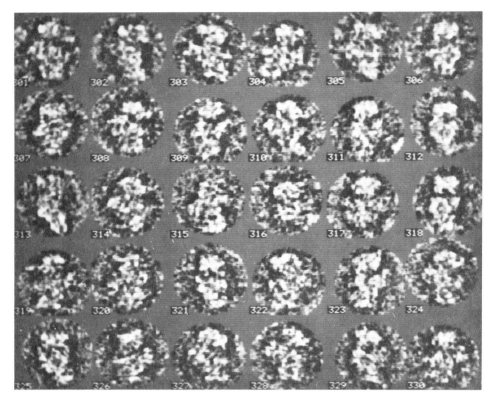

Figure 2. From a set of 914 original images of 30S ribosomal subunits of Bacillus stearothermophilus /17/, an arbitrarily chosen sequence of 30 images is shown after applying the alignment procedure described in the text with respect to several references. These images illustrate the noisiness and the heterogeneiety of the set of original images.

EIGENVALUE EIGENVECTOR DATA COMPRESSION

Given a set of aligned images, it is possible to express each of the images as a linear combination of predominant, independent "eigenimages" extracted from the set. The predominant eigenimages are the eigenvectors of the eigenvector eigenvalue problem described below. A full set of n images, each containing p pixels, can be represented by a single n x p matrix X. The matrix X has as its general element the measured density values x(i,j) at row number i and column j. Row number i contains the densities of all p pixels of image number i. The data set may be seen as a "cloud" of n points in a p-dimensional "image-space" (p-space) or as a cloud of p points in a n-dimensional "pixel-space" (n-space). We can extract the most characteristic properties of the data set X by solving the eigenvector equation:

$$X' . N . X . M . U = U . \qquad (1)$$

under the orthonormalization constraint:

$$U' . M . U = I \qquad (2).$$

In these formulae X' is the transposed matrix of X ; M is a p × p diagonal matrix containing weights ("metric" of the image-space); N is a n × n diagonal weight matrix for assigning different weights to different images (metric of pixel-space /16/), is the diagonal eigenvalue matrix and I is the unit matrix. The eigenvectors u, the columns of the eigenvector matrix U, are sorted according to the size of the corresponding eigenvalue.

If the metric matrices are taken to be unit matrices (which may be left out) we obtain the eigenvector equation for principal component analysis (cf. /10/), also known as the Karhunen-Loève transform in the field of image processing /cf. 18/. In correspondence analysis /9,10/, the diagonal elements of the metric matrices N and M contain as their diagonal elements:

$$n_i = \left(\sum_{j=1}^{p} x_{i,j}\right)^{-1} \qquad m_j = \left(\sum_{i=1}^{n} x_{i,j}\right)^{-1} \qquad (3)$$

With these metrics, the first eigenvector will necessarily point to the centre of mass of the data cloud; this eigenvector is predictable and may be ignored. In any subspace excluding the first eigenvector, the analysis is automatically relative to the centre of mass of the cloud which, in turn, allows for a straightforward simultaneous representation of "cases" and "variates" (2,16), or images and pixels in our nomenclature. With the generalized metrics M and N, the square distance in p-space between two images is given by:

$$d_{i,i'}^2 = \sum_{j=1}^{p} m_j \cdot (n_i \cdot x_{i,j} - n_{i'} \cdot x_{i',j})^2 \qquad (4),$$

which is the Chi-square distance for the metrics given by equation (3) /10,16/. The cross-correlation between image i and i' associated with this distance measure is:

$$vc_{i,i'} = \sum_{j=1}^{p} m_j \cdot n_i \cdot x_{i,j} \cdot n_{i'} \cdot x_{i',j} \qquad (5),$$

which is the general term of the symmetric variance-covariance matrix:

$$VC = N \cdot X \cdot M \cdot X' \cdot N \qquad (6)$$

The eigenvectors U obtained from equation (1) under the constraint (eq. 2), form a new basis of the image space which now is adapted to the shape of the cloud. The first eigenvector describes the largest variance of the cloud; the second, perpendicular to the first, describes the direction of the largest remaining variance etc. The matrix C containing the coordinates ("factorial coordinates" /16/) of all images relative to the unit vectors U of the new basis of the image hyper-space, is calculated by the weighted projection:

$$C = N \cdot X \cdot M \cdot U \qquad (7).$$

It is interesting to note that:

$$\begin{aligned} C \cdot C' &= N \cdot X \cdot M \cdot U \cdot U' \cdot M \cdot X' \cdot N \\ &= N \cdot X \cdot M \cdot X' \cdot N = VC \end{aligned} \qquad (8).$$

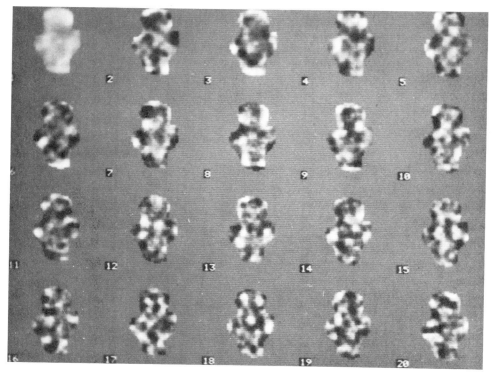

Figure 3a: The first 20 eigenimages (eigenvectors) of an analysis of 914 individual images of Bacillus stearothermophilus 30S ribosomal subunits /17/. The first eigenimage (with chi-square metrics) points at the centre of mass of the full data set and was ignored during the classification procedure. The eigenimages are orthogonal to each other (in the sense of formula 2) and describe a decreasing amount of the interimage variance of the data set. The variance described by each of the eigenimages is shown in Figure 3b. The mask within which the images were analysed is clearly visible in each eigenimage.

```
 2   6.356 %  I  ****************************************************************
 3   3.445 %  I  *********************************
 4   2.333 %  I  **********************
 5   1.630 %  I  ***************
 6   1.518 %  I  **************
 7   1.350 %  I  *************
 8   1.295 %  I  ************
 9   1.195 %  I  ***********
10   1.162 %  I  **********
11   1.099 %  I  *********
12   1.074 %  I  *********
13   1.009 %  I  ********
14   0.987 %  I  ********
15   0.971 %  I  ********
16   0.942 %  I  ********
17   0.917 %  I  ********
18   0.910 %  I  ********
19   0.889 %  I  ********
20   0.877 %  I  ********
```

Figure 3b:

Eigenvalue spectrum of second and higher eigenvalues associated with the eigenimages shown in Figure 3a. The first eigenvector, which points to the centre of mass of the data cloud was ignored. The 2nd through 20th eigenimages together describe 29.96 % of the variance of the data cloud relative to its centre of mass.

Equation (8) implies that the Euclidean cross-correlation as well as the corresponding Euclidean distance between image i and image i' in the coordinate space (factorial coordinate space or factor space) is identical to the generalised cross-correlation (eq. 5) and the corresponding generalised square distance (eq. 4) between images in our original data space.

Up to this point in the calculations, no actual reduction of the amount of data has taken place; we still have n eigenimages of dimension p and the number of coordinates per image is n. The data reduction occurs when we restrict ourselves to the eigenimages corresponding to the largest eigenvalues, thereby disregarding the higher ones which describe less and less of the variance in the data set and in our case contain merely noise. We typically use the first four to twenty eigenvectors which means that the images are now described by four to twenty numbers instead of the original 64x64 = 4096 density values. As an example a typical set of eigenimages is shown in Figure 3a. The corresponding eigenvalues are shown in Figure 3b expressed as percentage of the total variance relative to the centre of mass of the data set.

The data reduction obtained considerably facilitates the understanding of the information present in the data set in terms of classes or general trends. However, most of the (artificial) intelligence of the procedure lies in the next phase of the procedure: the decision making or classification phase.

CLASSIFICATION

Since we want to group similar images together into classes, in spite of the high noise levels present in the data, we exploit the noise reduction and data compression achieved by the eigenvector analysis and classify the images in the compact factor space. In this space, we apply a hierarchical ascendant classification (HAC) scheme /19/ which means that images (or classes) are merged two-by-two to form larger classes until, for example, a predetermined number of classes has been reached. The merging of images and/or classes into larger classes is based on a variance oriented proximity measure.

The variance criterion used, is based on a theorem that can be traced back to Huyghens /cf 20/. It states that for every possible partitioning of a collection of points in an Euclidean space into classes, the total variance of the set equals the sum of the inter-class variance and the total intra-class variance of the classes. The inter-class variance is defined as the variance ("second central moment" or "inertia") of the centers of mass of all classes relative to the center of mass of the total data set. The total intra-class variance is the sum of the internal variances of each of the classes, that is, the variance of each class relative to its own center of mass. An optimal partition, for a given number of classes, can now be defined as that partition in which the inter-class variance is maximal and the total intra-class variance is (thus) minimal. These classes are then as compact as possible in terms of their internal variance, while being at the same time as far apart as possible. To find this optimal partition, however, we have to check out an n! "infinity" of possibilities! The best solution achievable for larger data sets is therefore a local minimum which need not necessarily be the optimal partition defined above.

A good first approximation to such a local minimum can be calculated by the HAC using the "minimum added intra-class variance" as merging criterion. Two classes are merged if the increase of the total intra-class variance associated with this merging is the minimum possible increase at that level of the classification. The added intra-class variance (AIV) of the partition when merging two classes (or images) i and i', having the masses w(i) and w(i'):

$$\text{Add.Var.}_{i,i'} = (w_i \cdot w_{i'}) / (w_i + w_{i'}) \cdot d_{i,i'}^2 \qquad (9),$$

where $d(i,i')$ is the Euclidean distance in factor space, which was shown to correspond to the generalised distance in the original data space. At the starting point of the HAC procedure, each image has a mass $w(i)$ equal to $1/n(i)$ /19,16/.

As mentioned before, the partitions obtained through the HAC procedure are necessarily suboptimal. The fact that we take two classes (or images) together at a certain level of the HAC procedure binds the members of the two classes together forever. After an initial partitioning with the HAC scheme we therefore enhance the partition obtained by an iterative migration post-processor. Each original image is taken out of the class to which it is allocated and the AIV-proximity relative to all other classes is calculated. If the image is found to fit better in a different class after all, the image is placed into that other class and the intra-class variance of its original class and of its new class are recalculated; the process is then repeated with the next original image. Once all images have been checked and possibly reclassified, the whole partition has been changed so much that we have to restart the procedure. After a few iterations, no further migrations of images between classes occur: the process has stabilized. The reclassification refinement was found to decrease the total intra-class variance of the partition by values aroud 10%. Typically between 20% and 40% of the images changed their original HAC class membership during the reclassification.

A number of original images are notoriously bad in the sense that they are not typical within the data set. They may be misaligned particles or impurities in biochemical preparation etc., and we would like to exclude then from the data set by some automatic decision mechanism. Atypical images are often not well represented by the first few eigenvectors: these images will lie relatively close to the origin in the compact factorial coordinate space and will thus also lie close to each other. In the subsequent classification procedure, they end in the same class due to their proximity, although these images may have very little in common! This effect leads to "bad classes" (see Figure 4) which can, however, be pinpointed from their statistical behaviour during the summing from the original image data.

Another reason for (automatically) ignoring original images is that some images, although well represented by the first few eigenimages, have a very large AIV at the time they are merged into a class. In other words, these images do not resemble other images or classes very much. By ignoring (a predetermined number of) the last images to be merged into a class in the HAC procedure, this type of "bad" image may be excluded from the classification. The percentage of images to be ignored in this manner may be determined from a histogram of the AIV's of the images at the time they are being merged into a class. As an indicator of the quality of a specific class, we also find it useful to sort the classes by their compactness measured in terms of intra-class variance per class member. For further details on the classification procedures see /16/.

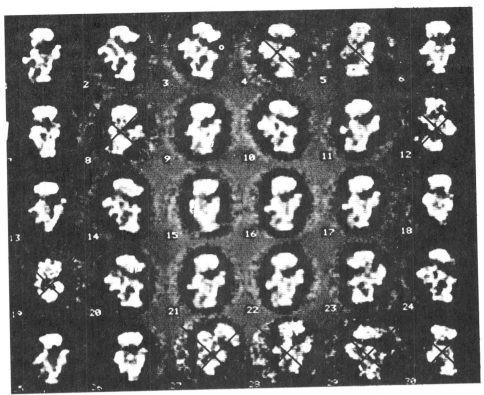

Figure 4. Result of a classification based on the image coordinates relative to the axes 2-20. The bad classes (see text) are crossed out. Together the good classes contain about 85% of all original images.

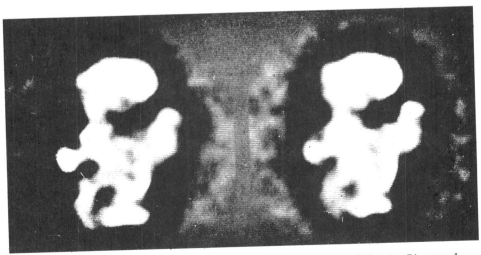

Figure 5. Two of the better classes, classes #1 and #17, in Figure 4, magnified to show the similarities between these two independent class averages. The resolution in the good classes is 17 Angstrom /17/.

The increase in SNR between the original images and the final class-averages can be very large: where no details can be seen directly in the very noisy originals, the class averages often show a wealth of structural detail. The resolution obtained in the good classes of this particular analysis is around 17 Angstrom, as determined by the Fourier ring correlation method /17/.

DISCUSSION

One fundamental problem with the alignment procedure needs some further attention: we can only align an image relative to a reference if that image looks like the reference. In a mixed population of images, this implies that we have to use the various class-averages as references, which are not available in the early stages of the analysis! The alignments are therefore alternated by classifications to find new references and the whole procedure is of an iterative nature. In aligning the set of images relative to a number of references, we proceed with that aligned version of each image which has the highest correlation coefficient relative to one of the references. The references themselves are also subjected to this "multi-reference alignment" procedure /21/ so as to assure a good total alignment of the data set. The whole iterative procedure of multi-reference alignments, eigenvector analysis and subsequent classifications is rather complex in organization, and an efficient and transparent computer image processing software system such as IMAGIC /12/ is indispensable.

Once the various characteristic views of the molecule in the electron microscopical preparation have been found, we try to fit these projections as well as possible into a three-dimensional reconstruction to get to the final goal of the study: the determination of the 3D structure of the macromolecule under investigation. This is a problem on which we are currently concentrating our attention /22,23,24/.

The image processing approach described in this paper relies heavily on computer resources. Several weeks of a dedicated computer (VAX 780) were needed to fully process the ribosome data presented here. The use of efficient algorithms at all levels of the analysis is important. Nevertheless, the new methods are evolving into a rapid and reliable technique to determine low-resolution (15-40 Angstrom) molecular structure in three dimensions. The approach is being used on various large (more than 200.000 MW) macromolecular assemblies.

ACKNOWLEDGEMENTS

I thank Dr. George Harauz and Vesna Fijala for their contributions to the work. Dr. Marina Stöffler-Meilicke provided the electron micrographs of the 30S Ribosomal subunit used in the example.

REFERENCES

1. D.J. DeRosier and A. Klug, Nature 217 (1968) 130-134.
2. P.N.T. Unwin and R. Henderson, J. Mol. Biol. 94 (1975) 425-440.
3. J. Baldwin and R. Henderson, these proceedings.
4. T.L. Blundell and L.N. Johnson, Protein Crystallography, (1976) Academic Press, New York.
5. J.Frank., W. Goldfarb, D. Eisenberg, and T.S. Baker, Ultramicroscopy 3 (1978) 283-290.
6. M. Steinkilberg und H.J. Schramm, Hoppe-Seyler's Z. Physiol. Chem. 361 (1980) 1363-1369.
7. J. Frank, A. Verschoor and M. Boublik, Science 214 (1981) 1353-1355.
8. W.O. Saxton and J. Frank, Ultramicroscopy 2 (1977) 219-227.
9. J.-P. Benzécri, L'Analyse des Données, Vol 2: L'Analyse des Correspodances, (1980) Dunod, Paris.
10. L. Lebart, A. Morineau and N. Tabard, Techniques de la Description Statistique (Dunod, Paris, 1977).
11. M. van Heel and J. Frank, Ultramicroscopy 6 (1981) 187-194.
12. M. van Heel and W. Keegstra, Ultramicroscopy 7 (1981) 113-130.
13. J. Frank, A. Verschoor and M. Boublik, J.Mol.Biol. 161 (1982) 107-137.
14. M.M.C. Bijlholt, M.G. van Heel and E.F.J. van Bruggen, J. Mol. Biol. (1982) 161, 139-153.
16. M. van Heel, Ultramicroscopy 13 (1984) 165-184.
17. M. van Heel and M. Stöffler-Meilicke, EMBO Journal 4 (1985) 2389-2395.
18. A. Rosenfeld and A.C. Kak, Digital Picture Processing, Vol 1, 2nd ed., Academic press (1982): Chap. 5-10.
19. F. Benzécri, Les Cahiers de l'Analyse des Données 5 (1980) 311.
20. J.-P. Benzécri, M. Danech Pejouh, T. Moussa and J.-P. Romeder, Les Cahiers de l'Analyse des Données 4 (1977) 369-406.
21. M. van Heel, J.P. Bretaudière and J. Frank, Proc. of 10th Int. Congr. on Elec. Mic., Hamburg 1982, Vol. 1, 563-564.
22. M. van Heel, Proc. of 8th Eur. Congr. on Elec. Mic., Budapest 1984, Volume 2, p 1347-1348.
23. G. Harauz and M. van Heel, in: Pattern Recognition in Practice II, Eds. Gelsema and Kanal, North-Holland (1985), in press.
24. G. Harauz and M. van Heel, submitted to OPTIK.

STRUCTURE ANALYSIS OF BACTERIORHODOPSIN BY ELECTRON CRYSTALLOGRAPHY

J.M. Baldwin, T.A. Ceska, R.M. Glaeser* and R. Henderson

MRC Laboratory of Molecular Biology, Hills Road
Cambridge CB2 2QH, England
* Permanent address: Biophysics Department
University of California, Berkeley, California 94720

INTRODUCTION

Bacteriorhodopsin (bR) is the protein found in the purple membrane of halobacteria, and it functions as a light-driven proton pump. It comprises a polypeptide chain of 248 amino acids and a light-absorbing chromophore retinal attached to one of the lysine residues. The protein is present in the purple membrane together with lipid molecules and forms a very well ordered two-dimensional array (Oesterhelt and Stoeckenius, 1971; Blaurock and Stoeckenius, 1971).

Several crystal forms of bacteriorhodopsin have been produced, as summarised in Table 1. Of these, three two-dimensional forms diffract in all directions to atomic resolution and are suitable for analysis by electron microscopy and diffraction. These three crystal forms are currently being analysed and are illustrated in Figure 1 where the projection structures are presented at 6Å resolution. From native p3 and p22$_1$2$_1$ data, 7Å resolution three-dimensional structures have been determined (Henderson and Unwin, 1975; Leifer and Henderson, 1983). For the deoxycholate treated (DOC) form, in which two thirds of the lipid content of the natural membranes has been removed, a preliminary 6Å three-dimensional map has recently been calculated (Tsygannick, unpublished) which shows essentially the same structure for the protein.

Three-dimensional diffraction intensities have been measured to 3Å resolution (Baldwin and Henderson, 1984) in the native p3 and p22$_1$2$_1$ forms. A little of this data is shown in Figure 2. Measurable diffraction intensity extends not only in the plane of the membrane (see Figure 4a) but also perpendicular to it (see also Figure 4b where an electron diffraction pattern from a 60° tilted membrane is shown). Table 2 gives an overall summary of progress on the three crystal forms and Figure 3 presents an averaged three-dimensional model based on the available native p3 and p22$_1$2$_1$ data (Baldwin, unpublished). It is clear from Figure 3 that the structure is composed of seven rod-shaped features which we presume to be α-helical sections of the polypeptide oriented perpendicular to the plane of the membrane. Further progress in understanding the structure and function of bR depends on the measurement of phases in three dimensions to 3Å resolution.

TABLE 1. CRYSTAL FORMS OF BACTERIORHODOPSIN

Description	Space group	N[†]	Cell dimensions Å	Degree of order Å	Reference
2-Dimensional					
In vivo form	p3	1	62.45	2.5	Henderson and Unwin, 1975
Orthorhombic	p22₁2₁	1	57.5;75.0	3.0	Leifer and Henderson, 1983
Acid form	p3	1	59.	better than 7	Henderson et al., 1984
DOC[*] form	p3	1	57.3	better than 4	Glaeser et al., 1985
3-Dimensional					
Hexagonal needles	P1	7	54;54; 230.	4;10	Michel, 1982
Cubes		6.7	220.	20	Michel, 1982
Membrane stacks	P312 or P321	1	62.45; 100.	2.5;50	Henderson and Shotton, 1980

† N = number of molecules in asymmetric unit
* DOC = lipid depleted form produced by treatment with deoxycholate.

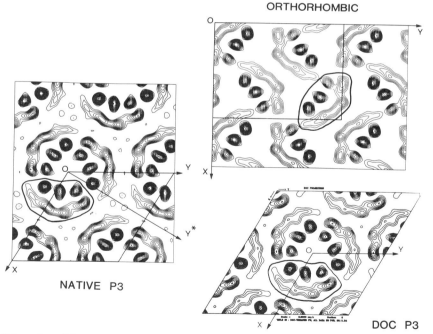

Fig. 1. 6Å projection maps of the 2-D crystal forms. An envelope is drawn round the bR monomer in each structure. The trimers of bR molecules are packed more closely together in the DOC p3 form than in the native p3 form as a boundary layer of lipid molecules has been removed.

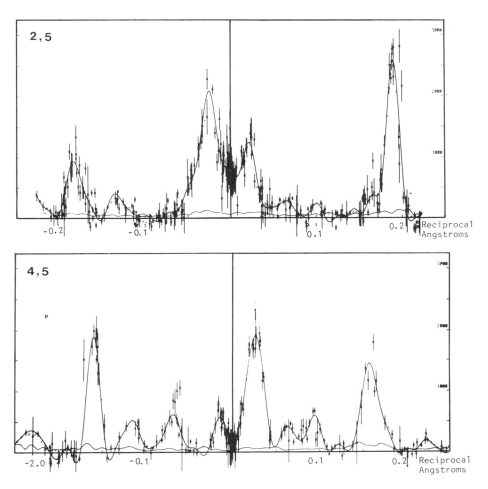

Fig. 2. Examples of merged electron diffraction intensities for two lattice lines in the p3 space group. The error bar on each observation is experimentally estimated from Friedel pair comparisons. The lower curve in each panel is the rms error in the best curve fitted to the observations.

TABLE 2. CURRENT STATE OF STRUCTURE ANALYSIS

Crystal	Native p3	Orthorhombic $p22_12_1$	DOC p3
7A			
Projection amps	✓	✓	✓
phases	✓	✓	✓
3-D amps	✓	✓	—
phases	✓	✓	—
3.5A			
Projection amps	✓	✓	—
phases	✓	—	—
3-D amps	✓	✓	—
phases	—	—	—

103

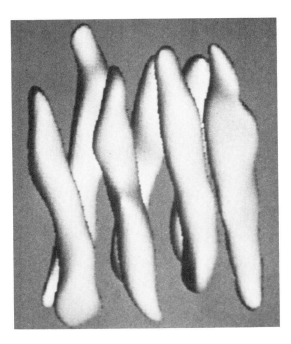

Fig. 3. A three-dimensional model produced by molecular averaging of the native p3 and the orthorhombic $p22_12_1$ data at 6Å resolution. The top of the model is at the cytoplasmic surface of the membrane. Connections between the seven rod-shaped features are not seen yet.

In the past, we have tried two approaches based on the use of the diffraction intensities. These were model-building and molecular averaging with phase extension (Henderson et al., 1983). Neither of these methods proved to be very powerful, so we have recently turned to (a) the use of isomorphous replacement with heavy atom derivatives and (b) high resolution imaging. There are papers already published (Katre et al., 1984; Hayward and Stroud, 1981) in which both these methods have been explored with little positive assurance that much could be achieved. Both methods have problems which is why we were initially deterred from undertaking them. However, as will be seen from the results presented below, it does look as though progress will be possible. One three-dimensional set of diffraction data on a phosphotungstic acid (PTA) derivative has now been collected using a liquid-nitrogen cooled cold stage on a Philips EM400. High resolution images have been successfully recorded on microscopes with stable cryo-stages (in collaboration with Fritz Zemlin in Berlin and others). Such images from untilted specimens have been measured and processed to 3.5Å resolution, giving a 3.5Å projection structure (see Figure 12 later). A paper describing this work has recently been prepared (Henderson et al., 1986).

USE OF ISOMORPHOUS REPLACEMENT

The amplitude for electron scattering does not increase linearly with atomic number as it does for X-ray scattering. Instead the increase is smaller, being approximately proportional to $Z^3/4$. This means that heavy atom derivatives of protein crystals with only a single heavy atom per protein molecule are unlikely to be as useful for the phasing of electron

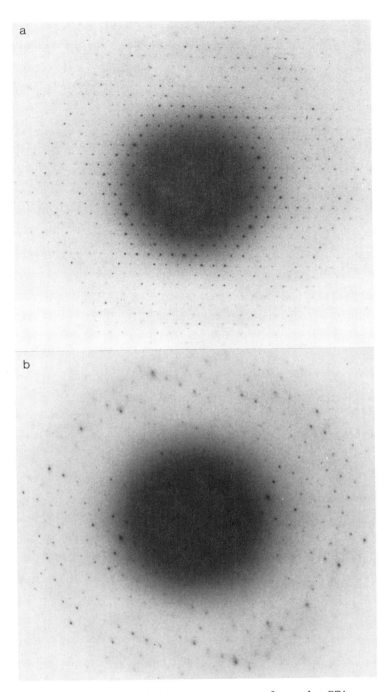

Fig. 4. Electron diffraction patterns from the PTA derivative. a) Untilted membrane. b) Membrane tilted at 60 degrees; the reflections are slightly less sharp perpendicular to the tilt axis. This is because the membranes are not perfectly flat.

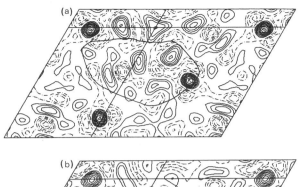
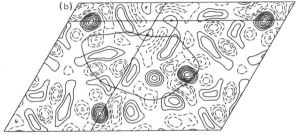

Fig. 5. PTA derivative difference maps at 6Å resolution. a) Projection map; several peaks can be seen within the protein boundary. b) Section of 3-D map near the cytoplasmic surface of the membrane, showing one of the major sites located in 3-D.

diffraction patterns as they are for X-ray diffraction. In particular the size of the changes in diffraction intensity will be smaller. A heavy atom derivative based on uranium atoms for electron diffraction is only likely to be as powerful as an iodine derivative would be in X-ray diffraction. Non-isomorphism, the main limitation encountered with X-ray diffraction methods is likely to be a greater problem for electron diffraction.

In the case of bacteriorhodopsin, we felt that the α-helical structure is so clear that rather poor phases would still be adequate to produce an interpretable high resolution map. Our approach in making heavy atom derivatives has relied on two premises. First, since single heavy atoms give small intensity changes we have been using heavy atom clusters. Second, since very good 6Å 3-dimensional phases are already available for initial determination of the locations of the heavy atom sites, we have been able to work with derivatives that have many sites and therefore larger intensity differences. We initially tried embedding purple membrane in high concentrations of heavy atom compounds. In general, positively charged heavy atom compounds destroyed the crystallinity, whereas negatively charged molecules preserved the membrane crystallinity and gave intensity changes in electron diffraction patterns.

Our evaluation of potential heavy atom derivatives involved densitometry of the electron diffraction pattern and extraction of the spot intensities (Baldwin and Henderson, 1984) followed by comparison of the derivative data with a 3-dimensional native data set to determine tilt axis, tilt angle, scale factor, R-factor, $\Delta F/F$, and the correlation between the ΔF values for symmetry related reflections. Difference Fourier maps were also calculated between derivative and native data. Of more than 20 heavy atom compounds tried, the one that gave the largest reproducible changes was

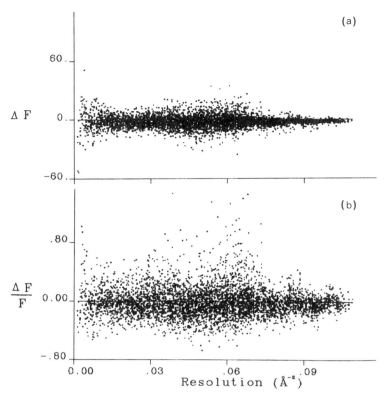

Fig. 6. Plots of ΔF versus resolution for the PTA derivative. The phasing power, scaling and some indications of isomorphism of this derivative can be estimated from the graphs of a) ΔF and b) $\Delta F/F$ as a function of resolution. The overall mean value of $|\Delta F|/F$ is 0.15. ΔF is largest at intermediate resolutions (5.0-3.5Å) and tapers off at the highest resolutions. The small size of ΔF and $\Delta F/F$ at the highest resolution suggests that the phasing power of this data may fall at 3Å resolution.

phosphotungstic acid (PTA). PTA is normally used as a negative stain for electron microscopy.

A 3-dimensional data set for the PTA derivative data set was collected and processed, and examples of diffraction patterns from untilted and 60° tilted membranes are shown in Figure 4. The entire PTA derivative data set consisted of 96 diffraction patterns from membranes tilted by 60°, 19 diffraction patterns from membranes tilted by 15° and 5 patterns from untilted specimens. After merging of this data set and scaling against the native data set a 3-dimensional difference map was calculated. The projection difference map is shown in Figure 5a and a section of the 3-dimensional map is shown in Figure 5b. The section is near the cytoplasmic surface of the membrane and shows the location perpendicular to the membrane of the large peak initially seen in projection. Nearly all of the PTA sites are on one or other of the membrane surfaces as is expected for a membrane which is embedded in, but not necessarily covalently reacted with, the heavy atom cluster.

Plots of ΔF and $\Delta F/F$ as a function of resolution are shown in Figures 6a and 6b. The differences, ΔF, tend to be largest in the intermediate

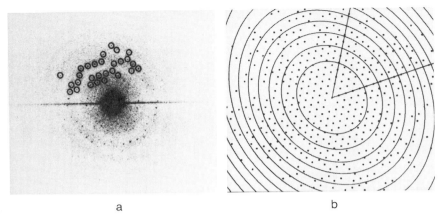

Fig. 7. a) Optical diffraction pattern from typical low-temperature image taken in Berlin. Some of the higher resolution diffraction spots have been circled for clarity. b) Computer generated plot for the same image showing all diffraction spots detected above the noise level. The zeroes of the contrast transfer function are indicated by lines; the image was underfocussed by 3300Å with some astigmatism.

resolution range. Two observations can be made from these plots. (a) The PTA derivative is isomorphous with the native data to 3Å resolution. (b) The differences taper off at the highest resolution and therefore the phasing power of the derivative will be weak there. Using the known phases to 6Å resolution it will be possible to refine the heavy atom positions and occupancies of this derivative. Preliminary attempts at this refinement indicate that five major sites and a few minor sites will be required to model the derivative accurately. Since we have only one derivative and the space group is non-centrosymmetric it is difficult to evaluate the phasing power beyond the known resolution. A second derivative data set will be needed before the heavy atom approach can be fully assessed.

HIGH RESOLUTION IMAGING

The problems with imaging at high resolution have recently been analysed quantitatively (Henderson and Glaeser, 1985). Image contrast is lost in comparison with what would be expected if the imaging was working according to theoretical calculations. At 4Å resolution, using any organic or biological specimen, the image contrast in the best images is only 2 to 3% of theoretical. The loss of image contrast can be attributed to several factors which are listed below in order of decreasing importance.
 (a) Specimen or image movement during the initial exposure to electrons as the specimen is being destroyed by the beam. This is also precisely the time during which the image contrast should be greatest.
 (b) Inelastic electron scattering which degrades the images.
 (c) Chromatic aberrations in the electron optics which gives rise to reduced contrast since the electrons have a finite energy spread.
 (d) Radiation damage causing weaker diffraction at the end of the exposure.
 (e) Poor resolution of film emulsions used in recording.
 (f) Stray magnetic fields in the vicinity of the microscope.
If some or all of these effects could be eliminated, the quality of the images would be greatly improved, and resolutions equal to that obtained by electron diffraction would be obtained. However, the first and most important problem, that of specimen movement, has not yet been solved.

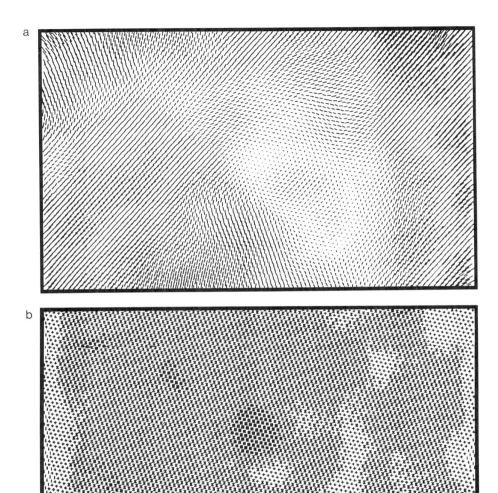

Fig. 8. Distortion maps of one of the images. In the computing procedure, a small reference area taken from the centre of the image is matched against each part of the whole image by means of a cross-correlation calculation. a) Vector displacements between the correlation peaks and the positions predicted based on a perfect lattice. Vectors are plotted at 20x actual size. In the centre of the image the displacement is zero since this is the reference area. b) Heights of correlation peaks at the actual lattice points; darker symbols indicate higher correlation.

Some of the other problems would be greatly reduced by the use of a high voltage electron microscope with better magnetic lenses, but without the most important problem being taken care of, these improvements would be of limited usefulness.

7	6	6	6	7	6	7	7	7		6	7	6	7	6	6	7	5	7
8	8	8	7	8	6	8	9	8		7	6	7	6	7	7	6	9	7
7	8	9	9	8	7	8	7	6		7	7	8	6	8	6	8	6	7
8	8	9	8	10	10	7	6	8		7	7	7	7	11	7	6	8	8
6	9	8	8	10	7	8	9	5		7	8	8	13	(35)	7	6	6	7
7	9	8	8	9	8	6	8	6		6	8	7	7	10	8	7	5	7
8	8	7	9	8	7	10	6	7		9	9	9	7	6	8	6	6	7
7	5	8	8	8	7	7	6	7		8	7	6	6	7	9	7	6	7
7	8	8	7	8	6	8	8	6		7	8	8	6	8	8	9	6	6

Fig. 9. Comparison of spots in the Fourier transform of an image before and after correction of the image to remove distortions. The summation of the intensities of 62 selected spots in the resolution range 5.5 to 3.5Å before correction (left) gives a peak which is just detectable, but after correction (right) the peak is quite strong, demonstrating the presence of high resolution information in the image of the crystal.

In spite of these problems, the use of large two-dimensional crystals and a cryo-microscope, together with the computer processing of large areas of each micrograph, can result in the restoration of enough of the lost signal-to-noise ratio that the diffraction spots out to 3.5Å resolution can be detected. This has now been done in projection for the native p3 form in collaboration with groups in Berlin, Heidelberg and Berkeley (Henderson et al., 1986). Some of the results of this work are described here.

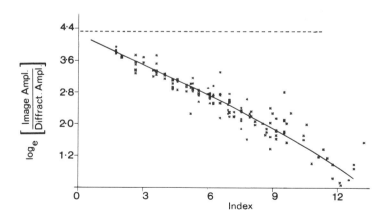

Fig. 10. Ratio of amplitude of a spot determined from the image to amplitude of the same spot from electron diffraction, plotted as a function of resolution for one of the images. The contrast at 4.0Å resolution is approximately 30 times smaller than it should be in theory (see dashed line). At low resolution the contrast is close to that expected. On the horizontal scale, an index of 12 corresponds to 4.5Å resolution.

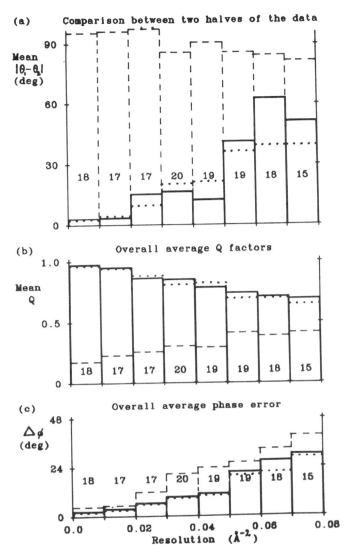

Fig. 11. Statistical analysis of phase data from 12 images.
a) Mean differences between estimates of the phases from two halves of the data treated separately are shown as a function of resolution by the solid lines. Also shown are the results of the application of an identical analysis to two sets of model data: 1) using random phases: the disagreements are 90 degrees as expected (dashed line); 2) using simulated phases having the same error distribution as that which we believe exists in the data: the disagreements are similar to those observed (dotted line). b) Q factor of resultant phase for each diffraction spot plotted as a function of resolution. The Q factor is defined as the ratio of the magnitude of a resultant vector to the sum of the magnitude of the individual vectors. Each of our vectors had direction given by the observed phase and magnitude related to the reliability of the phase measurements. c) Estimated phase error on the whole data set as a function of resolution. In b) and c) the solid lines are for the actual data, and the dashed and dotted lines are for the model data sets described in a).

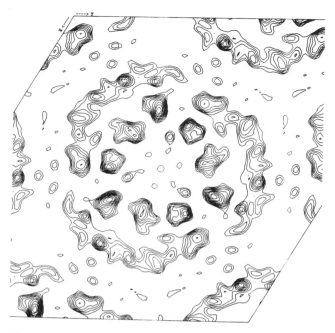

Fig. 12. 3.5Å projection structure of bR, based on 143 Fourier terms determined from 12 untilted high resolution images. Figure of merit weighting derived from the statistical analysis was used to weight down the less reliable measurements. The average figure of merit at 3.5Å is 0.85.
(From Henderson et al. 1986).

Figure 7 shows an optical diffraction pattern of an image taken near liquid helium temperature by Drs F. Zemlin and E. Beckman in Berlin, together with a computer simulation of the spots detected and the zeroes of the contrast transfer function for the same image. Figure 8 shows the application of a real space correlation method to treat small distortions in the positions of the molecules which are of the order of 1 part in 400 in magnitude. This corresponds to distortions from the ideal lattice positions of about 25Å over an image width of about 8000Å. Clearly to enable 3Å resolution data to be determined treatment for such distortions is essential. Figure 9 shows a display of the averaged intensity in the Fourier transform of 62 spots in the resolution range 5.5 to 3.5Å before and after treatment of the image for these distortions. The high resolution Fourier components are clearly seen after processing, even though they have only 2-3% of the theoretically expected contrast. A quantitative comparison of image and electron diffraction amplitudes is presented for one image in Figure 10. Clearly, if the initial images had contrast that was 30 times higher at the highest resolution, the results of image analysis would be superb and it would be easy to determine the high resolution structure. The improvement of these images is an important and useful goal.

Image analysis of 12 images taken on cryo-microscopes has provided enough phase information for a statistical analysis of the results to be made. Figure 11 presents Q-factor, figure of merit and phase error analyses. Figure 12 shows the 3.5Å projection map produced using the image phases. In the analysis of factors which require consideration in the

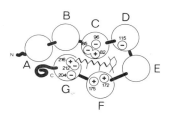

Fig. 13. Proposed model showing some of the features of the bR structure that may be important in its role as a proton pump. (Redrawn from Engelman et al. 1980).

image analysis (Henderson et al., 1986) several factors previously overlooked (Hayward and Stroud, 1981) were found. The most important of these was the need to make a correction to the image phases due to beam misalignment in the microscope (Zemlin et al., 1978; Zemlin, 1979; Smith et al., 1983).

PROSPECTS FOR EXAMINATION OF PHOTOCHEMICAL INTERMEDIATES

Our current model of the polypeptide arrangement in bacteriorhodopsin and the position of the retinal is shown schematically in Figure 13. We believe this to be the model with the best agreement to published data. The seven helices are shown connected as suggested by Engelman et al. 1980. The chromophore retinal, which is responsible for the purple colour of bR is attached to Lys 216 of the protein, and it is shown here in the all-trans conformation extending towards the position determined for it by neutron diffraction experiments (Jubb et al., 1984; Sieff et al., 1985). Charged amino acids known to be in the interior of the protein face each other towards the centre of the molecule, possibly forming a hydrophilic proton channel.

When a quantum of light is absorbed by the retinal group, the protein goes through a series of intermediates before returning to its unexcited state. During this cycle a proton is released on the outside of the membrane and taken up from inside the cell. At least four intermediate structures with distinct visible absorption spectra have been detected and all of these are likely to differ slightly in protein conformation, retinal isomerisation and protonation of ionisable groups. Some of these intermediates can be isolated in stable form at low temperature. Since it is now easy to carry out electron diffraction at low temperature, we have begun by examining the M intermediate, which absorbs maximally at 412 nm instead of at 568 nm like unexcited bR. This is the most accessible and probably the most interesting intermediate since the protons are released at this stage in the cycle of proton pumping. M probably has the retinal in the 13 cis conformation. Analysis of the data, recorded at liquid nitrogen temperature (Glaeser et al., 1986) has provided both a difference map in projection (Figure 14) and an estimate that the structural changes between bR and M are small, (equivalent to one or two side chains moving less than 5Å in addition to the isomerisation of the retinal). In the future, it should be possible to study other photochemical intermediates and to

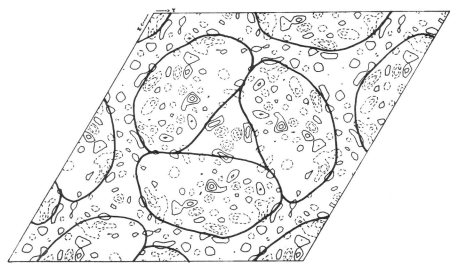

Fig. 14. Projection difference density map at 3.5Å resolution between the M intermediate and normal bR, showing small features mostly inside the molecular boundary. (From Glaeser et al. 1986).

complete the analysis of the M intermediate in three-dimensions once high resolution three-dimensional phases are obtained.

CONCLUSION

Three crystal forms of bacteriorhodopsin are presently analysed at 6Å resolution in three dimensions. Progress towards 3Å resolution has been made by the determination of the projection structure of the native crystal form by high resolution imaging and preliminary analysis of a heavy atom derivative. Future plans include the analysis of high resolution images of tilted membranes and the collection of a second derivative, which together will enable the determination of the three dimensional structure. Low temperature analysis of photochemical intermediates has demonstrated that small changes are detectable and should be interpretable.

REFERENCES

Baldwin, J.M., and Henderson, R. 1984, Measurement and evaluation of electron diffraction patterns from two-dimensional crystals, Ultramicroscopy 14:319-336.

Blaurock, A.E., and Stoeckenius, W. 1971, Structure of purple membrane, Nature New Biol., 233:152-155.

Engelman, D.M., Henderson, R., McLachlan, A.D., and Wallace, B.A. 1980, Path of the polypeptide in bacteriorhodopsin, PNAS, 77:2023-2027.

Glaeser, R.M., Baldwin, J.M., Ceska T.A., and Henderson, R., Electron diffraction analysis of the M412 intermediate of bacteriorhodopsin, 1985, Biophysical J., in press.

Glaeser, R.M., Jubb, J.S., and Henderson, R. 1985, Structural comparison of native and deoxycholate-treated purple membrane, Biophysical J., in press.

Hayward, S.B., and Stroud, R.M., 1981, Projected structure of purple membrane determined to 3.7Å resolution by low temperature electron microscopy, J. Mol. Biol., 151:491-517.

Henderson, R., and Glaeser, R.M., 1985, Quantitative analysis of image contrast in electron micrographs of beam-sensitive crystals, Ultramicroscopy 16:139-150.

Henderson, R., and Shotton, D.M., 1980, Crystallisation of purple membrane in three-dimensions, J. Mol. Biol., 139:99-109.

Henderson, R., and Unwin, P.N.T., 1975, Three-dimensional model of purple membrane obtained by electron microscopy, Nature 257:28-32.

Henderson, R., Baldwin, J.M., Agard, D.A., and Leifer, D., 1983, High resolution structural analysis of purple membrane, Proc. 41st EMSA, Ed. G.W. Bailey, pp. 418-421.

Henderson, R., Baldwin, J.M, Downing, K.H., Lepault, J., and Zemlin, F. 1986, Structure of purple membrane from Halobacterium halobium. Recording, measurement and evaluation of electron micrographs at 3.5Å resolution, Ultramicroscopy, in press.

Jubb, J.S., Worcester, D.L., Crespi, H.L., and Zaccai, G., 1984, Retinal location in purple membrane of Halobacterium halobium: a neutron diffraction study of membranes labelled in vivo with deuterated retinal, EMBO Journal, 3:1455-1461.

Katre, N.V., Finer-Moore, J., Stroud, R.M., and Hayward, S.B., 1984, Location of an extrinsic label in the primary and tertiary structure of bacteriorhodopsin, Biophysical Journal, 46:195-204.

Leifer, D., and Henderson, R., 1983, Three-dimensional structure of orthorhombic purple membrane at 6Å resolution, J. Mol. Biol. 163:451-466.

Michel, H., 1982, Characterisation and crystal packing of three-dimensional bacteriorhodopsin crystals, EMBO J. 1:1267-1271.

Oesterhelt, D., and Stoeckenius, W., 1971, Rhodopsin-like protein from the purple membrane of Halobacterium halobium, Nature New Biol., 233:149-152.

Seiff, F., Wallat, I., Ermann, P., and Heyn, M.P., 1985, A neutron diffraction study on the location of the polyene chain of retinal in bacteriorhodopsin, Proc. Nat. Acad.Sci. USA, 82:3227-3231.

Smith, D.J., Saxton, W.O., O'Keefe, M.A., Wood, G.J., and Stubbs, W.M., 1983, The importance of beam alignment and crystal tilt in high resolution electron microscopy, Ultramicroscopy, 11:263-282.

Zemlin, F., 1979, A practical procedure for alignment of a high resolution electron microscope, Ultramicroscopy, 4:241-245.

Zemlin, F., Weiss, K., Schiske, P., Kunath, W., and Hermann, K.H., 1978, Coma-free alignment of high resolution electron microscopes with the aid of optical diffractograms, Ultramicroscopy, 3:49-60.

NEUTRON DIFFRACTION : CONTRIBUTION TO HIGH AND LOW RESOLUTION

CRYSTALLOGRAPHY

B. Jacrot

European Molecular Biology Laboratory
c/o I.L.L.
156X, F-38042 Grenoble Cedex

The contribution of neutron diffraction to crystallography in molecular biology is essentially due to the difference of the scattering properties of H and D nuclei (1) ($-0.374 \cdot 10^{-12}$ cm for H and $+0.667 \cdot 10^{-12}$ cm for D). This characteristic can be used in high resolution work and in low resolution work.

I. HIGH RESOLUTION

The applications of neutron diffraction at high resolution have been recently reviewed by KOSSIAKOFF (2), and the results of investigations on myoglobin, trypsin, lysozyme, crambin and ribonuclease A are described in some details in the proceedings of a meeting hold at Brookhaven (3). Several informations are obtained from neutron diffraction, which are not, or are less accurately, obtained from X-ray diffraction. The main ones are the following :

- the coordination of a proton playing a key role in the active site. This has been done for trypsin (KOSSIAKOFF and SPENCER (4)) and for lysozyme (see MASON et al. in (3)).

- hydration. The normal strategy is that X-rays provide a first information on the oxygen. Neutrons then show the position of the proton. This position is obtained in a more reliable way if experiments are done both with crystal in H_2O and in D_2O, as the interpretation of a feature of the density map is than easier. This strategy has been well demonstrated by SAVAGE on vitamin B12. (Thesis work unpublished). There is certainly more reliability in assigning densities to water molecules from combined X-rays and neutron diffraction than from X rays alone.

- H bonds

- H-D exchange. This is an information not given by X rays but only by neutron diffraction and NMR. It has been interpreted in all proteins analysed by neutron diffraction to give information on accessibility to solvent and to infer from this accessibility something on protein dynamics (see for instance KOSSIAKOFF (5)). So far the experiments were done on samples soaked for very long periods in D_2O, and give the probability of exchange of a given proton at equilibrium. Recently in a study on lysozyme by MASON et al. (6) it has been possible to study the kinetics of exchange of amide protons. To do this, samples were soaked in D_2O at various pH (at pH 4.2 the exchange goes more slowly than at pH 8), various temperature and for period ranging from 3 hours up to 89 days, so it was possible to distinguish 4 classes of kinetics (very fast, fast, slow, very slow) and to assign each of the amide protons to one of these classes. There are no very obvious simple correlations between the kinetics and B factors or secondary structure features. The results are now analysed in detail to correlate the exchange to local motions or global displacement of part of the protein. Neutron data can be collected in reasonable time (a few weeks) and the method could be used for proteins too big for NMR, e.g. the heamoglobin or an Fab fragment of immunoglobulin.

If hydration can be studied by neutron diffraction, there is also some interest in localizing other molecules which can be substituted to water and see how they interact with the polypeptide chain. This had recently been done by LEHMANN and his colleagues (7) who have studied ethanol-lysozyme interactions. Deuterated ethanol has an average scattering density of $510^{-14} cm/Å^3$ very different from that of water (which is nearly zero), and is easily identified even if partially disordered. The figure 1 shows an ethanol molecule observed from a Fourier difference map, with the backbone of the polypeptide chain in contact with it. Thirteen molecules have been clearly identified with the following main conclusions :

The structure of the lysozymes is not modified with the 25% ethanol used in that study; the interaction between ethanol and the protein are essentially of hydrophobic character.

With position sensitive detectors and high flux reactors the time to collect the data on a native small protein up to ~2.5Å is of the order of a day. So the use of neutron as a final stage of refinement in protein crystallography of small or medium size proteins should be considered seriously.

II. LOW RESOLUTION

When data are collected at low resolution (<10Å), what matters is the average scattering density over rather large volumes. At the limit of very low resolution, this average density is (in $10^{-13} cm Å^3$ unit) 0.18 for a protein and ~0.32 for a nucleic acid, it is -0.05 for H_2O and 0.64 for D_2O (see 1 for further details). So it is possible by using crystals soaked in solvent made of various mixtures of H_2O and D_2O to obtain density maps in which the contribution of the protein and nucleic acid components in e.g. a virus can be modified in a controlled manner. This is quite useful for nucleic acid protein complex, which do not diffract to high resolution.

The method was first used for the nucleosome (8) showing in particular that the histone core in situ is at 16Å resolution identical to the histone octamer previously studied by electron microscopy. This result is of interest in the context of the controversy on the structure of the histone core.

The complex between tRNA$_{ASP}$ and tRNA synthetase from yeast also does not diffract to high resolution. Neutron diffraction could be used to localize and orientate the tRNA molecules in the unit cell (9).

The method can also be useful even with molecules diffracting to high resolution. Such is the case of the matrix porin which with detergents give crystal diffracting to high resolution but for which isomorphous derivatives have not been obtained. Neutron diffraction gives the position and the orientation of the molecule (a trimer) in the unit cell as well as the organisation of the detergent. Moreover is has provided (10) a structure of the trimer at 16Å. It is possible that this will be enough as a starting point to get a high resolution map using X-ray data, non-crystallographic symmetries and phase extension.

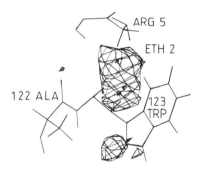

Figure 1

The contours show an ethanol (CD_3CD_2OH) molecule observed by Fourier difference in a crystal of lysozyme soaked in 25% ethanol solution. The molecule is on highly occupied site, in an hydrophobic pocket between an alanine and a tryptophan residue. (From Lehmann et al.)

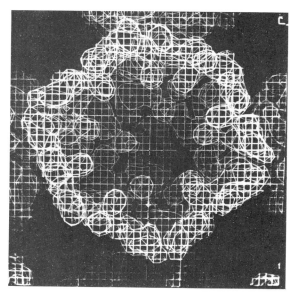

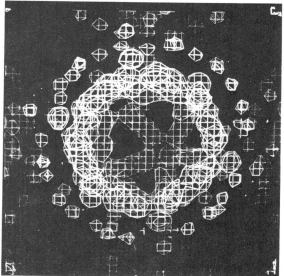

Figure 2

Maps of the protein part (Top) and of the RNA part (bottom) of Satellite Necrosis Virus (STNV) obtained from neutron diffraction respectively in 68% and 40% D_2O. The protein capsid agrees perfectly well with the high resolution structure obtained from X rays. The RNA is found mostly clustered on the 5 fold axis between the protruding part of the proteins and along the edge of the icosahaedron. (From Bentley et al.)

The case of plant viruses is different. X-ray diffraction at high resolution is possible, but with the disorder of the internal part of the virion, it does not provide information on the organization of the nucleic acid (RNA). At low resolution there are features everywhere in an X-ray map, but the interpretation is difficult. With neutron diffraction this map can be analysed in terms of an RNA map and protein map. In addition it is easier to collect the very low resolution peaks with neutrons (because $\lambda \sim 10 \text{Å}$ is used) than with X-rays, so the quality of the neutron map at low resolution ($\sim 16 \text{Å}$) is of higher quality; the method has been used for Satellite Tobacco Necrosis Virus (STNV) (11) and Tomato Bushy Stunt Virus (TBSV) (12).

With STNV, one finds RNA density clustered on the 5 fold axis (see figure 2) between the protruding protein subunits and along the edges of the icosahaedron. These last density features are nicely fitted with A-RNA. Altogether the RNA density corresponds to about 70% of the viral genome. This means that about 70% of the RNA is organized with the icosahaedron symmetry, at least within the 16Å resolution used in those experiments.

With TBSV it is found that the protein density is split into two domains. The outer one corresponding to the part known (at atomic resolution) from X-ray crystallography. The inner part, invisible in X-ray map is separated from the first one by a gap of some 30Å, filled with RNA with a very low protein density. These results agree with previous analysis based on neutron small angle scattering (13). The diffraction experiment allows a determination of the connectivity between the two domains and suggests models for the RNA folding.

The method of low resolution crystallography is now well established and its usefulness demonstrated by the above examples. The method does not require crystals of very large size, and data have been collected with samples of linear size of the order of a fraction of millimeter. With the operation of a dedicated diffractometer with a position sensitive detector at the Institut Laue Langevin, objects like ribosomal subunits could eventually be studied.

REFERENCES

1. B. Jacrot, Reports on progress in physics, 39:911-953 (1976).
2. A.A. Kossiakoff, Annual review of Biochemistry, 54:1195-1227 (1985).
3. "Neutrons in biology" edited by B.P. Schoenborn, Plenum Press (1984).
4. A.A. Kossiakoff and S.A. Spencer, Biochemistry 20:6462-6474 (1981).
5. A.A. Kossiakoff, Nature 296:713-721 (1982).
6. S. Mason, G. McIntyre, R. Stanfield, G. Bentley. Personal communication.
7. M.S. Lehman, S.A. Mason and G.J. McIntyre, Biochemistry, 24:5862-5869 (1985).
8. G.A. Bentley, A. Lewit-Bentley, J.T. Finch, A.O. Podjarny and M. Roth, J. Mol. Biol. 176, 55-75 (1984).
9. A. Podjarny, B. Rees, J.C. Thierry, J. Cavarelli, J.C. Jésior, M. Roth, A. Lewit-Bentley, R. Kahn, B. Lorber, J.P. Ebel, R. Giegé and D. Moras. Personal communication.
10. M. Garavito, M. Zulauf and P. Timmins. Personal communication.
11. B. Bentley, A. Bentley, L. Liljas, M. Roth, O. Skoblund and T. Unge. Personal communication.
12. D. Wild, P. Timmins and J. Witz. Personal communication.
13. C. Chauvin, J. Witz and B. Jacrot, J. Mol. Biol. 124, 641-651 (1978).

3 GRAPHICS AND STRUCTURE ANALYSIS

ELECTRON DENSITY FITTING WITH COMPUTER GRAPHICS:

A REVIEW AND A GLIMPSE

>T. Alwyn Jones
>
>Department of Molecular Biology
>
>Box 590, Biomedical Centre, S-751 24 Uppsala, Sweden

Introduction

The organizing committee has asked me to review the use of computer graphics in electron density fitting. Meeting them half way I shall give a brief description of this and then describe some recent developments that I have made in Uppsala.

Anyone who has used Beevers-Lipson strips to calculate a Fourier, realizes that a computer is a useful tool for crystallographers. Similarly, anyone who has used a pen and ruler to draw a molecule is quickly converted to computer graphics. From the beginning of the 1960´s therefore, crystallographers have been attracted to computer controlled pens and light beams.

By 1965, Levinthal had started drawing molecules on cathode ray tubes (see the review by Marshal et al., 1974 for a list of early systems). Many of these programs were interested in graphics as an aid in solving the protein folding problem. Pioneers should set their sights high.

The first protein model, of myoglobin, was built in 1959 out of a jungle of thin metal rods. Kendrew et al. (1960) used a scale of 5 cms/Ångström. Coloured paper clips were fixed to the vertical rods to mark density values. A more useful scale of 2 cms to the Ångström was later introduced by Kendrew. These wire models allowed bond rotations which could then be fixed by tightening screws. Together with the optical device designed by Richards (1968) this became the protein crystallographer´s plaything for many years. The Richards Box used a semi-silvered mirror to overlap the electron density (drawn as contours on plastic sheets) and the wire model.

One of the first attempts at making an electronic Richard´s box came from Oxford. David Barry and Tony North wrote a program that could display electron density with atoms superimposed. It ran on a Ferranti Argus and could draw only a limited number of vectors. It was used to look at substrate binding to lysozyme (Barry and North, 1971). Interestingly both Barry and North later constructed their own systems. Barry, working with Marshal in St. Louis produced the MMS-X, and North in Leeds made a PDP11/45 based system.

Figure 1 shows some of the systems developed from about 1973. The

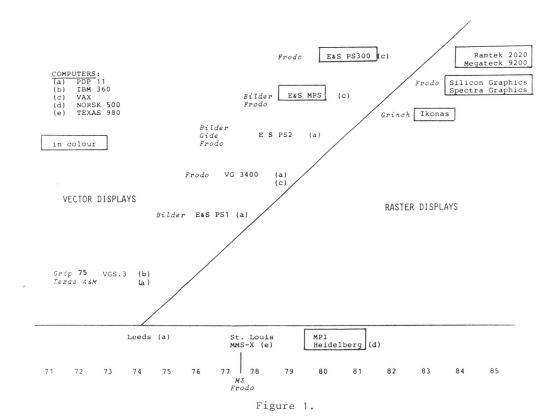

Figure 1.

1970´s showed a steady improvement in hardware which allowed one to draw and rotate enough vector to give a believable representation of the electron density. In 1973, Vector General produced their VG3 product that became the basis for a number of programs. The deluxe system was produced at Chapel Hill, N. Carolina in Brooks Laboratory and was known as Grip 75. The display was directly run by a PDP/11, which was used as a smart controller connected to an IBM 360. The program was written in PL/1 on the IBM 360 and the system was extensively used before the sudden spread of displays at the end of the 70´s (see Wright, 1982 for a list of projects). This list includes what is probably the first protein model to be completely built on a display (Tsernoglou et al., 1977). The program does many of the things we now take for granted, such as dihedral rotations, single atom and fragment movements, and model regularisation.

A number of completely mini-computer based VG3 systems were built. One should mention in particular the system at Texas A and M (Morimoto and Meyer, 1976) which was one of the first to interface to the Protein Data Bank (Bernstein et al., 1977).

By this time (mid-1970´s), the St. Louis group around Marshal and Barry had constructed MMS-X. This was controlled by a Texas 980 mini-computer, came with a well designed control box (2 joysticks, dials, function keys) and an architecture that was a delight to program. A number of programs were written for it. In particular, M3 written by Collin Broughton from Edmonton (Sielecki et al., 1982) has been extensively used by James´ group since 1980.

Although the early Evans and Sutherland equipment, LS1, was used to look at proteins, the first fitting program came with the Picture System series. Diamond at MRC, Cambridge, England wrote a program BILDER for the PS1 (Diamond 1980, 1982). Controlled by a PDP11/45, Diamond built a little masterpiece in 32K words. (Only someone who has tried to write a large program for the PDP/11 can appreciate this.) The user interface consists of a number of graphic menus (called pages), where items can be picked by a pen and tablet. Diamond used his experience from an earlier model building program (Diamond, 1966) to create a novel method for building the initial structure into a map. After building a standard conformation at one page, guide points can be chosen at another page to place a given atom at any other particular place in space (hopefully in density). The program attempts to achieve this by rigid body rotation, translation of the fragment, dihedral rotations of the flexible torsion angles, and by some flexibility of the amino acid tau angle. It was modified for use with the later model PS2 and at NIH for the VAX controlled Multi Picture System (MPS). NIH, MRC and Harvard in particular have used the program extensively.

Other PS2 based systems were designed in LaJolla by Dempsey (I know of no published description but it was used to build cytochrome C peroxidase, mentioned by Funzel et al., 1984) and in Drenth´s laboratory in Groningen (Brandenburg et al., 1981). One of the nice features of the Groningen program, GUIDE, is its ability to handle multiple sets of protein coordinates. It has been extensively used to correct errors during refinement, and to build a number of new structures.

Vector General´s next product, VG 3400, came in 1976, and John Gassmann inMartinsried bought the very first one. By a series of events that I will not go into, I was told to do something with it. It came equipped with only dials, but after talking with Diamond we rapidly bought a tablet. It was controlled by a PDP/11 which had a fast link to a Siemens 4004 mainframe computer. The Siemens was at this time heavily used by Huber´s group to refine structures using real space refinement (Diamond, 1971; Deisenhofer and Steigemann, 1975). What was most needed was a program to help in refinement, i.e. to remove the errors that could not be corrected by the existing refinement programs. This was a major consideration in the design of a program that was later called Frodo.

In Frodo, there are multiple user interfaces; only one is on the display, the rest are at a terminal. The display interface is kept as simple as possible, and designed on the what-you-see-is-what-you-get principle. The program is described in a number of publications (Jones, 1978; 1982). After leaving Munich, I ported it to a PS2 in 1979, and MMSX in 1980. Tickle (Birkbeck, London) made lots of improvements on my PS2 version which I incorporated with my own changes in a new PS2 version in 1980. This was put on a VAX/MPS system by Bush (Merk, New Jersey). The extensions I made for our VAX/VG3400 system at Uppsala were incorporated in some of the other systems (particularly by Bush). For my work on satellite tobacco necrosis virus (Jones and Liljas, 1984) I needed access to electron density rather than pre-contoured density. I therefore wrote on-line contouring (like M3) and interactive real space fitting. The MMS-X version (with map files to save space on the Texas 980) has been much used by the Rossmann, Matthews (Oregon) and Hendrickson groups.

By 1980, colour displays became available on the MPS, although few made good use of it in fitting programs. A further turn of the technological spiral came with the PS300 series. With a built-in microcomputer, this machine could draw lots of coloured vectors and promised to reduce the load on the controlling VAX. One of the first crystallography groups to get one was Quiocho´s at Rice University. They decided to port Frodo, using the MPS and VG programs as a start. This conversion was made by Pflugrath and Saper and extended by Sack. A conversion to one of the first useful raster displays (Silicon Graphics Iris 2400) has been made by Oatley at LaJolla.

The spread of graphics has quickened dramatically in the last five years for economic reasons - both displays and computers are now cheaper and more powerful. For example, a black and white VG 3400 cost about $70,000 in 1976, while a useable coloured PS330 costs about $60,000 today and does much more (although more memory, a fast link and display buffer increase the price by more than another $20,000). Similarly a micro VAX 2 is a blessing compared to a PDP11/40.

A different approach to building models has been suggested by Greer (1974). This involves describing the density as a skeleton of linked points. The aim was to use this to fully automate the building process. Another approach (starting with a skeleton) uses rule based artificial intelligence methods (Johnson, 1980). Neither have been successful in interpreting a new map. The North Carolina group has developed a graphics program Grinch that displays a skeleton on a raster machine (or more recently a PS300). (Since I know of no publication on the system, the following description may not be too accurate.) The level of the skeleton can be controlled by a dial to pick out the highest features in the map. These can then be coloured and used as a basis for building a model.

Some Recent Developments to Frodo in Uppsala

The present implementations of Frodo suffer from two main drawbacks. Firstly, unskilled users can create stereochemically incorrect (or un-favourable) conformations because of ignorance particularly when building a structure at non-atomic resolution in a poorly phased electron density map. Secondly, the representation we use for the electron density contains too much information so that when trying to interpret a map the user becomes overwhelmed. For this reason I have always suggested that mini-maps be used for solving the folding of the chain. A model can then be built by using guide points, measured from the mini-map, to fit secondary structure templates (e.g. α-helices, β-strands and/or standard turns) and filling in the rest with 2 or 3 guide points per residue (method 4 in Jones, 1982). Very few new structures have been built with Frodo without prior extensive study of mini-maps. (I know of only one, our own work on retinol binding protein, Newcomer et al., 1984). However, I think the tools described below will finally make mini-maps unnecessary.

There are at present 274 sets of protein coordinates deposited at the Protein Data Bank. Why not use this information (the results of thousands of man-years work) in our modelling? I have written a set of tools that pick out the best match between an imput set of $C\alpha$-coordinates and a data base of structures. This is done in real time at the graphics terminal using a very fast matching algorithm that works with diagonal plots.

Command DGNL allows one to define a connected piece of chain along which the position of $C\alpha$ atoms can be specified if so desired. Otherwise the program places $C\alpha$'s at the correct spacing. The resulting piece of chain is compared with all equivalent length pieces in a data base to find the ten best fits. These are transformed and can be displayed in turn as a stripped poly-alanine chain. Any of them can be accepted and put into the Frodo coordinate data set.

DGLP works in a similar way but is used to search for chain fragments where only the positions of a few $C\alpha$'s in the fragment are known. This command is meant for modelling related proteins.

Tests carried out on retinol binding protein suggest that all parts of a new protein already exist in the Protein Data Bank. I could build RBP from satellite tobacco necrosis virus, alcohol dehydrogenase and carbonic anhydrase C, such that the r.m.s. deviation for all main chain atoms was 1.0 Å after regularisation. The protein was built at the display with no attempt to find the optimum set of bits and pieces and with no consideration of the linking points. Since most people are not experts on protein conformation, it seems reasonable to suggest that if you build your first model from correct pieces, you will get a better model.

A few years ago I tried to put skeletonisation (Greer, 1974) into our VG3400 version of Frodo. It was hopeless (partly because of the black and white screen). I learned my lesson and have a new implementation for the PS300 designed in collaboration with Sören Thirup. The electron density has to be skeletonised, and its output processed. The result is a set of points with a connectivity where each point has a status (assigned depending on connectivity length) of main chain or side chain. These programs have been written by Thirup.

The "bones" atoms can be displayed with Frodo and are treated as normal atoms but with some important differences. As such, they can be moved around and bonds can be made or broken. However, the new connectivity can be saved. We allow each bone atom to have one of three possible statuses - M, S, or U - and the user can redefine the values given by the preprocessing program. The U status is used to define the main chain that you find acceptable (Useful) while M and S stand for main and side chains.

Our new version of Frodo has various atomic colouring modes, such that the MSU atom types can have different colours. It is therefore easy to see their current status and their position in the electron density.

Any combination of MSU can be displayed so that you can get a global view of your current main chain definition whenever you want.

The skeleton you see depends on the base value you define in the preparation program. If it is too low, there will be too many bonds and if it is too high there will be too few. Our experience to date suggests using approximately 1.3 times the standard deviation of the map. This is higher than the value we normally use for density contouring (which is approximately 1 standard deviation).

Once you are happy with your interpretation (or sooner) you can use DGNL to match the skeleton to suitable pieces from the data base. The side chains are automatically added in a standard conformation. These can be matched to the skeleton with BFIT (written by Thirup), and then to the density with RSR (Jones and Liljas, 1984). Our aim is to go from a suggested skeleton interpretation to an optimally fitted molecule, fully automatically. There are other ideas in the pipeline for matching the skeleton to the sequences, but you will have to wait a few months. Happy fitting!

References

Barry, C.D. and North, A.C.T., 1971, Cold Spring Harbour Symp. Quant. Biology, XXXVI, 577.
Bernstein, F.C., Koetzle, T.F., Williams, G.J.B., Meyer, E.F., Brice, M.D., Rodgers, J.R., Kennard, O., Shimanouchi, T. and Tasumi, M., 1977, J. Mol. Biol. 112 :535-542.

Brandenburg, N.P., Dempsey, S., Dijkstra, B.W., Lijk, L.J. and Hol, W.G.J., 1981, J. Appl. Cryst. 14 :274-279.
Deisenhofer, J. and Steigemann, W., 1975, Acta Cryst. B31 :238-250.
Diamond, R., 1982 ,Computational Crystallography. Edited by D. Sayre, p. 318-325. Clarendon Press: Oxford.
Diamond, R., 1980, Computing in Crystallography, 21.01-21.19, Indian Academy of Sciences.
Diamond, R., 1971, Acta Cryst. A27 :435-452.
Diamond, R., 1966, Acta Cryst. 21 :253-266.
Funzel, B.C., Poulos, T.L. and Kraut, J., 1984, J. Biol. Chem. 259 : 13027-13036.
Greer, J., 1974, J. Mol. Biol. 82 :279-302.
Johnson, C.K., 1980, Computing in Crystallography, 28.01-28.16, Indian Academy of Sciences.
Jones, T.A. and Liljas, L., 1984, Acta Cryst. A40: 50-57.
Jones, T.A., 1982, in Sayre, D. (ed.) Computational Crystallography, p. 303-317. Clarendon Press: Oxford.
Jones, T.A., 1978, J. Appl. Cryst. 11 :268-272.
Kendrew, J.C., Dickerson, R.E., Strandberg, B.E., Hart, R.G., Davies, D.R., Phillips, D.C., and Shore, V.C., 1960 ,Nature 185 :422-427.
Marshal, G.R., Basshard, H.E., and Ellis, R.A., 1974, Computer Representation and Manipulation of Chemical Information. Edited by W.T. Wipke, S.R. Hyde, S.R. Heller, R.J. Feldmann and E. Hyde, pp. 203-237. New York: John Wiley.
Morimoto, C.N. and Meyer, E.F., 1976, Crystallographic Computing Techniques. Edited by F.R. Ahmed, pp. 488-496. Copenhagen: Munksgaard.
Newcomer, M.E., Jones, T.A., Åqvist, J., Sundelin, J., Eriksson, U., Rask, L., and Peterson, P.A., 1984, EMBO J. vol. 3, no. 7 1451-1454.
Richards, F.M., 1968, J. Mol. Biol. 37 :225-230.
Sielecki, A.R., James, M.N.G., and Broughton, C.G., 1982, Computational Crystallography. Edited by D. Sayre, pp. 409-420. Clarendon Press: Oxford.
Tsernoglou, D., Petsko, G.A., McQueen, J.E., and Hermans, J., 1977, Science 197: 1378-1381.
Wright, W.V., 1982, Computational Crystallography. Edited by D. Sayre, pp. 294-302. Clarendon Press: Oxford.

COMPUTER GRAPHICS IN THE STUDY OF MACROMOLECULAR INTERACTIONS

Arthur J. Olson, John A. Tainer, and Elizabeth D. Getzoff

Research Institute of Scripps Clinic, Molecular Biology Department
La Jolla, California 92037

COMPUTER GRAPHICS IN MOLECULAR MODELING

The major purpose of molecular graphics is that of modeling, that is the representation of structure and/or process for the purposes of evaluation and understanding. Developments in computation and computer graphics have enabled the production and use of powerful tools for molecular modeling, far surpassing the physical models originally used for the purpose. Hardware advances in processor and display technologies have made possible real-time manipulation of three-dimensional models of increasing complexity and "realism". Algorithmic developments are producing a variety of useful, interpretable representations and intelligible renderings for construction and display of models. Because of these advances, it becomes most important to emphasize the distinction between model and primary observation. The computer generated image or simulation may look compelling, but seeing is not believing. Looking is the name of the game. A model is only as good as the information used to construct it.

If a model is only a reformulation of existing information, what then is its utility? Computer graphic molecular modeling supports three general activities: synthesis, analysis, and communication. In each of these activities it is the high bandwidth of information transmission and the three-dimensional context of the visual images that give utility to the computer graphic representation. In addition, the possibility for real time manipulation and modification make these representations flexible and extensible.

In the activity of synthesis, the modeling is used to build or extend existing models by combining information and knowlege from a variety of sources for its construction. The piecing together of molecular fragments from a chemical fragment database or the fitting of a protein structure into a three-dimensional electron density map typifies this modeling activity. Here, real-time interaction is a key ingredient in integrating the information at hand.

Analysis requires selective display of experimental and/or computational results in a comprehensible framework. Here, interpretation and evaluation are the critical operations and human pattern recognition is the dominant process. Display and comparison of any number of macromolecular properties such as chemical composition, connectivity, molecular shape, electrostatic properties, or mobility characteristics all fall into the domain of modeling analysis. As more macromolecular structures become available for examination and comparison, and as more techniques are developed for analysis, new patterns will emerge to help classify, organize, and interpret the wealth of information contained in any single macromolecular structure. This activity then feeds back to the synthesis activities which can construct models for the next level of biomolecular organization.

The third major activity of modeling is that of communication. Communication is critical in science if ideas are to compete and take hold. Computer graphics provides a powerful medium for conveying complex, three-dimensional relationships in the structure and function of biological molecules. It is imperative that this information is comprehensible not only to the structural scientist but also to the larger body of scientists whose expertise can add data and knowledge to increase overall understanding. The model as communicator also serves the important function of bridging the gap between science and the broader community. By making complex scientific results more accessible to the public, we can not only interest bright, young people in looking deeper into the area, but we can also demystify science to the public in general which supports this work.

TOOLS FOR COMPUTER GRAPHICS MOLECULAR MODELING

Software is the key ingredient in producing useful tools for molecular modeling. Both algorithms and programs must be designed with the application firmly in mind. Specification of an algorithm to calculate a property, render an object, or perform a task, depends upon a host of criteria including relevance to the modeling operations, generality, accuracy, intelligibility and interpretability of the results and computational expense.

As an example relevant to the modeling of macromolecular interactions we look at some algorithms for calculating and rendering macromolecular surfaces. The simplest representation is the van der Waals surface which produces a "CPK-type" model of the molecule. Production of the surface involves surrounding each atom with a sphere drawn at its van der Waals radius. Some computation is needed to remove pieces of the surface that lie inside neighboring atomic spheres. This algorithm has the advantage of being computationally inexpensive and of producing surface images that resemble a physical model that most chemists are familiar with. In addition, real-time bond manipulation with surface following is easily achieved if non-nearest neighbor overlaps are ignored. For displaying complete macromolecules or complexes of macromolecules, however, the representation and the rendering have several drawbacks. Firstly, much of the surface calculated in this manner is internal to a macromolecule and represents inaccessible interstices. Thus, for instance, a surface area calculation using this algorithm would grossly overestimate the amount surface at the solvent interface. In terms of the images or renderings that this surface presents, there are additional considerations for transparent surfaces or vector representations which do not use hidden line or hidden surface algorithms (e.g., dot or mesh surfaces used in interactive modeling). The van der Waals rendering is not intelligible because of the large number of overlapping surfaces in the display. In addition, even with hidden surface rendering a CPK model of a very large structure tends to confuse interpretation of surface shape because of the large number of lines created by the spherical intersections. Thus, the van der Waals surface serves best for calculation and rendering of small molecules or of small fragments of larger molecules.

Algorithms which incorporate solvent accessibility as a criterion overcome some of the objections of the van der Waals surface. Added volume algorithms, as implemented by a number of workers (1.2), expand each van der Waals sphere by the radius of the sphere representing a solvent molecule. This technique effectively eliminates any interstitial surfaces that cannot be accessed by the solvent. Because it resembles the van der Waals surface only using larger radii, the computational time for creating this added volume surface is relatively fast. Since the molecular surface is pushed out it is difficult, however, to assess complimentarity of two surfaces and rendering has similar drawbacks to the van der Waals surface in that sphere intersection lines can disrupt and confuse a smooth view of surface shape.

The solvent accessible molecular surface algorithm implemented by Connolly, first in numerical form and subsequently in analytical form (3), overcomes the drawbacks of the expanded volume surface but at the cost of computational expense.

The Connolly molecular surface produces a surface "skin" surrounding a macromolecule, constructed by its interaction with the probe sphere. The skin is smooth and continuous, composed of convex patches at the areas of contact between probe and single atoms of the target, and re-entrant patches composed of inward-facing pieces of the probe sphere when it contacts two or more target atoms simultaneously (see Fig. 1). This algorithm has the advantage of producing a description that has physical meaning with regard to macromolecular interactions, thus enabling accurate calculation of properties such as surface area, volume, and shape characteristics. From a modeling point of view, the fact that the surface is smooth and continuous and has both convex and re-entrant parts makes it effective in showing surface topography and in examining shape complimentarity at macromolecular interfaces. The disadvantage of this surface is that it is computationally expensive, requiring about an hour to calculate a typical protein surface on a VAX-11/750.

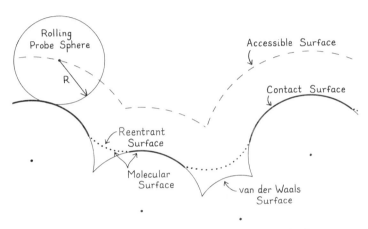

Figure 1. Different surface definitions. A slice through several solvent-exposed atoms of a molecule delineates the van der Waals surface on which a rolling probe sphere traces out the accessible, contact, and molecular surfaces (with component reentrant and contact pieces) described in the text.

In contrast to algorithm design, program development should be concerned with the human factors associated with utilization. An elegant algorithm may be of little value if the program that implements it is unused or unusable. Thus, issues such as software portability, the person-machine interface, flexibility, and extensibility come into play, especially in producing interactive modeling programs. It is not within the scope of this paper to address these issues in detail. However, it is worthwhile to briefly mention our experience.

Our approach to the problem of producing interactive software for molecular modeling has been one of functional modularity. Thus, GRAMPS (4) was developed to provide a real-time interactive language for three-dimensional picture editing, manipulation and animation. Layered on top of GRAMPS we developed GRANNY (5) as a molecular modeling package. Separating the graphics functionality from the molecular modeling has several advantages. Firstly, the necessary machine dependencies of the graphics interface are isolated from the molecular modeling software, providing a measure of portability. Secondly, the functionality of each system may be used separately or simultaneously to provide a large degree of flexibility. Thus, for instance, a novice molecular modeling user could access all of the functionality he needs by interacting with GRANNY alone, while an advanced user with special unanticipated needs can access GRANNY for standard tasks and GRAMPS to construct and manipulate non-standard objects. Thirdly, the capability of molecular animation is enhanced by this modular structure. Animation is principally a graphics task, but the construction of the molecular "actors" is a task for molecular modeling. Finally, the program structure gives a path for extending functionality by developing other programs that communicate with GRAMPS and/or GRANNY. For instance, a module for automatic construction of crystallographic symmetry can be developed without modifying the existing programs.

MODELING MACROMOLECULAR INTERACTIONS

Our approach to understanding macromolecular interactions is based upon the computer graphic analysis of crystallographically determined protein structures. Specific areas of interest include modeling molecular shape, electrostatic forces, solvent structure, and patterns of mobility. Ideally, computer graphics allow the integration of such diverse molecular properties into a model of protein structural and functional interactions which suggests specific and useful experiments. One can envision a typical macromolecular interaction as consisting of at least three stages: 1) precollision orientation by electrostatic forces, 2) local recognition involving the interaction of specific residues, and 3) the induced fit accomplished by local structural rearrangements of both molecules. Here we outline our current approaches for understanding and analyzing these three stages of interaction. By developing as many different ways as possible to examine protein structure, we hope to learn how to better approach the problems of modeling unknown complexes. The most useful approaches are those which simplify or skeletonize the information so that the degree of complexity is reduced to an interpretable level without inducing too much error in the parameter of interest. The goal of modeling is to effectively display properties of interest so as to allow recognition of patterns in complex data, rather than to develop ways to see all the details simultaneously. Thus, it is often more informative to model change or even rate of change in a parameter than it is to examine the parameter directly.

Molecular surface shape plays a critical role in molecular recognition. Analysis of the few known macromolecular complexes and the interfaces of oligomeric proteins indicates that these stable complexes are characterized by a high degree of shape complementarity. It is therefore important to develop ways to analyze and simplify shape so that this parameter can be used to predict molecular interactions. Molecular shape can be modeled from either local or global perspectives; both are useful for evaluating the influence of shape on docking.

The easiest way to examine detailed shape directly is to interactively display the molecular dot surface implemented by Connolly. This surface clearly depicts the shape of the molecular boundary as seen by interacting solvent molecules. We have examined the molecular dot surfaces of a number of high resolution proteins interactively using the graphics language GRAMPS (4) and the molecular modeling program GRANNY (5). We find that local shape clearly influences the interactions of small molecules with proteins. For example, in known crystal structures, water molecules lie along grooves in the molecular surface and metal ion binding sites found in isomorphous "heavy atom" derivatives occur in pre-existing pits or invaginations (6,7). These hydrophilic pockets range in size from small indentations, just large enough to encompass the ion, to deep invaginations up to 15 or more Å long (see Fig. 2). Local shape complementarity is also important at oligomeric macromolecular interfaces where it can be assessed by the volumes of gaps and interpenetrating surfaces, as well as the twist and degree of interdigitation of interacting surfaces. Dot surfaces defining the surface area of each subunit buried from solvent accessibility by the adjacent subunit are helpful in analyzing all of these characteristics.

Figure 2. Preexisting hydrophilic pockets which bind metal ions in Cu,Zn superoxide dismutase. The external molecular surface of superoxide dismutase is shown as a thin, hollow shell colored white on the outside and gray inside. A clipping plane cuts away the front surface of the protein to simultaneously reveal both inside and outside surface topography. The entire alpha carbon backbone is shown unclipped as a bent wire tube (dark gray). Two light colored spheres representing the copper (upper center) and zinc (upper right) can be seen inside the external surface underneath the active site. Bonds and atoms (white spheres and connections) are shown for the side chains of Arg 141 (top left) and Asn 63 (upper right) which close off the two deep pockets from opening into tunnels. The programs AMS (8) and MCS (9) were used for calculating the surface and the raster image.

Information on molecular surface indentations can be skeletonized by calculating and displaying the trajectory of the rolling probe sphere when in contact with two atoms (3). Increasing the size of the probe sphere decreases the detail with which shape is sampled by both the dot surface (10) and the probe trajectory. While bigger probes can be used for examining accessibility to molecules larger than water, their utility is limited by the errors resulting from poorly defined positions of long side chains and by their successively poorer representation of the more complex shapes of larger interacting molecules. The shape complementarity at molecular interfaces has been approached more successfully by constructing sets of spheres that touch the molecular surface at two points and have centers on the normal vectors from the surface points. The spheres thus generated outside the larger molecule are mapped onto those generated inside the smaller molecule to identify complementarity (11). More global information on overall shape has been obtained by approximating proteins as ellipsoids (12). This method can be successfully applied to structures at low resolution using only alpha carbon positions. Higher order quadratic terms might be added to the equation of the ellipsoid to better approximate local shape. We are now working on this approach in collaboration with Nelson Max.

Modeling changes in shape can often be more useful than looking at shape directly. One method is to examine the average change in the direction of the surface normal vectors over different sized regions. We have also examined the change in surface area by displaying the derivative of the molecular surface area with respect to position as a vector on each atom. By examining the surface of superoxide dismutase in this way, there appear to be specific patterns of vector magnitude and direction associated with differently shaped regions.

LOCAL MOBILITY AND INDUCED FIT

In addition to dependence on the shape of the uncomplexed molecules, complementarity at molecular interfaces is influenced by flexibility and induced fit. Ideally, this would be examined in many examples where the structures are known for both the complexed and individual molecules; unfortunately, few such complexes are known. The metal ion binding sites described above act as a database for simple protein-small molecule systems, which involve very little induced fit, i.e. preexisting pockets are found in the uncomplexed proteins. In order to evaluate induced fit in protein-protein interactions, we have used computer graphics modeling to design an experiment using anti-peptide antibodies representing sites in the protein myohemerythrin (13) and for retrospective studies on 12 other proteins (14) (see Fig. 3).

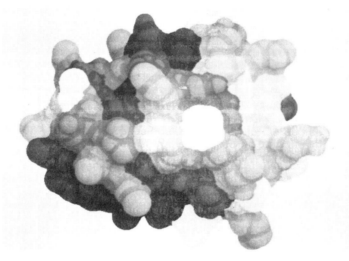

Figure 3. Computer graphics view showing cytochrome c solid external molecular surface. The external surface was calculated using M. L. Connolly's program RAMS (8) and coded by the average main-chain temperature factors using a radiating body brightness scale (highest white to lowest black). Except for areas in crystal contacts, the antigenic determinants are associated with the most highly mobile regions. Adapted from ref. 14.

The selection of peptides from the myohemerythrin sequence was made using computer graphics to dissect the protein based upon main-chain temperature factors averaged by residue, since the main-chain mobility reflects the global conformational variability of the protein. Mapping the results onto the structure, we found that anti-peptide antibodies against contiguous, highly mobile regions react strongly with the native protein, while anti-peptide antibodies from well-ordered regions do not. In addition, simultaneous mapping of antigenity and mobility parameters onto the protein sequence and surface respectively using computer graphics identified a relationship between mobility and antigenicity of contiguous determinants in other proteins. Thus, it may be useful to develop ways to combine the parameters of molecular shape and mobility. For example, one attractive approach is to generate spheres bounded by touching two surface points as described above, but allow the spheres to interpenetrate the surfaces formed by more mobile atoms. Such plastic surface shapes may also be useful for evaluating electrostatic forces which often depend upon the distance separation from mobile charged side chains.

PRECOLLISION ORIENTATION BY ELECTROSTATIC FORCES

To define the possible role of electrostatic forces in orienting molecules prior to their actual collision, we have developed a new method for both the analysis and color-coded computer graphics display of the three-dimensional electrostatic vector field surrounding a macromolecule (15) (see Fig. 4). We calculate both the electrostatic potential and field from partial charges (16) assigned to each of the atoms of the protein (17), thus enabling the contributions of specific residues to be identified. Using this approach, we are currently examining the electrostatic molecular surface and electrostatic field of Cu,Zn superoxide dismutase (15), the variable region of phosphorylcholine-bound mouse immunoglobulin Fab M603 (18), and the electron transfer partners plastocyanin and cytochrome c (19), as well as electrostatic interactions at inter-domain and inter-subunit interfaces of superoxide dismutase, immunoglobulins, and the coat protein of tomato bushy stunt virus (20).

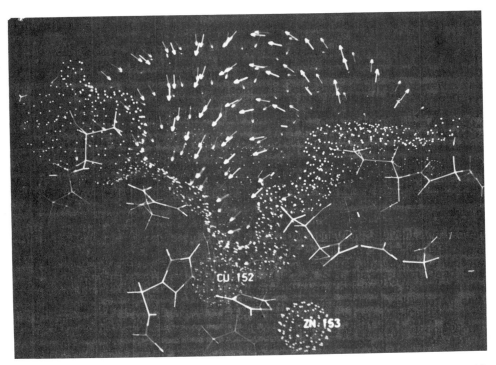

Figure 4. A slice through the deep active-site channel of Cu,Zn superoxide dismutase with electrostatic field vectors. The active-site channel is seen in cross section. The vectors were calculated at points outside the molecular boundary, on concentric spheres of up to 14 Å radius centered on the axial water ligand of the copper ion. The skeletal model Cu ligands and active-site Arg 141 (left center) can be seen along with the Cu (bottom center) in the most positive part of the active channel. The electrostatic field or electrostatic potential gradient (the magnitude and direction of maximum change in the potential), was calculated by evaluating the partial derivative of the potential with respect to each of the three coordinate axes (15). By convention, electrostatic field vectors indicate the direction a positive charge would move in the field; we reversed this convention here to show the directional force on the negatively charged superoxide substrate.

While there is considerable evidence supporting the importance of electrostatic forces in macromolecular interactions, characterization of their roles in sufficient detail to allow the design of specific experiments has been more difficult to obtain. Examination of the electrostatic potential surfaces of macromolecular complexes indicates that electrostatic complementarity is often a feature of stable complexes, but the poorly defined dielectric constant in regions inside and around macromolecules complicates the calculations. A useful accomplishment would be the development of an automated docking algorithm based upon electrostatic forces. In approaching this goal, our current analysis has focused upon the electrostatic fields surrounding macromolecules, as their directionality is important in determining electrostatic orienting effects. The fields also have the advantage of being relatively insensitive to the value of the dielectric. Furthermore, dissection of the electrostatic field allows the relative influences of individual residues to be determined and to be tested experimentally. Electrostatic field vectors displayed graphically can also be used to model the docking of two molecules based solely upon electrostatic parameters (see Fig. 5). Finally, the interpretation of electrostatic and other parameters can be greatly aided by their animation (21). Problems that have yet to be addressed include the effects of ordered solvent molecules and the movements of water molecules and amino acid side chains induced as molecules approach each other.

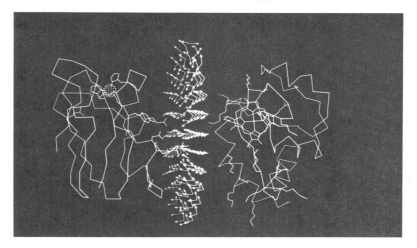

Figure 5. Disk of superimposed electrostatic field vectors for plastocyanin and cytochrome c. A separation of 12 Å between the interacting molecular surfaces of each docked complex was produced by a 6 Å translation of each molecule along a line joining their centers of mass. A set of points separated by about 2 Å and lying on concentric circles, were selected to form a disk 20 Å in diameter midway between the two molecules and normal to the intermolecular axis described above. Disks of electrostatic field vectors originating from these points were calculated from the partial charges of each molecule. When both molecules were displayed together with their associated electrostatic field disks (color coded by field magnitude), the superimposed disks formed pairs of arrows, one resulting from each molecule. Cytochrome c and its disk of electrostatic field vectors were then rotated around the intermolecular axis to identify the rotational positions in which the directions and magnitudes of the electrostatic field vectors belonging to the two molecules were as closely matched as possible.

References

1. Richards, F. M., 1977, "Areas, volumes, packing, and protein structure", *Annu. Rev. Biophys. and Bioeng.* **6**:151-176.
2. Pearl, L. H., and Honegger, A., 1983, "Generation of molecular surfaces for graphic display", *J. Mol. Graphics* **1**:9-12.
3. Connolly, M. L., 1983, "Solvent-accessible surfaces of proteins and nucleic acids", *Science* **221**:709-713.

4. O'Donnell, T. J., and Olson, A. J., 1981, "GRAMPS - A graphics language interpreter for real-time, interactive, three-dimensional picture editing and animation", *Computer Graphics* **15**:133.
5. Connolly, M. L., and Olson, A. J., 1985, "GRANNY. A companion to GRAMPS for the real-time manipulation of macromolecular modes", *Computers & Chem.* **9**:1-6.
6. Tainer, J. A., Getzoff, E. D., Connolly, M. L., Olson, A. J., 1983, "Topography of protein surfaces", *Fed. Proc.* **42**:1998.
7. Getzoff, E. D., and Tainer, J. A., 1985, "Superoxide dismutase as a model ion channel", *In* Ion Channel Reconstitution (C. Miller, ed.), Plenum, New York.
8. Connolly, M. L., 1983, "Analytical molecular surface calculation", *J. Appl. Crystallogr.* **16**:548-558.
9. Connolly, M. L., 1985, "Depth-buffer algorithms for molecular modeling", *J. Mol. Graphics* **3**:19-24.
10. Pique, M., 1982, "What does a protein look like", computer animated film, University of North Carolina.
11. Kuntz, I. D., Blaney, J. M., Oatley, S. J., Langridge, R., and Ferrin, T. E., 1982, "A geometric approach to macromolecule-ligand interactions," *J. Mol. Biol.* **161**:269-288.
12. Taylor, W. R., Thornton, J. M., and Turnell, W. G., 1983, "An ellipsoidal approximation of protein shape", *J. Mol. Graphics* **1**:30-39.
13. Tainer, J. A., Getzoff, E. D., Alexander, H., Houghten, R. A., Olson, A. J., Lerner, R. A., and Hendrickson, W. A., 1984, "The reactivity of anti-peptide antibodies is a function of the atomic mobility of sites in a protein", *Nature* **312**:127-133.
14. Tainer, J. A., Getzoff, E. D., Paterson, Y., Olson, A. J., and Lerner, R. A., 1985, "The atomic mobility component of protein antigencity", *Annual Review Immunol.* **3**:501-535.
15. Getzoff, E. D., Tainer, J. A., Weiner, P. K., Kollman, P. A., Richardson, J. S., and Richardson, D. C., 1983, "Electrostatic recognition between superoxide and copper, zinc superoxide dismutase", *Nature* **306**:287-290.
16. Weiner, P. K., and Kollman, P. A., 1981, "AMBER: Assisted model building with energy refinement. A general program for modeling molecules and their interactions", *J. of Computational Chem.* **2**:287-303.
17. Tainer, J. A., Getzoff, E. D., Richardson, J. S., and Richardson, D. C., 1983, "Structure and mechanism of copper, zinc superoxide dismutase", *Nature* **306**:284-287.
18. Getzoff, E. D., Tainer, J. A. and Lerner, R. A., 1985, "The chemistry of antigen-antibody union", *In* Immune Regulation, (M. Feldmann, ed.) Humana Press, Clifton, NJ, in press.
19. Freeman, H. C., Getzoff, E. D., and Tainer, J. A., 1984, "Electrostatic effects in metalloprotein electron transfer: A computer graphics study of the interaction between plastocyanin and cytochrome *c*", *Proceedings of the Division of Coordination and Metal-Organic Chemistry of the Royal Australian Chemical Institute*, in press.
20. Getzoff, E. D., Tainer, J. A., and Olson, A. J., 1985, "Recognition and interactions controlling the assemblies of β barrel domains", *Biophys. J.* **49** in press.
21. Olson, A. J., Getzoff, E. D. and Tainer, J. A., 1984, "Terms of entrapment: Structure and function of superoxide dismutase", computer animated film, copyright Research Institute of Scripps Clinic.

PREDICTION OF PROTEIN STRUCTURE FROM AMINO ACID SEQUENCE

Michael J.E. Sternberg

Department of Crystallography
Birkbeck College, Malet Street
London, WC1E 7HX

INTRODUCTION

Renaturation experiments show that a folded active protein can be unfolded and inactivated and then this process can be reversed to yield the active folded molecule. Thus it is considered that the amino acid sequence and the protein environment determine the three-dimensional conformation of a protein. Accordingly workers are trying to predict theoretically the three-dimensional structure of a protein from its sequence. Prediction will be useful as protein crystallography is at best time-consuming (typically at least 10 man years) and sometimes impossible because of unsuitable or unavailable crystals. In contrast, amino acid sequence information, today often derived from the nucleic acid sequence, is becoming available for many proteins. Recently, the advent of protein engineering has increased the requirements to model the relationship between protein sequence and conformation (eg. Fersht et al, 1985). This paper reviews available methods of structure prediction and introduces developments using data bases and logic programming computer languages.

PREDICTION OF SECONDARY OR LOCAL STRUCTURE

Many algorithms have been developed to predict the local conformation of the chain from local sequence. The rationale for this work on secondary structure prediction stems in part from experimental work that showed synthetic homopolymers preferentially adopt certain conformations. In addition there is the notion that α-helices and perhaps also β-strands and β-turns might act as nucleation sites during folding in vivo. The justification for this work is that better than random predictions generally can be obtained.

In general the methods can be classified into two types. The empirical schemes use parameters obtained from analysis of known sequence and structures. The propensities for formation of a given structure of the different residue types, and sometimes pairs or triplets of residue types, are used for prediction. The other approach is based on stereochemistry. Three commonly used methods are the empirical approaches of Chou and Fasman (1974) and of Robson and co-workers (Garnier et al, 1978) and the stereochemical method of Lim (1974).

Recently Kabsch and Sander (1983) have assessed the accuracy of these three methods of prediction and found the percentage of residues correctly predicted α, β or otherwise for the methods were:

Chou and Fasman	50%
Robson	56%
Lim	59%

The limited success in part stems from the failure to include longer range interactions in the algorithms.

Two groups (Taylor and Thornton, 1983; Cohen et al, 1983) have reported improvements in secondary structure predictions. They both begin by considering one of the classes of protein structures - the α/β class. Taylor and Thornton develop a template of the β-strand/α-helix/β-strand motif that occurs with typical number of residues in and between each secondary structure. This motif then modifies a prediction based on the Robson approach. Improvements of about 10% are observed. Cohen et al (1983) employ the notion that a segment of chain can be predicted as either a definite or a possible α-helix, β-strand or turn based on the pattern of residues. The global view of an α/β protein is then used to decide which possible segments should be made definite and which not. Their results are better than if a similar series of rough assignments were made from one of the standard methods.

PREDICTION OF TERTIARY STRUCTURE BY ENERGY CALCULATIONS

Based on the conclusion from renaturation experiments, the most direct method, in principle, of predicting tertiary structure is by searching for the conformation of lowest free energy of the protein/solvent system.

One major problem in this approach is that the contributions to the free energy cannot be adequately modelled. Many workers (e.g. Weiner et al, 1984) have values for bond geometry deformations and for van der Waals interactions. Electrostatic effects, in particular salt bridges and hydrogen bonds, are harder to model because of the need to include the complex dielectric response of the protein/solvent system. The hydrophobic effect, which is a major effect in folding the protein, involves considerations of the burial of non-polar groups and is often not modelled. In addition, the conformational entropy of the protein chain is not directly included and is considered by attempting to explore the energy surface.

An indication of the limited understanding of these interactions is that on minimising the energy of the molecule starting from the X-ray conformation, shifts of up to 1Å on α-carbon atoms are often observed (e.g. Weiner et al, 1984) which is far greater than the experimental error in the system.

Recently, attention has been given to the selection of accurate methods to calculate the electrostatic potential within and around a protein. A recent study (Rogers et al, 1985) has used a variety of different treatments to model the experimental redox potential change (65mV) in cytochrome C_{551} that occurs on titration. Inaccurate values were obtained from the simple methods such as uniform dielectric (500mV) and distance-dependent dielectric (213mV) whilst fair agreement (90mV) was obtained when an algorithm was used that modelled the shape and different dielectrics of the protein/solvent system.

The second problem is that even the smallest proteins can adopt a myriad of conformations. Minimisation techniques may not be able to search the entire conformation space and instead locate a global, rather than a local, minimum.

TERTIARY STRUCTURE PREDICTION BY HOMOLOGY

The evolutionary mechanisms of gene duplication followed by divergence leads to families of related proteins that share similar amino acid sequences and three-dimensional structures. The extent of sequence homology can be around 15% identity with additionally many of the substitutions being between chemically similar residues. Thus the standard homology searches can identify these relationships. More conserved than sequence is the three-dimensional structure. The pairing arrangements of α-helices and β-strands is largely unaltered although there can be shifts of a few Angstroms in the exact packing geometry. The major conformational changes between homologous proteins occur in the loop regions connecting the regular secondary structures. In these loops there are substantial sequence insertions and deletions.

These observations provide a method to predict an unknown structure based on the crystal conformation of an homologous protein. The approach is
(1) start with the co-ordinates of the X-ray structure
(2) then replace the amino acid residues to those of the unknown structure based on the sequence alignment
(3) make insertions and deletions of residues in the loops connecting the regular secondary structures.
(4) then adjust the model by rotations about main-chain and side-chain dihedral angles to yield a stereochemically sensible model in which there are no close contacts beween atoms, the hydrophobic core is tightly packed, and all buried polar and charged atoms make the required hydrogen bonds and salt bridges.

This method has been used for many years. In 1969 a physical wire model of α-lactalbumin was built based on the X-ray structure of hen lysozyme (Browne et al, 1969). However, the recent developments of interactive molecular graphics have enabled the method to be performed far more easily and accurately. In programs for fitting molecular models to electron density maps (e.g. FRODO, Jones, 1978) the residue replacements are performed simply and the interatomic distances monitored as the dihedral angles varied. Many molecules have been modelled using this method including renin (Blundell et al, 1983) and the HLA-DR proximal domain (Travers et al, 1984). Such predicted structures provide good models to consider detailed protein/ligand interactions (e.g. renin with inhibitors) and thus provide predictive information of great value to drug design, protein engineering and vaccine design.

The major difficulty, and the source of the major errors, occurs in modelling the loops in which insertions and deletions occur. Recently analysis of the known protein structures have been conducted to find rules for modelling. The short connections between two antiparallel, adjacent β-strands has been found to have distinct conformational preferences (Sibanda and Thornton, 1985). Two-residue turns adopt the I' or the II' conformation rather than the I or II arrangement which is more usually observed in other sharp turns in proteins. Even the 5-residue connection has one preferred arrangement. These motifs can be used as trial structures during model-building. In addition to using empirical guidelines, methods are being developed to explore the power of energy calculations to model loops.

TERTIARY STRUCTURE PREDICTION BY TEMPLATES

There are several examples where a set of proteins or protein domains that do not have homologous sequences adopt a similar three-dimensional fold in terms of the local packing of α-helices and β-strands. For example the nucleotide-binding domain of several proteins consists of a parallel β-sheet flanked by α-helices. It is an open question whether the cause of this structural similarity is due to divergence from a common ancestor or convergence to a stereochemically and/or functionally favourable motif.

In the absence of sequence homology, a potentially powerful method of structure prediction is to recognise the presence of a tertiary motif in the sequence being examined. One method is to identify certain key residues that form a template dictating the motif. For example a sub-structure of the nucleotide binding region often consists of

β-strand 1 / α-helix / β-strand 2

The connection between β1 and the α contains a sequence pattern

Gly - X - Gly - X - X - Gly.

These patterns occur because of requirements to bind the nucleotide (Wierenga and Hol, 1983). This sequence pattern occurs in the ATP-binding site of several homologous oncogene products such as V-src and a variety of other proteins - the epidermal growth factor receptor and mammalian cAMP-dependent protein kinase. Accordingly a three-dimensional model for the ATP-binding site of these unknown structures (Fig. 1) can be constructed from the known co-ordinates of a protein with the Gly - X - Gly - X - X - Gly sequence (Sternberg and Taylor, 1984).

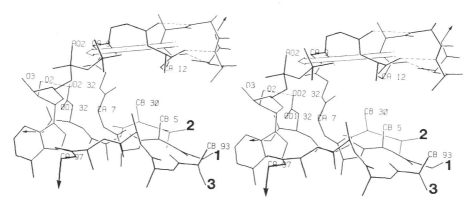

Fig. 1 - A stereo diagram of the proposed ATP-binding site of V-src showing the polypeptide backbone and Cβ atoms together with ATP (from Sternberg and Taylor, 1984).

TERTIARY STRUCTURE PRDICTION BY α/β DOCKING

The three-dimensional structures of most globular proteins can, as a first approximation, be described by the stacking of α-helices and of β-strands. Indeed, about half of the known structures belong to the α/β, β/β or α/β class of proteins in which the packing of the secondary structures conforms to fairly well-understood principles. Thus one approach to predict protein structure involves starting with the α-helices and/or β-strands and packing them together to predict the fold of the appropriate class of molecule.

In the development of these approaches (Cohen et al, 1979; Cohen et al, 1980; Cohen et al, 1982) the correct assignment of secondary structures is used so that the algorithm can concentrate on the tertiary packing. The approach is called the combinational mehtod in which first all possible packing of α-helices and β-strands are generated which are consistent with geometric principles and the locations of hydrophobic docking patches on the surfaces of the secondary structures. Then filters are applied to eliminate structures based on considerations of the available length of connections between secondary structures, the absence of steric clashes and the appropriate topological constraints. In general around 10^8 pairings are generated which is filtered down to between 10 and 10^3 possible folds one of which is close to the native fold. Typical root mean square deviations between predicted and observed Cα positions for the secondary structures are between 2 and 6Å.

The concepts used in this approach have been applied to predict unknown structures. The pattern of hydrophobic residues in the β-sheets of β/β structures were central to the proposal for a fold for the cell-surface antigen thy-1, (Cohen et al, 1981). More speculative was an attempt to model the structures of α- and β-interferons (Sternberg and Cohen, 1982). Secondary structure predictions suggest that there may be 4 α-helices central to the tertiary structure. The α/α packing algorithm was then used to dock these predicted α-helices into a four-fold α-helical bundle.

DATA BASES

Comparative analyses of protein conformations are revealing many empirical rules and preferences which are of help to structure prediction. However, such analyses are time consuming and the information gathered becomes outdated as new structures are solved and existing structures refined. In addition, workers wishing to perform certain mutation experiments or involved in drug design are likely to have specific queries which were never previously imagined. Currently protein co-ordinates are obtained from the Brookhaven Data Bank (Bernstein et al, 1977) in a serial form and analysis of proteins requires programming to extract the required information. At Birkbeck, in collaboration with Leeds University, we have just begun to arrange the protein co-ordinates in a suitable form so that an on-line interrogation could for example, list the side-chain dihedral angles of all Arg-Asp salt bridges when the Arg is in an α-helix and the Asp in a β-sheet. This is being tackled by use of a commercial relational data base which organises the data into several tables which can be searched and cross-correlated. Previous work has used relational data bases for input to molecular graphics (Morffew et al, 1983). In this proposed study the tables will span three levels of information: (1) details of each protein including the resolution and other details of the crystallographic analysis; (2) details at the residue level with sequence, secondary structure assignment, exposure to solvent, dihedral angles, and which other residues are close in space; (3) details at the atom level with co-ordinates and suggestions of which atoms form hydrogen bonds and salt bridges. The data base query language will enable users to ask detailed questions relevant to their specific interests.

REASONING ABOUT PROTEIN TOPOLOGY USING PROLOG

Proteins exhibit complex topological features such as a motif formed from four or five β-strands known as a Greek-key (Richardson, 1981). The proposed relational data base approach, although able to extract and correlate information rapidly, would be unable to reason about topology. Indeed computer based reasoning about topological features has proved difficult in imperative programming languages such as FORTRAN. In part the difficulty arises because such languages require the programmer to define both the logical specification of the task and the control information needed to execute it as an algorithm. In contrast, logic programming languages (e.g. PROLOG) require only the specification in logic. Control of program execution is left to the language's own problem solver. Accordingly the use of PROLOG to represent and reason about topological features in protein structures has been explored (M.J.E. Sternberg, W.R.Taylor, J.Nyakairu, J.P.Fox and C.J. Rawlings; work in progress with the Imperial Cancer Research Fund).

In outline, the features of PROLOG enable the explicit representation of logical relationships amongst objects in the program database. For example the geneology of the British royal family might be encoded in the database thus:

    philip is-father-of charles
    charles is-father-of william
    elizabeth is-mother-of charles

Rules of inference can derive additional relationships from the program database. For example:

    A is-grandfather-of B   IF
        A is-father-of X AND X is-father-of B.

A, B and X are variables that become filled from the information in the database that satisfies this rule (see Clocksin and Mellish, 1981).

The topology of β-sheet proteins was expressed as PROLOG facts. Most of the information was extracted automatically from the Brookhaven database of protein structures (Bernstein et al, 1977). These facts take the form:

    strand(b) follows   strand(a).
    strand(b) is-adjacent-to strand(a).
    strand(a) is-antiparallel-to strand(b).

However additional features had to be extracted from the papers describing the structures in order to adequately specify the protein topology.

Rules have been constructed that define the structure of hairpins, meanders, greek keys and jelly-rolls (see figure 2). These rules reflect the hierarchical origin of topological motifs. A user may interrogate the program database using the rules to establish the presence of these motifs in a given protein and to determine the component strands.

This program was implemented on a DEC system-20 at the Imperial Cancer Research Fund, London using DEC-10 PROLOG. A search for all greek keys in an eight stranded structure using a compiled program requires of the order of 50 msecs of CPU time. Methods of reducing the search time are being explored.

This approach is being extended to proteins that have α-helices and is being combined with analysis of the relationships between amino acid sequence and protein topology.

PROLOG has recently gained popularity as an implementation language for expert systems. Although this program cannot be described as an expert system for a variety of reasons, this study has demonstrated the potential of programming in PROLOG to reason about topology. The expression of structural definitions is natural, fast and easy to examine. Programs based on these principles could be used as front ends to protein structure databases.

```
greek_key ( [3,1,1] , strands ( [ A, B, C, D ] )) :-
    succeeds (B, A),
    adjacent (A, D),
    are_antiparallel (A, D),
    meander (strands( [B,C,D] )),
    A \== C.
```

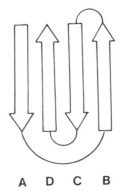

A   D   C   B

Figure 2: A PROLOG clause to define a four-stranded Greek-key. The clause reads that there is a Greek key with strand order 311 formed from strands A,B,C and D if (1) strand B succeeds strand A (ie. B is the next sequential strand), (2) strand D is adjacent to strand A, (3) strand D is antiparallel to strand A, (4) strands B,C and D form a meander (ie. strand B is adjacent and antiparallel to strand C and strand C is adjacent and antiparallel to strand D), and (5) strand A is different from strand C.

CONCLUSION

It is unlikely that over the next few years there will be a general solution to the problem of predicting protein structure from sequence. However the use of the information being obtained from protein crystallography for model building by homology or using sequence templates is providing predictions of use not only for biochemical research but for drug design and in studies on protein engineering. Over the next few years major developments will occur in the use of sequence and structural databases to obtain rules for prediction.

ACKNOWLEDGEMENTS

I would like to thank all my colleagues in the Crystallography Department at Birkbeck for all the stimulating discussions on this field. Particular thanks to Professor Tom Blundell and Drs. Willie Taylor and Janet Thornton for their advice. My work is supported by a Royal Society University Research Fellowship.

REFERENCES

Bernstein,P.C, Koetzle,T., William,G.J.B, Meyer,E.Jr, Brice,M.D., Rodgers,J.R., Kennard,O., Shimanouchi,T., and Tasumi,M., 1977, J. Mol. Biol., 112:535-542.
Blundell,T.L., Sibanda,B.L., and Pearl,L.H., 1983, Nature, 304:273-275.
Browne,W.J., North,A.C.T., Phillips,D.C., Brew,K., Vanaman,T.C., and Hill,RC., 1969, J. Mol. Biol., 42:65-86.
Chou,P.Y. and Fasman,G.D., 1974, Biochemistry, 13:211-222.
Clocksin,W.F., and Mellish,C.S., 1981, "Programming in Prolog", Springer Verlag.
Cohen,F. E., Richmond,T.J., and Richards,F.M., 1979, J. Mol. Biol., 132:275-288.
Cohen,F.E., Sternberg,M.J.E., and Taylor,W.R., 1980, Nature, 285:378-382.
Cohen,F.E, Novotny,J., Sternberg,M.J.E., Campbell,D.G. and Williams, A.F., 1981, Biochem. J., 195:31-40.
Cohen,F.E., Sternberg,M.J.E., and Taylor,W.K. 1982, J. Mol. Biol.
Cohen,F.E., Abarbanel,R.M., Kuntz,I.D., and Fletterick,R.J., 1983, Biochemistry, 21:4894-4904.
Fersht,A.R., Shi,J-P., Knill-Jones,J., Lowe,D.M., Wilkinson,A.J., Blow, D.M., Brick,P., Carter,P., Waye,M.M.Y., and Winter,G., 1985, Nature, 314:235-238.
Garnier,J., Osguthorpe,D.J., and Robson,B., 1978, J. Mol. Biol. 120:97-120.
Glover,I., Haneef,I., Pitts,J., Woods,S., Moss,D., Tickle,I. and Blundell,T.L., 1983, Biophysics, 22:293-304.
Jones,T.A., 1978, J. Appl. Crystallogr., 11:268-272.
Kabsch,W., and Sander,C., 1983, FEBS Letters, 155:179-182.
Lim, V.I., 1974, J. Mol. Biol., 88:873-894
Morffew,AJ., Todd,S.J.P., and Snelgrove,M.J., 1983, Computer and Chemistry, 7:9-16.
Old,R.W. and Primrose, S.B., "Principles of Gene Manipulation", 3rd edition, Blackwell Scientific, Oxford.
Richardson,J.S., 1981, Av. Protein Chem. 34:167-339.
Rogers,N.K., Moore,G.B., and Sternberg,M.J.E., 1985, J. Mol. Biol. 182:613-616.
Sibanda,B.L., and Thornton,J.M., 1985, Nature, 316:170-174.
Sternberg,M.J.E., and Cohen,F.E., 1982, Int. J. Biolog. Macromol., 4:137-144.
Sternberg,M.J.E., and Taylor,W.R., 1984, FEBS Letters, 175:387-392.

Taylor,W.R., and Thornton,J.M., 1983, *Nature*, 301:540-542.
Travers,P., Blundell,T.L., Sternberg,M.J.E., and Bodmer,WF., 1984, *Nature*, 310:235-238.
Weiner,S.J., Kollman,P.A., Case,D.A., Singh,U.C., Ghio,C.,Alagona,G., Profeta,S., and Weiner,P., 1984, *J. Am. Chem. Soc.*, 106:765-784.
Wierenga,R.K., and Hol,W.G., 1983, *Nature*, 302:842-844.

4 WATER AND DYNAMICS

X-RAY ANALYSIS OF POLYPEPTIDE HORMONES AT ≤1Å RESOLUTION:
ANISOTROPIC THERMAL MOTION AND SECONDARY STRUCTURE OF PANCREATIC POLYPEPTIDE
AND DEAMINO-OXYTOCIN

A.C. Treharne, S.P. Wood, I.J. Tickle, J.E. Pitts, J. Husain,
I.D. Glover, S. Cooper and T.L. Blundell

Laboratory of Molecular Biology
Department of Crystallography
Birkbeck College, University of London
Malet Street, London WC1E 7HX, UK

INTRODUCTION

Polypeptide hormones are synthesised as larger precursors and stored in endocrine cells from which they are released into the circulation in response to peptide factors, metabolite levels and other agents. At target tissues they bind specific cell surface receptors which give rise to intracellular secondary effects such as changes in the levels of cAMP, tyrosine phosphorylation, etc. Their degradation is rapid and is often mediated by internalisation of the hormone-receptor complex. The role of the conformation of polypeptide hormones during this complex series of events may help in our understanding of the various processes at the molecular level.

Analyses of conformation, flexibility and dynamics present a range of challenging problems both experimentally and conceptually. Small polypeptide hormones of less than 30 amino acids will generally be flexible, and their conformations may depend on their concentration, on the solvent and on other molecules in solution (Schwyzer, 1973; Blundell, 1983; Blundell and Wood, 1982). For example, glucagon has little defined structure in dilute aqueous solutions but secondary structure is stabilized by self-association at high concentrations and by the presence of non-aqueous solvents and lipid micelles (see Blundell 1983 for review). The conformations of some small hormones such as oxytocin may be limited by disulphide bridges and consequently they may have better defined mainchain conformations, but even these are flexible and comprise several conformers in equilibrium in aqueous solutions (Meraldi et al, 1977). Many larger polypeptide hormones, such as insulin and probably glycoprotein hormones and growth hormones, have globular structures with hydrophobic cores (Blundell and Wood, 1982). Their preferred conformations are probably retained in aqueous solutions, in oligomeric forms, and in the crystalline state, although even insulin shows large changes under conditions of high salt concentrations.

In an attempt to understand the relationship of conformation to the molecular biology of polypeptide hormones, we have carried out a series of X-ray and biochemical studies (for reviews see Blundell and Humbel, 1980; Blundell and Wood, 1982; Pitts et al, 1983). Two polypeptide hormones, deamino-oxytocin and avian pancreatic polypeptide (aPP), crystallise in forms which are suitable for very high (≤1Å) resolution X-ray analysis. Here we review the analyses, we discuss the conformations and flexibilities of the hormones and attempt to relate them to receptor binding.

OXYTOCIN

The hormone oxytocin

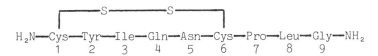

elicits smooth muscle contraction causing milk ejection and uterine contractions in mammals.

Although crystals of oxytocin were first reported by Pierce et al (1952), the crystal structure has proved elusive. Preliminary crystal data for the biologically active deamino-oxytocin (Ferrier et al, 1965) were reported by Low and Chen (1966) and an active 6-seleno deamino-oxytocin analogue was later purified and crystallised (Chiu et al, 1969). No further progress has been reported although other crystals of oxytocin have been described and structure analyses have been completed for the COOH-terminal peptide (Rudko et al, 1971).

Deamino-oxytocin, synthesised by Dr. V.J. Hruby, was crystallised in forms closely related to those of Low and Chen (1966). Crystal data were collected for deamino-oxytocin in spacegroup $P2_1$ to resolution ~1.0Å, and for air-dried deamino-oxytocin in spacegroup C2 to 1.2Å. Data were also collected for 6-seleno deamino-oxytocin, synthesised by Dr. H.R. Wyssbrod, which crystallised in C2 and diffracted to 1.9Å. The close isomorphism of the latter two data sets, enabled location of the sulphur positions from isomorphous (Se-S) and anomalous difference (S and Se) Pattersons, although there was some confusion due to disorder. A best Fourier was then calculated which gave poorly resolved electron density, but it was possible to improve the model by cycles of refinement and remodelling. In the later stages, restrained refinement of the non-hydrogen atoms anisotropically and hydrogens isotropically gave an agreement value (R) of 10.0% at 1.2Å resolution for the C2 crystal form.

The cell dimensions of the C2 and $P2_1$ crystal forms are closely related, so that the asymmetric unit of the $P2_1$ crystals contains two molecules related by a local two-fold axis (Wood et al, 1985). The resulting parameters from the C2 map were then used to initiate the refinement of the crystal form with spacegroup $P2_1$ with data to 1Å resolution after an appropriate origin shift to the coordinates. The final residual was 7% after several cycles of modelling from difference maps and refinement.

The conformations found in the C2 and $P2_1$ crystal structures are shown in Figures 1 and 2a. Each has two β turns; a type II turn between Tyr 2 and Asn 5 stabilized by two hydrogen bonds between the amide and carbonyl of both residues and a type III turn between Cys 6 and Gly 9 stabilized by a hydrogen bond between Cys 6 CO and Gly 9 NH. Table 1 gives the intramolecular hydrogen bonds while Table 2 gives the mainchain torsion angles for the conformations in the $P2_1$ crystal form.

Table 1. Intramolecular hydrogen bonds and angles for deamino-oxytocin (two molecules of the P2₁ crystal asymmetric unit A and B)

X	Y	r_{X-H}	r_{X-Y}	(X-H...Y Θ=(X...H-Y (degrees)
		(units in Å)		
molecule A				
N A2	O A5	1.93	2.97	161.26
O A2	N A5	1.90	2.95	162.61
O A6	N A9	2.70	3.55	135.97
molecule B				
N B2	O B5	2.05	3.09	162.75
O B2	N B5	1.82	2.85	156.85
O B6	N B9	2.54	3.49	146.16

Table 2 Torsion angles of the main chain and side chains of the two molecules of deamino-oxytocin in the P2₁ crystal asymmetric unit

residue	ϕ	ψ	ω	χ^1	χ^2	χ^3	χ^4
	(units in degrees)						
molecule A							
Cys A1		120	-171	-109/-178	-102/87	77/-87	
Tyr A2	-133	167	176	-50	87		
Ile A3	-54	120	179	-68	-79		
Gln A4	58	24	171	-66	-55	-48	
Asn A5	-151	75	=177	168	26		
Cys A6	-145	98	-176	-177	57/110		
Pro A7	-78	-8	179	24	-36	31	-15
Leu A8	-73	-40	-176	-59	174		
Gly A9	170	-177		0			
molecule B							
Cys B1		101	-172	178/-111	80/-93	-94/83	
Tyr B2	-126	164	-177	-55	98		
Ile B3	-65	125	-173	-169	163		
Gln B4	56	29	164	-64	-51	-47	
Asn B5	-158	66	177	-180	-43		
Cys B6	-128	98	-173	-179	113/45		
Pro B7	-73	-12	-179	29	-40	35	-18
Leu B8	-77	-33	-173	-67	175		
Gly B9	-176	168		-23			

The 20-membered ring is curved (see Figures 1 and 2a) with Tyr 2 and Asn 5 side-chains on the top side. The curved shape is completed by the acyclic tail of the molecule which is oriented away from the ring with a trans Cys 6-Pro 7 peptide. The top surface of the curve is also made up of Ile 3, the α- and β- carbon of Gln 4, the ring of Pro 7 and side-chain of Leu 8. The underside of the curve is devoid of side-chains, the terminal amide of Gly 9 and side-chain amide of Gln 4 being stationed at the extremities of the underside cleft. In the crystal there are at least two possible conformations for the disulphide bridge region which differ at the S1 position.

In the C2 cell one conformation results in a disulphide torsion angle of +76° with right-handed chirality and the other with a torsion angle of

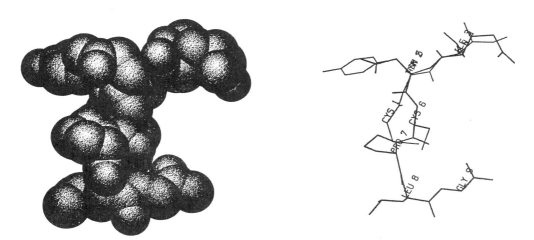

Fig. 1   Space-filling and stick model for deamino-oxytocin in space-group C2, showing the 'pleat' of the 20-membered ring and orientation of the 'tail' peptide.

-101° and left-handed chirality. The two S1 positions are about 2Å apart and equally populated. There is no observable mainchain disorder away from the bridge atoms, the thermal parameters of Cαs 1 and 6 being typical of other backbone atoms.

In the P2$_1$ cell the molecules are more ordered, with different chiral forms similar to those seen in the C2 cell and are predominantly populated (70/30) in each molecule. The torsion angles are +$\underline{73}$°, -74° and -$\underline{93}$°, +81°, the dominant isomer being underlined. Furthermore, the dominant conformers are associated with distinct Cα-β rotamers of the Ile 3 sidechain. Small differences are also apparent in the sidechains of Tyr 2 and Asn 5. Together these differences in structure between the two molecules of the P2$_1$ cell correlate well with those atoms showing higher thermal parameters in the C2 cell.

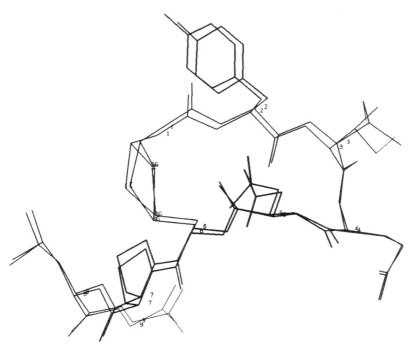

Fig. 2a   The two molecules of the P2$_1$ asymmetric unit shown superimposed to indicate the main differences located at the sulphur-1 and Ile 3 side-chain, with other conformational differences.

Figure 2b shows aniso-tropic thermal ellipsoids for the molecule of the C2 cell. All the hydrogen bonding potential of the molecules is satisfied either by intramolecular interactions or bonding of water and other oxytocin molecules. The details are described elsewhere (Wood et al, 1985).

Oxytocin and its deamino analogue have been the subject of many spectroscopic studies. NMR proton assignments (Brewster et al, 1973) and Raman spectroscopy (Tu et al, 1979) indicate that the β turn between residues 2 and 5 with a hydrogen bond between Tyr 2 CO and Asn 5 NH is a conformational feature for oxytocin in DMSO as it is in the crystals. Although the initial definition of the β-turn in DMSO was not explicit, the bulky corner residues at positions 3 and 4 were thought to favour a type I turn, which is preferred unless Gly or D Ala is located at residue i + 2. However, Glickson et al (1975) suggested that unique conformational constraints due to the 20-membered ring might stabilise a type II turn and this is confirmed by the NMR spectra of Ile 3 and the similarity of the spectra of oxytocin and Gly 4 oxytocin. This is the conformation observed in the crystals indicating that the conformation is retained in different environments. An additional hydrogen bond between Asn 5 CO and Tyr 2 NH has been indicated by NMR for deamino-oxytocin in solution and this is also observed in the crystal structure.

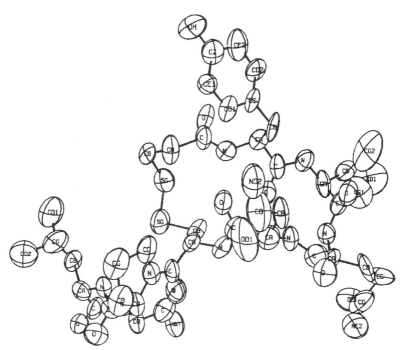

Fig. 2b   Anisotropic thermal ellipsoids for C2 structure of deamino-oxytocin. One disulphide chirality only is shown.

The chirality of the disulphide bridge in solution has also been the subject of much discussion (Urry et al, 1968; Wyssbrod et al, 1975). Circular dichroism and laser Raman spectroscopy indicate that most molecules in solution have dihedral angles close to ±90°, so that both conformers probably exist in equilibrium and the population of each conformer, in solution as in the crystal, depends on the analogue and the environment.

There is evidence for the maintenance in DMSO of the type III β-turn between residues 6 and 9 with a hydrogen bond between Cys 6 CO and Gly 9 NH, and the existence of a <u>trans</u> Cys 6 Pro 7 peptide, both of which are observed in the crystals. However, the tocic ring and acyclic tail were proposed to fold over in DMSO, so that a hydrogen bond is formed between Asn 5 amide carbonyl and Ile 8 NH. This appears to be replaced by intermolecular hydrogen bonds in the crystal structure which lead to a more open structure.

In fact, such an open conformation is consistent with the NMR in water (Hruby et al, 1983) where there is no evidence for such intramolecular hydrogen bonds. In any case it is likely that there is conformational flexibility between the tocic ring and the β-turn of the acyclic tail and this is consistent with conformational energy calculations, normal mode analysis and molecular dynamics calculations which indicate considerable flexibility (Treharne et al, 1985).

In conclusion the solution data taken together with the crystal structures indicate that the two β-turns are stable features of oxytocin and its analogues. Both crystal structure and solution data are consistent with flexibility of the disulphide and it is probable that both conformers exist, but in different proportions in different environments. Finally, there may be considerable flexibility in the spatial relationships between the tocic ring and acyclic tail, but it is probable that where intermolecular hydrogen bond donors and acceptors are available - in water and in the crystal, for example - there are no intramolecular hydrogen bonds between the two parts.

PANCREATIC POLYPEPTIDE

Pancreatic polypeptide (PP) is a 36 amino acid hormone which is synthesised in specific cells of the Islets of Langerhans and may act as a satiety factor. It is a member of a larger family of homologous hormones and neuroregulators.

Collection and processing of the X-ray data for aPP crystals (spacegroup C2) to 0.98Å resolution have been described previously (Blundell et al, 1981; Glover et al, 1983) and involved 53,000 intensity measurements which were processed and merged to give 17,058 independent relections with an agreement index, R_{sym} = 6.0%. An asymmetric unit of the crystal cell contains 300 non-hydrogen atoms. The solvent channels contains about 100 water molecules, 65 of which are relatively well-ordered. The model was refined at 0.98Å resolution using the program RESTRAIN (Haneef et al, 1985) in which bond lengths, bond angles and planarity of aromatic rings and peptides were restrained to target values defined from analyses of smaller peptides and amino acids. Refinement of the anisotropic atomic displacements was undertaken using six parameters of the mean-square vibration tensors for each non-hydrogen protein atom and the oxygens of water molecules. The observation-to-parameter ratio was approximately 5:1. At convergence the conventional agreement factor (R) was 15.3% calculated without omission of weak reflections. The resulting anisotropic thermal parameters were subjected to rigid group analysis (Schomaker and Trueblood, 1968) and to the rigid bond tests of Hirschfeld (1976). The aromatic side groups were treated as rigid bodies and the components of their TLS were refined directly against the X-ray data.

During least-squares refinement of TLS parameters appropriate libration corrections were made to calculated distances in the geometrical restraints. Hydrogen positions were included where they were defined by the polypeptide geometry, but those attached to atoms with large thermal motions especially in sidechains did not refine well.

The refinement confirmed the overall features of the aPP molecule defined at 1.4Å resolution (Blundell et al, 1981) and shown in Figure 3. These include a polyproline-like helix (residues 2-8) which is packed against an α-helix (residues 14-32) so that the proline residues at 2, 5 and 8 interdigitate between the sidechains of the α-helix to give a well-defined hydrophobic core. The COOH-terminal residues are relatively flexible even in the crystals. aPP forms symmetrical dimers involving hydrophobic interactions and in the presence of zinc ions, which each coordinate three aPP molecules, an extended oligomeric lattice is formed.

Examples of the thermal ellipsoids are given in Figure 4. Most atoms show clear evidence for anisotropy with the amplitudes usually being smaller in the directions of covalent bonds. The carbonyl atoms of the mainchain peptide groups are particularly anisotropic, with a strong correlation between the major axes of the motion and the nature of the hydrogen bonding.

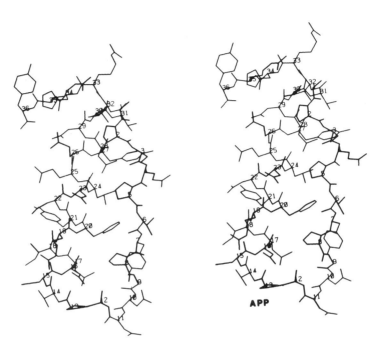

Fig. 3   Stereo representation of aPP monomer showing the α-helix and polyproline-like helix.

In general, thermal ellipsoids of residues exposed to the solvent have greater amplitudes than those packed against the interior of the molecule. The COOH-terminal residues have particularly high thermal parameters and are completely exposed to solvent. The hydrophilic, solvent-exposed sidechains of Glu 4 and Asp 6 show greater amplitudes than residues 2, 5, 7 and 8 which are more closely involved in the core of the protein. In several cases the anisotropy appears to be due to disorder between two alternative positions rather than harmonic thermal motion. For example, the hydroxyl of Ser 3 is constrained by close contacts to occupy two positions close to the dyad of the dimer, and the Cγ atoms of the prolines flip between conformations which place the Cγ on either side of the five-membered ring "best plane" so that there are far greater apparent amplitudes of motion perpendicular to the rings.

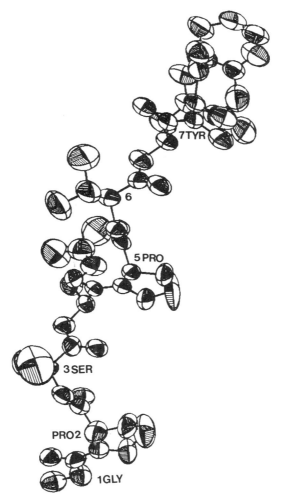

Fig. 4   Anisotropic thermal ellipsoids (with 50% probability) for the polyproline-like helix of aPP, showing residues 1 to 8.

Many sidechains show evidence of correlated motions or disorder. Rigid body TLS analysis (Schomaker and Trueblood, 1968) of the unrestrained thermal ellipsoids of the aromatic sidechains did not always give meaningful results, but we were able to define the librational motions more precisely by refining the components of the TLS tensors directly from the X-ray data. Tyr 7, Phe 20 and Tyr 27 each show an approximately isotropic rigid body translation associated with one dominant direction of libration with an amplitude in the range 4-6°. In each case the libration axis makes an angle of about 50° to the plane defined by $C\alpha$, $C\beta$ and $C\gamma$, suggesting motions more complex than simple librations about the sidechain bonds. Tyr 21 shows more complicated motion with anisotropic translation and significant libration about two axes (see figure 5).

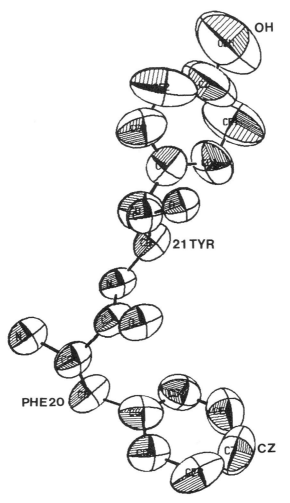

Fig. 5   Evidence of correlated side-to-side motion of side chains being depicted by residues Phe 20 and Tyr 21.

Intermolecular interactions appear to affect the amplitudes signficantly. Thus the thermal parameters of the zinc ligands are unusually low. As observed in previous structure analyses, the thermal parameters of the water molecules are strongly correlated with the thermal parameters of protein atoms to which they may be bonded and to the number of well-defined hydrogen bonds they make.

Circular dichroism and NMR studies indicate that a similar conformation is conserved in aqueous solution when the dimer is present and that the conformation is conserved when the monomer predominates at high dilutions.

CONFORMATION AND BIOLOGY

Pancreatic polypeptide is synthesised as a larger precursor from which it is cleaved. The high concentrations in the granules in the pancreatic islets indicate that the storage form is probably a dimer and, in the case of aPP, zinc linked oligomers may form (see Blundell et al, 1981 for a review). In this way pancreatic polypeptide resembles insulin and glucagon for which there is good evidence of oligomers - zinc insulin hexamers and glucagon trimers - in storage granules (Blundell and Wood, 1982). On the other hand, oxytocin is synthesised as part of a much larger precursor which appears to form a globular structure and after cleavage into neurophysin and oxytocin a stable complex exists in storage granules.

When released into the circulation the complexes will dissociate at the high dilutions and monomers will predominate in solution. Here the conformations probably retain some of their secondary structure but will undoubtedly be more flexible.

At the receptor it is probable that a defined conformation will be assumed. Eventually, complexes of the hormone and its receptor need to be isolated and characterised by NMR spectroscopy, X-ray analysis and other physico-chemical techniques. In the meantime, analysis of conformationally restricted analogues can be useful in defining the likely receptor conformer or conformers.

The first and most comprehensive model for oxytocin receptor binding, proposed by Walter and coworkers, was based on the solution conformation in DMSO as defined by NMR together with an assessment of the analogue assay data avilable for the uterine target (Walter et al, 1971). Ile 3, Gln 4, Pro 7 and Leu 8 were held to constitute the binding message while the proximity of Tyr 2 -OH and Asn 5 carboxamide were important for activation. In this model the 20-membered tocic ring was seen as essentially planar with an almost featureless underside and a topside dominated by Tyr 2 and Asn 5 in a cleft between ring and acyclic tail. Some features of this model may be correct but the X-ray analysis shows that the ring is pleated, like part of a β-sheet . This is probably maintained in solution and at the receptor.

Using rigid analogues, Hruby (1981) and Wood et al (1985) have extended these ideas to a "dynamic" model for oxytocin activity. 1-Penicillamine derivatives of oxytocin, which have the two hydrogens at the uterine receptor. Circular dichroism, Raman and NMR spectroscopy show that they have much reduced flexibility in the ring moiety, a preferred disulphide torsion angle of between 110° and 115° (Hruby et al, 1983). The crystal structures have shown the existence of two conformers of differing disulphide bridge chirality in both deamino-oxytocin and its 6-seleno-analogue. Wood et al (1985) propose that conformers with a right-handed chirality bind to the receptor most strongly but a conformational change to the opposite chirality is required to elicit a biological response. Thus 1-penicillamine derivatives with a more rigid, right-handed disulphide are antagonists. The high receptor affinity and biological response of deamino-oxytocins may then be the result of lowering the energy barrier between conformers, possibly due to the replacement of the charged terminal amino group by the smaller, neutral hydrogen atom.

The model does not imply a necessary interaction of a disulphide with the receptor as small conformational changes occur in the mainchain of the tocic ring and larger differences in the conformation of the sidechains of Tyr 2 and Ile 3 which are correlated with the conformation of the disulphide. Thus the model is quite consistent with the activity of derivatives in which the sulphur atoms of the cystine are replaced by methylene groups or selenium atoms.

At the receptor, water will be excluded from much of the surface of the hormone. Experimental data from analogues indicate that hydrophobic and hydrogen bonded interactions - possibly through the free mainchain CO and NH groups as well as side chains - are likely to dominate the receptor interaction. We have suggested that the molecule binds as a neutral species so that the deprotonated amino terminus can form a hydrogen bond to the receptor. The ability to form such a hydrogen bond without deprotonation would explain the enhanced activity of analogues in which the amino terminus is replaced by an hydroxyl group. If there were a charge on the receptor neither these analogues nor deamino-oxytocin would be expected to bind to the receptor strongly. The receptor binding may also require the formation of the two hydrogen bonds between residues 2 and 5 as found in deamino-oxytocin and analogues with glycine at residue 4. All of these observations indicate that the conformation of the neutral deamino-oxytocin in an environment which excludes water from much of the hormone - as in the crystal - might reveal important aspects of the conformation at the receptor.

For aPP few data are available concerning receptors or receptor binding. However, the X-ray studies described here indicate that if the receptor presents a hydrophobic surface it is likely that the globular α-helical/polyproline containing conformer formed in the chain will be favoured. The variation of this surface involving both the α- and polyproline-helices may account for the species specificity of the pancreatic polypeptide family. However, conserved COOH-terminal residues - 32-36 - appear to be essential for activity; they are also free of the globular core and very flexible. It is therefore possible that once the globular region is bound to a receptor in a species specific fashion that the conserved region is induced to fit the receptor in a species general fashion and thus lead to activation.

Clearly models for receptor binding and activation are rather speculative at the present time. Nevertheless, it is likely that the conformations defined by X-ray analysis and the inherent flexibility will play roles in biological activity.

ACKNOWLEDGEMENTS

We thank the SERC and MRC for financial support. We are grateful to Drs. V.J. Hruby and H.R. Wyssbrod for synthesis of deamino-oxytocins. We thank Glenda Dryer for typing the manuscript.

References

Blundell, T.L. (1983) in Glucagon (ed. P.J.Lefevre, Springer-Verlag, Berlin) pp37-56
Blundell, T.L. and Humbel, R.E. (1980) Nature 287 781-787
Blundell, T.L. and Wood, S.P. (1982) Ann.Rev.Biochem. 51 123-154
Blundell, T.L., Pitts, J.E., Tickle, I.J., Wood, S.P. and Wu, W-C. (1981) Proc.Natl.Acad.Sci. USA 78 4175-4180
Brewster, A.I.R. and Hruby, V.J. (1973) Proc.Natl.Acad.Sci. USA 70 3806-3809
Chiu, C.C., Schwartz, I.L. and Walter, R. (1969) Science 163 925-926
Ferrier, B.M., Jarvis, D. and Du Vigneaud, V. (1965) J.Biol.Chem. 240 4264-4266
Glickson, J.D. (1975) in Peptides: Chemistry, Structure and Biology (eds. Walter, R. and Meienhofer, J.) Ann Arbor Science Publ., Michigan pp787-802
Glover, I.D., Haneef, I., Pitts, J.E., Wood, S.P., Moss, D.S., Tickle, I.J. and Blundell, T.L. (1983) Biopolymers 22 293-303
Haneef, I., Moss, D.S., Stanford, M.J. and Borkakoti, N. (1985) Acta Cryst. A41 426-433

Hirschfield, F.L. (1976) Acta Cryst. A32 239-244
Hruby, V.J. (1981) in Topics in Molecular Pharmacology (eds. Burgen, A.S.V. and Roberts, G.C.K.) North Holland, Amsterdam pp100-126
Hruby, V.J., Rockway, T.W., Viswanatha, V. and Chan, W.Y. (1983) Int.J.Pept.Protein Res. 21 24-34
Low, B.W. and Chen, C.C.H. (1966) Science 151 1552-1553
Meraldi, J.P., Hruby, V.J. and Brewster, A.I.R. (1977) Proc.Natl.Acad.Sci. USA 74 1373-1377
Pierce, J.G., Gordon, S. and DU Vigneaud, V. (1952) J.Biol.Chem. 199 929-930
Pitts, J.E., Wood, S.P. and Blundell, T.L. (1983) CRC Crit.Rev.Biochem. 13 141-213
Rudko, A.D. Lovell, F.M. and Low, B.W. (1971) Nature New Biology 232 18-19
Schomaker, V. and Trueblood, K.n. (1968) Acta Cryst. B24 63-76
Schwyzer, R.J. (1973) J.Mondial. Pharm. 3 254-260
Smyth, D.G. (1970) Biochem.Biophys.Acta 200 395-403
Treharne, A.C., Haneef, I. and Blundell, T.L. (1985) this volume
Tu, A.T., Lee, J., Deb, K.K. and Hruby, V.J. (1979) J.Biol.Chem. 254 3272-3278
Urry, D.W., Quadrifoglio, F., Walter, R. and Schwartz, I.L. (1968) Proc.Natl.Acad.Sci. USA 60 967-974
Walter, R., Schwartz, I.L., Darnell, J.H. and Urry, D.W. (1971) Proc.Natl.Acad.Sci. USA 68 1355-1359
Wood, S.P., Tickle, I.J., Treharne, A.C., Pitts, J.E., Mascarenhas, Y., Li, J-Y., Husain, J., Cooper, S., Blundell, T.L., Hruby, V.J., Wyssbrod, H.R., Buku, A. and Fischman, A.J. (1985) Science, in press
Wyssbrod, H.R., Ballardin, A., Gibbons, W.A., Roy, J., Schwartz, I.L. and Walter, R. (1975) in Peptides: Chemistry, Structure and Biology (eds. Walter, R. and Meienhofer, J.) Ann Arbor Science Publ., Michigan pp815-820

WATER AT BIOMOLECULE INTERFACES

Julia M. Goodfellow, P. Lynne Howell and Robert Elliott

Department of Crystallography
Birkbeck College
Malet Street
London WC1E 7HX

SUMMARY

The aims of this paper are (i) to review our knowledge of solvent structure obtained from crystallographic studies of proteins and nucleic acids and (ii) to discuss structural aspects of computer simulations intended to increase our understanding of aqueous hydration of macromolecules. I shall use results from both crystallography and computer simulations in order to present a unified view of the structure of hydrogen bonded water networks surrounding biomolecules.

INTRODUCTION

It is well known that the interactions of water with macromolecules affects a wide range of properties and functions including protein folding (Finney et al., 1980), protein dynamics (Finney and Poole, 1984), enzymatic activity (Rupley et al., 1980), protein secondary structure (Blundell et al., 1983) and nucleic acid structural transitions (Texter, 1978). Many techniques have been used to study hydration of proteins (Finney et al., 1983) including infra-red and Raman spectroscop (Finney and Poole, 1984), nmr spectroscopy (Piculell, 1985; Piculell and Halle, 1985), differential scanning calorimetry (Rupley et al., 1980, Poole, 1983), X-ray and neutron crystallography (Savage and Wlodawer, 1985) and computer simulations (Finney et al., 1985[a]).

A similar variety of techniques have been used to look at the water associated with nucleic acids especially as there were no oligonucleotide crystal structures until around 1978. These techniques include infra-red spectroscopy (Falk et al., 1963), gravimetric studies (Falk et al., 1962) and fibre diffraction (Leslie et al., 1980). The general conclusions (see reviews by Texter, 1978 and Saenger, 1984) are that a primary shell of twenty water molecules per nucleotide exist of which 11 or 12 are hydrogen bonded directly to the DNA with A-T base pairs having one or two more water molecules bound than G-C base pairs. This primary layer is impermeable to cations and does not freeze. A secondary shell of water molecules may also exist which includes some counter-ions. Solvent accessibility calculations (Alden and Kim, 1979) showed that the total surface accessibility of the A and B helical forms were similar even though the A form is regarded as the dehydrated form in contrast to the hydrated B form.

Edsall and Mackenzie (1979, 1983) have produced an excellent review of all aspects of protein hydration whereas theoretical aspects of both nucleic acid and protein hyration are covered in a review by Berndt and Kwiatowski (1985). Although it is clear from these reviews that a large amount of data is now available on all aspects of macromolecular hydration, the meaning of such commonly used concepts as bound water and hydrophobic hydration is not always clear in terms of molecular structure. Therefore, we shall concentrate on results from crystallography and simulations which are two techniques capable, in principle, of defining the molecular structure of water networks close to biomolecule interfaces.

CRYSTALLOGRAPHIC STUDIES

(A) <u>Small Crystal Hydrates</u>

Analysis of 97 neutron structures of crystal hydrates has lead to a detailed knowledge of the geometry of the water molecule hydrogen bond (Chiari and Ferraris, 1982). The hydrogen bond is a relatively weak interaction and therefore there are only relatively weak stereochemical constraints acting on solvent molecule networks. Thus, the O$\cdots$O and H$\cdots$O distances are found to occur over a wide range of 2.5 to 3.15Å (mean 2.81Å) and 1.52 to 2.26Å (mean 1.86Å) respectively. The hydrogen bond angle O—H$\cdots$O tends to be nearly linear with the majority of angles greater than 150°. The O$\cdots$O$\cdots$O angle exhibits a large range of values (68.9 to 147.8) around the expected approximately tetrahedral angle of 108°.

(B) <u>Medium Sized Crystal Hydrates</u>

There are 7 structures of medium sized biomolecules (50-300 atoms) determined to better than 1Å resolution by X-ray crystallography and, in four cases, by neutron crystallography as well (Savage and Wlodawer, 1985). The water structures seen in these crystals are more extensive than in the small hydrates. Therefore, one begins to see networks of hydrogen bonded water molecules in ordered regions of well-defined electron density. Moreover, disordered solvent regions may exist which makes solvent modelling from crystallographic data more difficult.

One of the first networks to be studied was that in α-cyclodextrin crystal hydrate (Saenger, 1979). In this crystal, hydrogen bonded water molecules and hydroxyl solute groups were seen to form ring structures with five or six sides. The directions of these hydrogen bonds (i.e. donor to acceptor) were either homodromic (unidirectional) or anti-dromic (changing direction). Saenger (1979) suggests that these ring structures together with chains of water molecules may well form the main water structures hydrating macromolecules. Pentagonal rings of water molecules and polar solute atoms were also seen in the dCpG proflavine complex in which there were a group of four edge-linked pentagons supporting the heterocycle stacking at 3.4Å apart (Neidle et al., 1980).

The vitamin B_{12} coenzyme structure has been studied by X-ray and neutron crystallography at about 0.9Å resolution. The solvent region is divided into two distinct regions separated by an acetone molecule with an occupancy of 0.7. These regions are (i) an ordered pocket in which there is one major network (0.7 occupancy) and one minor network (0.3 occupancy) and (ii) a disordered channel (Savage, 1983). The analysis of the disordered regions was started with the assignment of main sites with different occupancies (with those sites of similar occupancies forming a hydrogen bonded network) and proceeded to an assignment of 'continuous' sites which occupy the elongated regions of electron density (Savage, 1985). Thus, at this high resolution, it is possible to model disordered solvent regions in considerable detail.

(C) Protein Crystals

Finney (1979) reviewed the structural role of water molecule sites as determined by X-ray crystallography. These sites tended to be relatively well-ordered and in the first hydration shell around the protein. They could be categorized into four main groups: (i) internally bound to metal ions, charged groups (not involved in ion-pairs) and polar groups (not otherwise involved in hydrogen bonds), (ii) on the surface covering exposed polar groups and in single or multiple bridges across the surface, (iii) forming bridges between different molecules and (iv) in enzyme active sites. In a more recent review on hydrogen bonding in globular proteins, Baker and Hubbard (1984) have summarized water$\cdots$protein contacts in 13 well-defined X-ray structures. Average hydrogen bond distances and angles are given for water$\cdots$main-chain interactions but the water$\cdots$side-chain interactions tend to show very broad distributions such that it was not possible to establish preferred distances or angles. Presumably this is due partly to the weak stereochemical constraints on hydrogen bonded water networks and also to partial disorder in the solvent regions close to side-chains.

In the best refined protein structures (better than 1.5Å), it has been possible to model a large part of the solvent regions in terms of ordered sites, e.g. 77% for crambin (Teeter, 1984) and greater than 70% for 2 Zn insulin (Baker et al., 1985). Neutron studies on crambin and HEW lysozyme lead to similar solvent sites in the ordered surface regions but not in the more disordered solvent regions (Savage and Wlodawer, 1985). The Crambin structure at 0.945Å resolution shows chains of water oxygen atoms forming links between polar groups on the protein surface. Moreover a cluster of four pentagonal arrays exist in a hydrophobic cleft at an intermolecular contact region. These rings are also hydrogen bonded to polar protein atoms.

(D) Oligonucleotide Crystals

The structure of the dodecamer d(CGCGAATTCGCG) provided the first detailed information of hydration sites for a B-DNA helical fragment (Drew and Dickerson, 1981, Kopka et al., 1983). The most ordered water sites were found when crystallization took place either with high concentrations of MPD (60%) or at low temperature (16K). Under these conditions, both the phosphate groups and the base keto oxygen and amino groups were hydrated. Strings or filaments of water molecules, involving those bound to phosphates and bases, were seen connecting opposite phosphate backbones across the major groove. The most noticeable solvent structure was a 'spine' of hydration in the minor groove. This consists of a bidentate bridging of one water molecule to base atoms, on opposite strands, on adjacent base pair steps. This occurs mainly in the central AATT region. A further layer of water molecules bridging those waters associated with the bases forms a 'spine' filling the minor groove.

This B DNA structure has been further refined by Westhof et al. (1985) using a different refinement procedure to that originally used by Dickerson and colleagues. They also see the spine of hydration in the minor groove but conclude that the hydration of the major groove is not very ordered except on two base pairs and with only one pentagonal networks of three waters and two base pair atoms. The phosphates are associated with ordered water molecules sites which connect to other phosphates but rarely to polar atoms on the bases.

One of the first A-DNA oligonucleotide structures (d CCGG) by Conner et al., 1984) showed no spine of hydration in either groove nor an extensive hydration shell around base or sugar groups. However, the hydration of the phosphates was found to be more ordered than that seen in the B-DNA crystals. Chains of 1 to 3 water molecules linking adjacent phosphates were seen as were chains of four water molecules across the major groove.

The more recent A DNA crystal structures of d (GGBrUABrUACC) (Kennard, 1984) and d(GGTATACC) (Kennard et al., 1985) clearly show that A DNA oligonucleotides are as 'hydrated' as B DNA. In the major groove nearly all the functional groups are hydrated and a ribbon of hydration - made from pentagonal water networks - extends down this groove principally in the central TATA region. The phosphate backbone shows water molecules linking adjacent groups along one chain. In the minor groove, 16 out of 20 functional groups and over half the O1' sugar atoms have ordered water molecule sites associated with them.

In contrast to most of the A and B oligonucleotide structures, the Z-DNA crystal structures are at high resolution (better than 1.5Å) and counter-ions can also be seen in the solvent. In one of these structures d(5BrCG5BrCG5BrCG) (Westhof et al., 1985) ordered water molecules are seen in the major and minor grooves as well as between successive phosphate groups. The deep minor groove is filled with ordered water molecules linking phosphate oxygens and polar base atoms.

In summary, the oligonucleotide crystal structures of all three helical forms are showing ordered water molecule sites near polar groups and more extensive water networks, chains and rings especially in spatially confined regions such as the minor grooves of B-and Z-DNA and the major groove of A-DNA. The ordered water molecule sites seen by crystallography are not necessarily the most important from a thermodynamic viewpoint. It can be argued (Beveridge et al., 1984) that, in fact, disordered water sites contribute more to the stability of a conformation because of the increased entropic contribution compared with that from ordered sites.

COMPUTER SIMULATIONS

Computer simulations, both Monte Carlo and Molecular dynamics, were used to study liquid water (Wood, 1979) partly because experimental information on liquids was limited to radial distribution functions and partly to understand more about the thermodynamics of liquids and solutions. These techniques have been extended to study water at biomolecular interfaces for similar reasons i.e. to increase our understanding of both structural and energetic aspects of water···macromolecule interactions (Finney et al., 1985[a]). However, this extension from homogeneous water ensembles to complex heterogeneous solvent biomolecule interfaces is not trivial and involves the 'realistic' modelling of interatomic force fields and problems of equilibration and convergence (Finney et al., 1985[a]).

Not only can simulations be used to predict solvent structure in disordered or liquid regions but can also be used to study the energetics of water···water and water···solute interactions. Such simulations have lead to a better understanding of the hydrophobic interaction between inert solutes including the presence of the solvent-separated hydrophobic interaction (Ravishanker et al., 1982). More recently, techniques have been developed to calculate free energy changes associated with water (Sussman et al., 1984) and with changes in solvent structure related to different conformations of a solute (Mehrotra et al., 1984).

(A) Models for Water···Biomolecule Interactions

Currently, many models are available for liquid water usually based on a water···water dimer potential (Finney et al., 1985[b]). Similarly, modelling the properties of the atoms on a biomolecule depends on the values chosen for the partial atomic charges and for the non-bonded interaction parameters (Finney et al., 1985[a]). In order to understand the idiosyncrasies of different combinations of water and solute models (i.e. potential energy functions), Vovelle and Goodfellow (1985) undertook calculations on a very simple system consisting of one water molecule and one small solute molecule, that of urea (which has carbonyl and amino groups common to amino and nucleic acids). The results of energy minimization calculations showed that the primary hydration sites (i.e. local energy minima) of urea were determined mainly by the choice of water model. Some models lead to out of plane sites around the carbonyl, other models lead to water molecules in bridging positions between carbonyl and amino groups. Using partial atomic charges from various quantum mechanical calculations, we found changes in the magnitude of the binding energy of the water molecule to the polar solute atoms at each hydration site. Using different sets of non-bonded interaction coefficients lead to differences in the geometry of the hydrogen bond between water and solute.

(B) Crystal Hydrate Simulations

Crystal hydrates provide an obvious test for the structural predictions of any simulation. Initial studies on protein crystal hydrates (Hagler and Moult, 1978) showed little agreement with experimental data. Simulations of small crystal hydrates showed that the water molecule positions were very sensitive to the interatomic potential energy functions used to model water···water and water···solute interactions (Goodfellow et al., 1982). Moreover, cooperative (non-pair additive) effects appeared to be important at solvent biomolecule interfaces (Goodfellow, 1982). Problems of comparison of predicted and experimental solvent networks (i.e. in deciding which predicted network was closest to the experimental solvent network) lead to the introduction of various agreement factors in order to compare the environment of each water in simulated and experimental positions (Goodfellow, 1984, Vovelle et al., 1985).

As described in a previous section on the crystallographic studies of medium sized molecules, the solvent regions of B_{12} coenzyme have been modelled in great detail using X-ray and neutron data at around 0.9Å resolution. In order to look at the effects of different water models, we undertook five independent simulations on this well defined solvent crystal hydrate in order to compare the simulated results with experimental solvent model (Vovelle et al., 1985; Goodfellow et al., 1985). Five water models were chosen from three different classes, namely three point charge (TIPS2), four-point charge (ST2 and EMPWI) and polarisable electropole (PE) models. In summary, whatever parameter was used to judge the success of the simulation, EMPWI was best, the PE models were worst and the ST2 and TIPS2 models were intermediate (Table 1). The implication was that to achieve better agreement with experiment one has to improve on the EMPWI model. Analysis of the water dimer energy surfaces, from which the water models were derived, showed a similar trend in properties (e.g. the position and depth of minima, height of repulsive regions) with the PE and EMPWI models being at the extremes. To improve on the EMPWI model would be to further decrease the minimum and to increase the repulsions in the water···water interaction. However, this type of modification would lead to a totally unrealistic model for water. Thus, we are left with an

Table 1

Water Model	RMSD[a] (Å)	Agreement Factors[b] (Å)	
		I	II
PE(DO)	1.26±0.60	.79	.44
PE(QQ)	1.22±0.73	.65	.35
ST2	1.02±0.56	.33	.22
TIPS2	0.86±0.55	.37	.30
EMPWI	0.75±0.24	.26	.12

(a) Root mean square deviation between experimental and simulated water oxygen positions.

(b) Agreement factors I and II are the root mean square deviations, in nearest neighbour distances within a 3.75Å sphere around each water, between (i) experimental and simulated environments and (ii) between the four independently simulated asymmetric units.

impasse. If we use the present functional forms for the water models, we cannot improve the agreement with experiment and maintain a reasonable model for water itself. The way out may be to modify the currently used functional forms for potential energy functions.

(C) Simulations of Aqueous Solutions

Numerous simulations of aqueous solutions have been reported (see Beveridge et al., 1981 and for a review Berndt and Kwiatowski, 1985) for solutes as simple as methane (Darschevsky and Sarkisov, 1974) through glycyl zwitterions (Mezei et al., 1984) to complex solutions involving helical turns of DNA, water and counter-ions (Corongiu and Clementi, 1981). The results can be confusing with some authors finding for example that water close to apolar solute groups is composed of 'distorted defective pentagonal dodecahedral clathrate contributions' and others concluding that 'a qualitative description of the observed geometry... as solid-like and clathrate-like would be very misleading'. Simulation results are often presented in terms of probability density maps (Romano and Clementi, 1980), statistical state solvation sites (Mehrotra and Beveridge, 1980) or quasi-component distribution functions (Mezei and Beveridge, 1985) which are all methods of partitioning water molecule sites into association with specific solute atoms. One of the clearest studies on nucleic acid hydration by Beveridge et al., 1984 shows the most likely number of water molecules close to each functional group on sugars, dimethyl phosphate anion and the five bases separately.

We have undertaken simulation studies on six amino acid solutions (those of glycine, alanine, valine, leucine, threonine and serine zwitterions) as well as on a dinucleotide dCpG in the classical (fibre) B conformation. In each case, we have analyzed the results in order to look at the overall pattern of hydration and not just at the water molecules associated with an individual base. Although very different water models were used for the amino acid solutions compared with the dinucleotide solutions, we find that similar patterns can be seen. Apart from the water molecules within hydrogen bonding distance of solute polar atoms, we have included those others proximal to the solute within 6.0Å. Water molecules not directly hydrogen bonded to the solute but within 4.0Å of an apolar solute group (e.g. methyl) have been termed 'apolar' waters for ease of reference.

The first pattern or motif which is apparent is that of self-bridging loops of hydrogen bonded water molecules around each polar solute atom (Figures 1a and 3). The second arrangement is that of polar bridging chains which extend from one polar solute atom to another around an amino acid (Figure 1b) or along the phosphate sugar backbone of a dinucleotide (Figure 4). Finally we looked at the networks involving the 'apolar' water molecules. These consisted of closed loops of most frequently five water molecules (Figure 2). Pentagonal networks were also seen to be most common in the self-bridging loops and in the polar bridging chains. All rings or loops were distinctly non-planar unlike the cages of pentagonal and hexagonal networks found in clathrates (Jeffrey and McMullan, 1967).

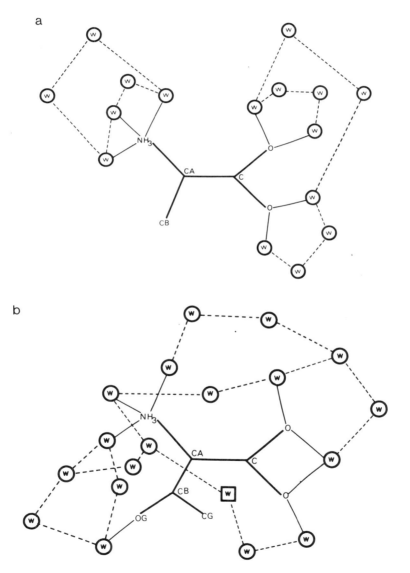

Figure 1 shows schematically the water networks (a) associated with each polar atom of alanine (self-bridging loops) and (b) between polar atoms of threonine (polar bridging chains).

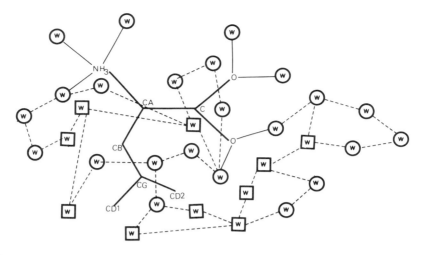

Figure 2 shows a schematic view of the water networks associated with the "apolar" waters around leucine.

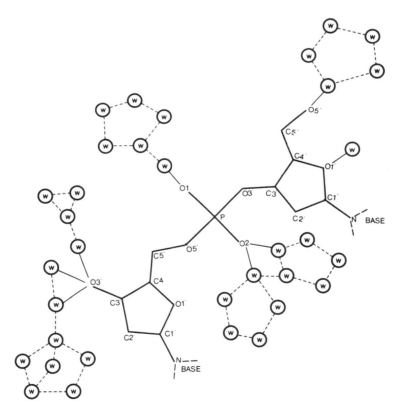

Figure 3 shows water networks associated with each polar or charged atom on the phosphate-sugar backbone (self-bridging loops).

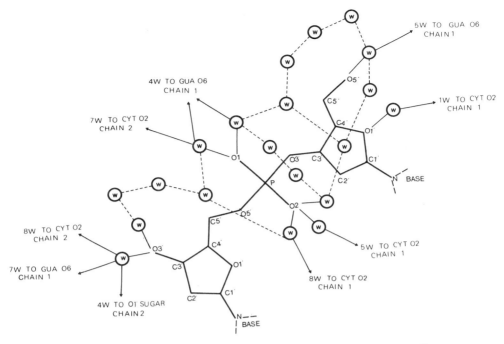

Figure 4 shows networks between polar groups on the phosphate-sugar backbone (polar-bridging chains).

CONCLUSIONS

X-ray and neutron crystallography at high resolution are leading to a molecular level picture of hydration which started with individual water molecule sites and is developing towards a view of hydrogen bonded water networks around sugars, proteins and oligonucleotides. Although computer simulation techniques are not yet capable of simulating the detailed molecular geometry of these solvent networks (as seen from simulations of crystal hydrates), we are beginning to understand the modifications to the potential energy functions necessary to improve our modelling of water···water and water···solute interactions. Moreover, simulations of amino acids and nucleotides in solution are showing similar patterns or motifs of hydration as those found from experimental crystallographic studies. These include self-bridging loops, polar bridging chains and rings of hydrogen bonded water molecules (most frequently consisting of five molecules). Non-planar pentagonal rings also occur in bulk water and high pressure ice. Planar pentagonal and hexagonal rings occur in clathrate hydrates. Thus rings and chains of hyrogen bonded water molecules appear to be common to most aqueous systems.

ACKNOWLEDGEMENTS

The authors would like to thank the SERC for project grant numbers GR/C/22417, GR/C/65704, GR/D/14372 and for a studentship to P.L.H.

REFERENCES

Alden, C.J. and Kim, S-H. 1979, J. Mol. Biol. 132 411.
Baker, E., Dodson, G., Hodgkin, D. and Hubbard, R. 1985, This volume.
Berndt, M. and Kwiatowski, J.S. 1985, in 'Theoretical Chemistry of Biological Systems', Ed. Naray-Szebo, G., Elsevier, Amsterdam.
Beveridge, D.L., Mezei, M., Mehvotra, P.K., Marchese, F.T., Thirumalai, V. and Ravi-shanker, G. 1981, Annals N.Y. Acad. Sci. 367: 198.
Blundell, T., Barlow, D., Borkakoti, N. and Thornton, J. 1983, Nature 306: 281.
Chiari, G. and Ferraris, G. 1982, Acta Cryst. B38: 2331.
Conner, B.N., Yoon, C., Dickerson, J.L. and Dickerson, R.E. 1984, J. Mol. Biol. 174: 663.
Corongiu, G. and Clementi, E. 1981, Biopolymers 20: 551.
Darschevsky, V.G. and Sarkisov, G.N. 1974, Mol. Phys. 27: 1271.
Drew, H.R. and Dickerson, R.E. 1981, J. Mol. Biol. 151: 535.
Edsall, J.T. and McKenzie, H.A. 1979, Adv. Biophys. 10: 137.
Edsall, J.T. and McKenzie, H.A. 1983, Adv. Biophys. 16: 53.
Falk, M., Hartman, K.A. and Lord, R.C. 1962, J. Amer. Chem. Soc. 84: 3843.
Falk, M., Hartman, K.A. Jr., and Lord, R.C. 1963, J. Amer. Chem. Soc. 85: 391.
Finney, J.L. 1979, in 'Water' Vol. 6, Ed. F. Franks, Chapter 2, p.47.
Finney, J.L. Gellatly, B.J., Golton, I.C. and Goodfellow, J.M. 1980, Biophys. J. 32: 17.
Finney, J.L., Goodfellow, J.M. and Poole, P.L. 1982, in 'Structural Molecular Biology', Eds. D.B. Davies, W. Saenger and S.S. Danyluk, Plenum Press, N.Y., p.387.
Finney, J.L., Goodfellow, J.M., Howell, P.L. and Vovelle, F. 1985a J. Biomol. Structure and Dynamics. In Press.
Finney, J.L. and Poole, P.L., Comments on Molecular and Cellular Biophysics Vol. II, 129.
Finney, J.L., Quinn, J.E. and Baum, J.O. 1985b, in Water Science Reviews, Ed. F. Franks, Vol. I, Cambridge University Press.
Goodfellow, J.M. 1982, Proc. Natl. Acad. Sci. USA, 79: 4977.
Goodfellow, J.M. 1984, J. Theor. Biol., 107: 261.
Goodfellow, J.M., Finney, J.L. and Barnes, P. 1982, Proc. Roy. Soc. B214: 213.
Goodfellow, J.M., Vovelle, F., Quinn, J.E., Savage, H.F.J. and Finney, J.L., 1985, in Colston Symposium on Water and Aqueous Systems, Bristol, April 1985.
Hagler, A.T. and Moult, J. 1978, Nature 272: 222.
Jeffrey, G.A. and McMullan, R.K. 1967, Progress in Inorganic Chem. 8: 43.
Kennard, O. 1984, Pure and Applied Chem. 56: 989.
Kennard, O. et al. 1985. In preparation.
Kopka, M.L., Fratini, A.V., Drew, H.R. and Dickerson, R.E. 1983, J. Mol. Biol. 163 129.
Leslie, A.G.W., Arnott, S., Chandrasekaran, R. and Ratliff, R.L. 1980, J. Mol. Biol. 143: 49.
Mehrotra, P.K. and Beveridge, D.L. 1980, J. Am. Chem. Soc. 102: 4287.
Mehrotra, P.K., Mezei, M. and Beveridge, D.L. 1984, Intl. J. of Quant. Chem., Symposium 11 301.

Mezei, M., Mehrotra, P.K. and Beveridge, D.L. 1984 , J. Biomol. Structure and Dynamics 2 :1.
Mezei, M. and Beveridge, D.L. 1985, in "Methods in Enzymology" (Ed. L. Packer). In Press.
Neidle, S., Berman, H. and Shieh, H.S. 1980, Nature 288 :129.
Piculell, L. 1985, PhD thesis, University of Lund, Section VI, pp.1-37.
Piculell, L. and Halle, B. 1985, submitted for publication.
Poole, P.L. 1983, PhD thesis, University of London.
Ravishanker, G., Mezei, M. and Beveridge, D.L. 1982 , Faraday Symp. Chem. Soc. 17 :79.
Romano, S. and Clementi, E. 1980, Intl. J. Quant. Chem. XVII, 1007.
Rupley, J.A., Yang, P-H. and Tollin, G. 1980, in 'Water in Polymers', Ed. Rowland, S.P., ACS, Washington DC, p.111.
Saenger, W. 1979, Nature 279 :343.
Saenger, W. 1984, in 'Principles of Nucleic Acid Structure', Springer-Verlag, New York, Chapter A, p.368.
Savage, H. 1983 , PhD thesis, University of London.
Savage, H. 1985 , Biophysical Journal. Submitted.
Savage, H. and Wlodawer, A. 1985, in Methods in Enzymology: 'Biomembranes: Protons and Water Structure and Translocation'. In Press.
Sussman, F., Goodfellow, J.M., Barnes, P. and Finney, J.L. 1984 , Chem. Phys. Letters 113 :372.
Teeter, M. 1984, Proc. Natl. Acad. Sci. USA 81 :6014.
Texter, J. 1978, Prog. Biophys. Mol. Biol. 33 :83.
Vovelle, F. and Goodfellow, J.M. 1985 , Submitted to J. Biomol. Struct. and Dynamics.
Vovelle, F., Goodfellow, J.M., Savage, H.F.J., Barnes, P. and Finney, J.L. 1985 , Euro. Biophys. J. 11 : 225.
Westhof, E., Prange, Th., Cheutier, B. amd Moras, D. 1985 , Biochimie. In Press.
Wood, D.W. 1979, in "Water: A Comprehensive Treatise', Ed. F. Franks, Plenum, New York, Vol. 6, p.279.

THE WATER STRUCTURE IN 2Zn INSULIN CRYSTALS

Ted Baker*, Eleanor and Guy Dodson, Dorothy Hodgkin and Rod Hubbard

Dept. of Chemistry
University of York
York, England

*Dept. of Biochemistry and Chemistry
Massey University
Palmerston North, New Zealand

INTRODUCTION

Proteins have evolved in an aqueous medium and they have exploited with advantage the bonding and steric properties of the water molecule in achieving their 3 dimensional folding, stability, control of substrate bonding and catalytic reactions. The study of water protein interactions has relied on spectroscopic and calorific methods which generally confirm the essential role water plays in protein structure and folding. The detailed description of the water-protein contacts and the dynamical behaviour of the protein surface atoms and the surrounding water molecules however is a much more challenging undertaking. It must be emphasized that the water molecules on protein surfaces are extremely mobile. They exchange sites very rapidly, even when well ordered by contacts to the protein, and move extensively along the connected networks across the protein. Where there are few specific sites with suitable H bond and other contacts the water molecules exist in a practically continuous population of networks (J Finney, et al., 1982).

Protein crystals form, very fortunately, an excellent system for studying water interactions of a protein surface. Since they contain typically 30 - 70 per cent of solvent these observations can be further used as a basis for describing the water protein behaviour in solution. The X-ray analysis of numerous protein crystals has revealed the presence of many water molecules attached to the sidechain and exposed peptide nitrogen and oxygen atoms (Watenpaugh et al., 1978, Blake et al., 1983, Baker, 1980, James and Selieki, 1983) essentially blanketing the molecule with solvent. Some water molecules prove to be completely buried within the protein, others filling cavities make strong contacts to both protein and solvent while again others make only one or two H bonds and form a loose covering surface. This rather simple picture correlates with such observations as the patterns of deuterium exchange, or the difficulty in removing a residual 5 per cent or so of the water from a protein as well as the other studies which have demonstrated that water needs to comprise about 30 per cent of a protein : water system (the so called "bound" water) before the molecule is functional. (Finney et al., 1982).

The principal drawbacks in the X-ray analysis of water structure are first that many of the solvent molecules are poorly defined, with apparent thermal parameters (B) greater than 50Å^2 (equivalent to a r.m.s. amplitude of vibration of about .2Å). Compared to most protein atoms these water molecules have low and error-prone electron density in the Fourier map. Secondly there are very few geometrical restrictions on the 3 dimensional arrangement of water molecules, which makes interpretation of the electron density more complicated. It is essential that analysis of solvent structure be very carefully and critically carried out. High resolution diffraction data ($d_{min} \leqslant 1.5$Å) are very desirable in such studies since it increases the contrast between well ordered and poorly ordered atoms; see especially Figure 3 below.

The Crystal Structure of 2Zn Insulin

The 2Zn insulin rhombohedral crystal contains one hexamer constructed of 3 dimers coordinated to 2 zinc ions which lie on a central 3 fold axis. The dimers contain a 2 fold axis which is perpendicular to the hexamers 3 fold axis (Dodson et al., 1979). The crystal unit cell is arranged approximately in a body centred cubic array. The hexamers are packed fairly tightly in rods along the crystal's 3 fold axis and are related to each other by 3 fold screw operations. The crystal contains about 30 per cent by volume of solvent, distributed in channels along the screw axes which extend through the whole crystal. These channels extend up to 10 - 16Å across and form the largest volumes of water structure in the crystal. (Figure 1) Around the circumference of the hexamer and joining the channels at the screw axes the water molecules form a layer broken only by occasional crystal contacts. There is an extensive and largely structured region of water at the hexamer centre. This is connected to the water molecules on the hexamer surfaces which come together top and bottom along the 3 fold axis.

The crystal structure has been refined using Fourier and restrained least squares procedures of Agawarl and Konnert and Hendrikson (Machin et al., 1981). The agreement factor $R = \Sigma ||Fo|-|Fc||/\Sigma|Fo|$ is .153 for all the 13371 data which extend to 1.5Å spacing. All the protein atoms have been positioned, although 8 of the 102 sidechains in the crystals asymmetric unit are disordered with the sidechains assuming two identifiable conformations; in some cases there are probably others at lower (<.33) occupancy. The solvent structure has been analysed in detail and 282 water molecules

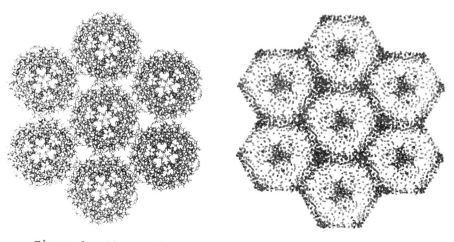

Figure 1. The complete structure of the R3 2Zn insulin crystal showing 7 hexamers viewed in the direction of the 3 fold axis. The protein atoms (a) and the water molecules (b) are shown separately.

(very nearly equal to the 285 calculated present from the density) have been located, distributed between 341 sites.

THE PROCEDURES FOR IDENTIFYING WATER MOLECULES

The refinement of the crystal structure has gone through several phases in which the details of the water molecules became increasingly clear. In these studies Fourier maps were used with coefficients Fo, Fo - Fc, 2Fo - Fc and Fc, together with systematic difference Fourier maps in which 1/8 of the structure was systematically removed. The appearance of the solvent's electron density in these different maps was an important guide in interpreting water structure. (Dodson, 1981). Figures 2 - 4 illustrate some regions of the water molecule structure in the Fo map calculated from the final atomic coordinates.

The well defined water molecules, generally attached to the protein are very clear features, with density ranging from 1.0 - 2.0 $\bar{e}/Å^3$. As refinement progressed, the lower levels of electron density (0.4 - 1.0 $\bar{e}/Å^3$) became clearer, usually retaining the same appearance in the various Fourier maps. The appearance of this lower density, spreading away from the well defined water, varies from spherical peaks, to relatively well defined threads, to regions of confused peaks in which the waters form interpenetrating and mobile networks. The occupancy of these sites was estimated by the character of the density and the presence of steric and H bonding contacts.

Least squares refinement of positional coordinates and B values of the water molecules were carried out. Typically water molecules had B values of 40 - 80 $Å^2$. Because of their low electron density and positional inaccuracy water molecules with B>50 $Å^2$ were systematically reviewed at fairly regular intervals during the refinement calculations. When during refinement the B value for a water continued to increase it was removed from the phasing and its density checked for several cycles.

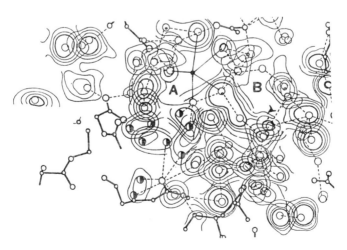

Figure 2(a) A section of the electron density about the 3 fold screw axis.
The density is contained as follows:
--- 0.0 - 0.3 $\bar{e}/Å^3$ in divisions of 0.1 $\bar{e}/Å^3$
— 0.4 - 0.6 " in divisions of 0.1 $\bar{e}/Å^3$
— .7 - " in divisions of 0.3 $\bar{e}/Å^3$

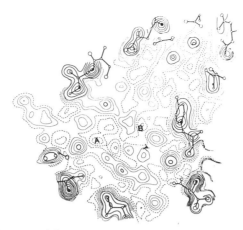

Figure 2(b) The electron density (sectioned through 6Å and on a larger scale) and its associated water structure around the two cavities A and B where the contour level is < .4 $\bar{e}/Å^3$. The contour levels start at 0.4 $\bar{e}/Å^3$ and follow the same scheme as in (a). The half filled in water molecules are in partially occupied sites, often mutually exclusive.

CHARACTERISTICS OF THE WATER STRUCTURE

The solvent electron density illustrated in Figures 2 and 3 shows the variability in the appearance of the water peaks, from the very well-ordered water (B < 30Å^2) attached to the protein, through linking waters of variable definition, to waters further out from the protein surface, usually much lower in density. Sometimes this density extends into continuous threads, apparently representing specific pathways of very mobile water molecules, but even in the centre of the largest water regions, ∼8Å from the protein surface, there is still evidence for definite water structure, mostly connected together by threads of low density.

Analyses and descriptions of the water structure are limited by the inadequacies in representing this electron density as individual peaks. These peaks are often diffuse or oddly shaped, suggesting that they do not represent a single water molecule site, but rather a number of close alternative positions; often these form networks. In some cases (eg. for some elongated peaks illustrated in Figure 2(b)) partial-occupancy sites have been introduced, but in others diffuseness has simply been taken up in the B values. We do believe that the density genuinely depicts water structure, but because of these complexities, statistics of water molecule contacts etc. must be treated with caution. Nevertheless, some observations can be made.

A summary of the probable hydrogen bonded contacts made by water molecules is given in Table 1. This is constructed for only the full-weight waters, taking potentially "good" hydrogen bonds (H---OW < 2.24Å or O(N)---OW < 3.25Å, - see below) into account. (A more liberal cut-off of 2.5Å and 3.5Å gives a distribution essentially the same in character but shifted to the right towards more contacts).

Table 1

H bonding by Insulin to fully occupied water sites.

(distance < 3.25Å)

Protein	H Bonds BavÅ²	0	1	2	3	4	5	6	7	Total
0	61	2	4	18	28	24	16	3	2	97
1	57			13	20	23	11	7	3	77
2	44			4	12	11	6	2		35
3	39				1	4	1			6
4	32					2				2
5	28						1			1
		2	4	35	61	64	35	12	5	218

Using these criteria, eight water molecules make contact only with protein, with 2, 3, 4 and, in one case 5 potential hydrogen bonds. On the other hand nearly 100 water molecules are hydrogen bonded only to other waters, mostly lying in the water-filled cavities between the protein molecules. A total of 77 waters make one hydrogen bonded contact with protein, while 44 make two or more. The average total number of hydrogen bonds is between 3 and 4, but a considerable number of waters have 5 or more, attributable to the diffuse nature of many of the water positions and to the undoubted existence of non-hydrogen bonded contacts in the range 3.0 - 3.25Å.

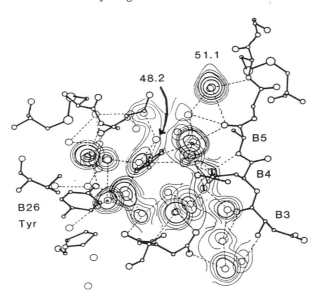

Figure 3 The electron density in the solvent region between the extended peptide chain $B_3 - B_6$ and the sidechains in an adjacent molecule. The water positions used in the structure factor calculations are shown by open circles; dotted lines indicate H bonding contact distances < 3.5Å. Note the marked variation in peak height in the water peaks, sometimes even when adjacent. The zig-zag character of the water chains along the peptide chain and stretching from B3 to B26 tyr OH is apparent.

Protein Hydrogen Bonding and the Role of Water

Water molecules cover essentially all of the exposed protein surface, and fill small pockets and crevices. In doing so they have an important role in satisfying the hydrogen bonding potential of polar groups on the protein. Thus in the insulin dimer only 8 of the main chain NH and C=O groups (out of 194) and 2 of the polar sidechain groups (out of 72) do not have identified hydrogen bond partners. Moreover, most of these apparently non-hydrogen bonded groups (7 of the 8 main chain and one sidechain) do in fact have potential hydrogen bond partners at suitable angles, but just outside the distance criteria we have used (O---H <2.5Å and O---O(N) < 3.5Å). Only the main chain C=O of residue B6.1* which projects into a non-polar pocket inaccessible to water, and the sidechain amide of Asn A21.2* which is on the surface but appears to have no attached waters, are exceptions to this general hydrogen bonding of all polar groups. Water molecules constitute the sole hydrogen bond partners for nearly 30 per cent of the main chain C=O and NH groups and 60 per cent of the polar sidechain groups.

Many of the H bonding contacts and geometry made by the surface atoms on insulin are very similar to those observed in other studies (Baker and Hubbard, 1984). There is one feature, the helix:water H bonding geometry that deserves mention. Here the C=O groups on the exposed sides of helices are often found to complete their H bonding potential through water molecules. Moreover our analysis on insulin and other refined protein structures show that not only are these H bonds geometrically favourable but that the α-helix offers a quite specific binding site for water molecules as first suggested by Bolin et al., (1982). This is illustrated by the plot of angles at the carbonyl O in terms of the in-plane (γ) and out of plane (β) components (Figure 4) which shows a fairly tight and separate distribution for the intra helical C=O---HN and C=O---Water bonds. (Artymiuk and Blake, 1981).

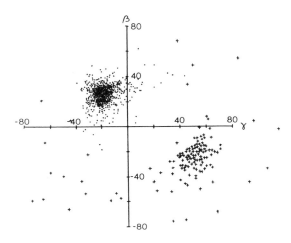

Figure 4 The distribution of the in-plane (γ) and out-of-plane (β) angles for water molecules (+) H bonded to helical carbonyl O. These are generally directed out in a relatively specific direction which is markedly different to the inward directed H bond made from the carbonyl O to the amide NH. (·) (Artymiuk and Blake, 1981)

---

*B6.1 indicates residue B6 in molecule 1, and A 21.2 residue A21 in molecule 2.

When the full hydrogen bond potential of the various groups is considered (taking each carbonyl, carboxyl and phenolic oxygen as having two hydrogen bonding sites, hydroxyl oxygens of Ser or Thr having three, etc.) coverage appears less complete. Out of 65 NH sites not used in protein hydrogen bonds, 57 are occupied by water molecules, and of 210 O sites available 163 are occupied by water. The main reason for the latter result, however, is that some 40 per cent of main chain C=O groups (37 out of 98) are unable to make more than one hydrogen bond. Most of these are on the inner, buried, side of helices, while others are sterically blocked by their presence in elements of β-structure, or simply by unfavourable packing of protein around them. This failure of many C=O groups to express their full hydrogen bond potential must be seen as an inevitable consequence of the internal folding in the protein. On the other hand, it is still true that where polar sidechains are concerned, nearly 90 per cent of the hydrogen bonding sites (131 out of 150) are used, mostly by water molecules.

As has been noted in other protein structures, oxygen ligands for bound water outnumber nitrogen by a factor of about 3 to 1 (3.3:1 insulin). In part this is due to the greater angular flexibility allowed by O, compared with NH, and in part to the observation that there are simply more O atoms present (150 O and 126 N in the insulin dimer) and that each oxygen offers at least two hydrogen bonding sites, compared with one for most nitrogens. Since almost all accessible hydrogen bonding sites do in fact bind water molecules the ligands used simply reflect the population of groups present on the protein surface. However we do note the (i) charged groups, especially COO$^-$, are highly hydrated and often support extensive water networks, whereas by comparison the amide sidechains of Asn and Gln have little ordered water attached, and (ii) main chain C=O and NH groups play an important role as ligands for the most firmly bound water.

Geometry of Protein-Water Contacts

Distributions of distances and angles for protein...water hydrogen bonds are shown in Figure 5. Although the numbers are small in this sample, the distributions are very similar to those seen in other proteins (Baker and Hubbard, 1984). For N-H---OW interactions the angle at the hydrogen is mostly in the range 145 - 170°, consistent with a close-to-linear hydrogen bond. For O---OW interactions the range of angles C-O---OW is much broader, a spread of about 45° over the range 100 - 145°, distributed around a median of about 125°. An interesting feature of the hydrogen bond distances, seen also in Baker and Hubbard (1984), is that while the distribution for H---OW (for NH---OW interactions) falls away towards 2.5Å, that for O---OW (for C-O---OW interactions) falls, then rises again towards 3.4Å. This is because the angle of 120 - 130° at the oxygen atom means that a "second shell" water molecule (ie. one hydrogen bonded to a CO-bound water) can itself be less than 3.5Å from the CO group, whereas the linear hydrogen bonding at the NH group means that a second whell water will normally not come within 2.5Å of the hydrogen (or 3.5Å of nitrogen). We have therefore used two distance criteria, 2.5Å (or 3.5Å), to identify possible hydrogen bonds (but inevitably including some second shell waters), and 2.25Å (or 3.25Å) to identify probable hydrogen bonds.

Ordering in the Water Structure

Despite the difficulties resulting from the correlation between B values and occupancies, the appearance of the electron density suggests that, for the full-weight waters at least, the B values give a reasonable indication of the mobility or ordering of the water molecules. We have thus attempted to correlate B values with structural observations.

There proves to be however only a very weak correlation between the

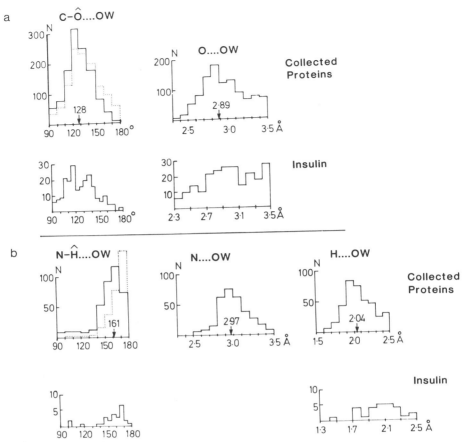

Figure 5 The histograms for the observed value of the angles and distances to water at the peptide O and N for collected proteins (Baker and Hubbard, 1984) and for 2Zn insulin.

water molecules and the B values of protein atoms to which they are attached. This is shown most dramatically for waters bound to main chain C=O groups where almost no correlation between B_{H2O} and $B_{protein}$ exists, and many C=O groups with low B values have waters with $B>60Å^2$ "bound to them.

Neither the total number of likely hydrogen bonds made by each water molecule (with protein and/or other waters) nor the total number of contacts with neighbouring atoms appear to correlate with the water B values (see Figure 6). The only structural correlation we have been able to make is with the number of good hydrogen bonds made with protein atoms. Thus the 25 best-ordered waters, as judged by their B values ($\bar{B}$ = 26.4) average 2.4 protein hydrogen bonds, the next 25 ($\bar{B}$ = 37.8) average 1.4 and beyond that the average number is very low (ca. 0.7). Viewed in another way, the average B value for waters drops from 61 for those making no protein hydrogen bonds, to 30 for those making 3 or more. There is some suggestion that waters attached to N-H groups may be better ordered than those attached to O ligands, since for water attached to a single nitrogen $\bar{B} = 53Å^2$ compared with $\bar{B} = 58Å^2$ for those bound to a single oxygen, while those bound to 1O + 1N have $\bar{B} = 42Å^2$ compared with $46Å^2$ for those bound to 2 O atoms. This could be explained by the greater angular restriction at NH groups, but the

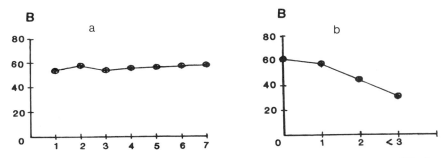

Figure 6 The distribution of the average thermal parameter (B) for the water molecules against (a) the number of contacts $< 3.5 \text{Å}$ and (b) the number of H bonds $< 3.25 \text{Å}$

populations are small and the difference may not be significant.

We therefore conclude that there is only a small number (about 40 for the insulin dimer) of "firmly" bound water molecules. These are waters which make two or more good hydrogen bonds with the protein (H---OW$< 2.25 \text{Å}$ or O(N)---OW$< 3.25 \text{Å}$). They represent almost all of the low-B water molecules in the overall population. They tend to occupy small pockets in the protein surface or in inter-subunit interfaces, and many play an important role in bridging between groups which cannot interact directly either because of distance or chemical character (eg. C=O and COO$^-$ groups). Most of them (70 per cent) have at least one main chain C=O or NH ligand, while the others bind to the zinc ion, or bridge between well-ordered sidechains or the chain termini. Interaction with just one protein group does not seem to be sufficient, in general, to provide a well-ordered water binding site. Presumably the wide range of hydrogen bond angles which are tolerated (section III.2) allows considerable looseness. This is no doubt why in insulin some well-ordered main chain C=O groups have only weakly-defined (high-B) waters attached, and why in most protein structure analyses water molecules are not seen around some 10 per cent of the polar groups.

The close interactions between water and sidechains are most important in defining water molecule positions. Thus changes in sequence will for example usually disturb the water structure as illustrated in Figure 7(a). Similarly disordered sidegroups can produce populations of partially weighted sites occupying space complementary to one of the sidegroup's conformations (Figure 7(b)).

Water - Water Interactions

Much of the water structure in the 2Zn insulin crystal is defined by interaction with polar groups on the protein surface, since about 50 per cent of the water molecules (121 out of 128 full-weight waters, and 49 out of 132 part-weight) contact at least one protein atom. The firmly bound waters act as anchoring points from which chains of weaker, less well ordered waters extend. These chains of water molecules, zig-zag or nearly straight, are often linked by continuous threads of density $\sim 0.4 \, \bar{e}/\text{Å}^3$ in height, representing pathways in which the less well ordered waters are found. (See Figures 2 and 3).

There are also frequent examples of water molecule sites where the local interactions provide some point of stability but not sufficient to prevent movement of the water to nearby alternative sites at less than H bonding distance. This can lead to networks of partially occupied sites, following still rather well defined paths. As the organisation of the

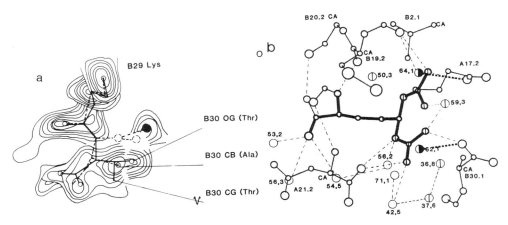

Figure 7(a) The difference electron density from human 2Zn insulin at residues B29 lys and B30 thr. Full lines represent the refined positions of the human insulin coordinates and dotted lines the equivalent positions for pig insulin (sequence B29 lys B30 ala). Two waters have been displaced by the arrival of the threonyl sidegroup in human insulin.

Figure 7(b) The residue B22 arginine (molecule 1) and its environment in 2Zn insulin. This residue is disordered and has been given two conformations related by a rotation about 150° around the CG-CD bond. The bars represent partially occupied atomic positions. The half filled in water molecules 64,1 and 62,1 share their positions with the guanidinium group.

water further from the protein comes looser these networks become less structured. In the extreme case the electron density suggests a continuum of water sites, which nonetheless are still evidently constrained to quite specific pathways.

Numerous rings of water molecules can be seen. There is one striking example of a 5-membered ring system covering an exposed non-polar sidechain (Val. A3.1), and anchored by protein atoms on either side (Figure 8). Other exposed non-polar groups however do not appear to have any characteristic pattern of waters associated with them, however. Three-membered rings are common throughout the water structure, but we also find 4-, 5-, 6- or more-membered rings in contrast to the finding in Crambin (Teeter, 1984).

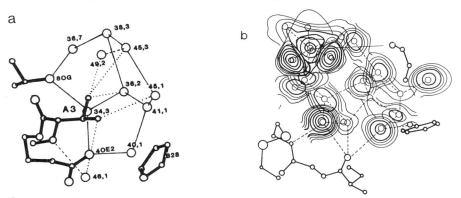

Figure 8 Ring structures in 2Zn insulin, (a) about the non-polar A3.1 valyl sidechain, (b) within the solvent volume.

Distributions of water....water distances are almost uniform over the range 2.3 - 3.5Å, although there is a rise towards 3.5Å, probably due to "second shell" effects. The angles at water molecules, show a very broad spread (Figure 9), not surprising because of the disordered nature of some of the sites, and the imprecision of their positions. There appear to be two distinct maxima, one at 55 -80°, the other at 90 - 115°, although following the latter there is a long spread to about 145°. The peak at 90 - 115° is as expected for tetrahedral geometry. The peak at 55 - 80° appears to arise from two effects, viz. the triangles of waters which are a common feature of our maps, and the effect of "second shell" waters (waters attached in turn to primary ligands of a given water molecule). There is indeed a slight reduction in this peak if only angles subtended by two "good hydrogen bonds" (both distances < 3.25Å) are considered.

Water Behaviour - Understood and Not Understood

Here we simply choose a few interesting examples of phenomena we have observed and are trying to understand.
(i) Why do some water molecules not interact more closely with protein groups? Examples are given by the NH groups of A2.2 and A3.2 (whose only potential hydrogen bond partners are waters at H---OW = 2.6Å and the C=O groups of B21.2 (only potential hydrogen bond partner is water at O----OW = 3.6Å). Taking the latter example, the water molecule, 53.2, has a low B value (27.4Å^2) and good density, and makes three hydrogen bonds in the present model (see Figure 10). It could easily be moved 0.6Å, without seriously altering those three interactions (in either distances or angles) or non-bonded contacts, to a position 3.0Å from B21.2, apparently giving a good hydrogen bond with B21.2 and completing a tetrahedron about the water molecule. Yet the density shows that this does not occur. Our present explanation is that the two protons of water 53,2 are directed towards B22.2 and water 56,3 (which in turn has its two protons directed towards B22.2 and A21.1 OD1). A simultaneous hydrogen bond with B21.2 is then not possible. The argument does not hold, however, if OD1 is replaced by ND2, which seems easily achieved. Similarly we can move the water molecules 21,3 and 23,1 to good hydrogen bond distances from A2.2NH and A3.2NH

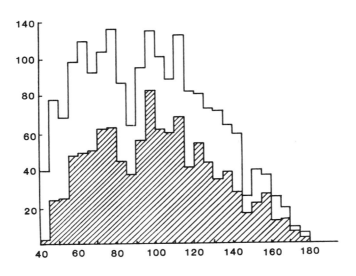

Figure 9 A histogram of the angles at the water molecule calculated with separation of d ≤ 3.25Å in hatched and d ≤ 3.5Å in open lines.

respectively, without bad contacts or disruptions of other hydrogen bonds, yet the density does not allow it. These are puzzles, and perhaps good candidates for simulation and energy calculations.

(ii) Can we correlate B values with water environments? The broad trend in B values can be related to the number of good hydrogen bonds with protein (see section III.3). In some cases individual B values can be explained nicely in structural terms. For example water 59,1 with 3 protein hydrogen bonds and one water has a lower B (12.2Å^2) than 46,1 (B = 27.5Å^2) with 4 protein hydrogen bonds. The former, however, is at the inner end of a deep pocket surrounded tetrahedrally by its 4 ligands, whereas the latter is less well buried, with a less tetrahedral geometry. Water 54,1 has only 3 hydrogen bonding ligands (2 protein, 1 water), in a trigonal plane, but has a low B value (29.9Å^2), - in this case non-bonded interactions contribute because movement in the axial direction is prohibited by too-close contacts with other protein atoms. Other B values are less readily understandable. Thus water 56,1 has a low B (= 31.1Å^2) despite no interaction with protein atoms (it makes 3 hydrogen bonds to other waters). Moreover it can be moved 0.5Å to make an additional hydrogen bond and complete a tetrahedron of waters around it. Why it is so relatively well-ordered and why it does not occupy this alternative position are puzzles. Similarly, water 52,1 has a relatively low B (= 40.0Å^2) despite contacts only with other waters, and only one of these < 3.25Å. While it is unwise to place too much reliance on individual B values, much could be learned from further analysis of these local water structures.

(iii) Why do apparently empty cavities appear in the water structure? We have found several instances where interconnected strings of water molecules leave a cavity, 6Å or so in diameter, with essentially no electron density (see Figure 2(b)). In this case, a water molecule placed at the centre of the cavity would have 10 contacts in the range 2.6 - 3.2Å with surrounding water molecules. Moreover 7 of them would be ≤ 3.0Å. We would suggest that while 2.6 - 3.2Å would be ideal for hydrogen bonding, there would simply be too many non-bonded contacts (for H_2O----H_2O the non-bonded contact distance is ~ 3.0Å). Similarly a water molecule in the cavity enclosed by waters 48,8; 50,2; 43,1; 50,1; 45,2; 50,6; 44,2; and 54,6 would make 7 contacts of 2.7 - 3.1Å (6 of them <3.0Å). It seems that the strings and

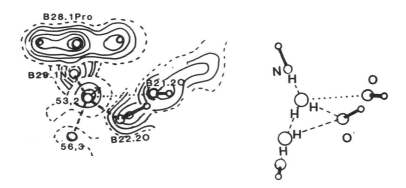

Figure 10 The H bonding pattern and contacts made by water 53,2. In (a) the electron density for the interacting atoms is shown; note the well defined density for 53,2. The arrow indicates the extent of the move which would give sensible H bonding contacts to B21.2 carbonyl O. In (b) the possible H positions are shown; the absence of an available H between 53,2 and B21.2 O would explain the longer contact distance of 3.6Å.

networks of water molecules defined by interaction with the protein surface result in some cavities which cannot be occupied at specific sites because of the large numbers of close contacts that would result. (Note that only 10 out of 218 full-weight waters in the present model have more than 4 contacts of < 3.0Å).

Conclusions

The principal conclusions from the study of the 2Zn insulin crystal are
1) Careful crystallographic refinement can reveal considerable details in the water structure, particularly when the solvent regions are not too large. There are intractable difficulties however in refining atomic positions, occupancies and thermal parameters.
2) There appears to be a rather extensive organisation of the solvent molecules in 2Zn insulin even though the water molecules themselves are usually very mobile.
3) Two or more H bonds to protein are needed to tie down attached water molecules; with fewer constraints the waters move rapidly about their mean positions.
4) The loosest waters, often parts of mobile dynamic networks still possess a definite directional character with their movements restricted to quite definite channels.
5) The failure of certain well defined water molecules to complete an apparently achievable H bonding pattern has yet to be explained. It suggests that other steric factors such as the position of the hydrogens, need to be accounted for in understanding H bond patterns (H Savage, private communication).
6) The water molecules frequently form ring structures; including the occasional 5 membered ring systems.
7) Finally the electron density of the water molecule structure at the insulin surface can serve as a framework of reference for techniques such as molecular dynamic simulations of solvent molecules. Although these approaches still need perfecting they may help us account for the H bonding behaviour of the better ordered water molecules as well as modelling the structure of the mobile water which trace out their apparently specific networks around the protein and water inferface.

References

Agarwal, R. C., 1978, Acta Cryst., A34:791.
Artymiuk, P. J. and Blake, C. C. F., 1981, J. Mol. Biol., 152:737.
Baker, E. N., 1980, J. Mol. Biol., 141:441.
Baker, E. N. and Hubbard, R. E., 1984, Prog. Biophys. molec. Biol., 44:97.
Blake, C. C. F., Pulford, W. C. A. and Artymiuk, P. J., 1983, J. Mol. Biol. 167:693.
Dodson, E. J., Dodson, G. G., Hodgkin, D. C. and Reynolds, C. D., 1979, Can. J. Biochem., 57:469.
Dodson, G. G., 1981, p95, in "Refinement and Protein Structures", P. Machin, J. W. Campbell, and M. Elder, eds., SERC, Daresbury Laboratory Workshop.
Goodfellow, J. M. Howell, L. P. and Elliot, R. This volume.
Finney, J. L., Goodfellow, J. M. and Poole, P. L., 1982, p387, in "Structural Molecular Biology : Methods and Applications", D. B. Davies, W. Saenger and S. S. Danylak, eds., Plenum Press, New York.
Hendrikson, W. A. and Konnert, J. H., 1980, 13.01 - 13.23, in "Computing in Crystallography", R. Diamond, S. Ramasechan and K. Venketasan, eds., Indian Institute of Science, Bangalore.
James, M. N. G. and Sielecki, A. R., 1983, J. Mol. Biol., 163:299.

Machin, P. A., Campbell, J. W. and Elder, M., 1981, "Refinement of Protein structures", SERC, Daresbury Laboratory.
Teeter, M., 1984. PNAS(USA), 81:6014
Watenpaugh, K. D., Marglus, T. N., Sieker, L. C. and Jensen, L. H., 1978. J. Mol. Biol., 122:175.

THE STRUCTURE OF WATER AROUND A MACROCYCLIC RECEPTOR: A MONTE CARLO STUDY OF THE HYDRATION OF 18-CROWN-6 IN DIFFERENT CONFORMATIONS

G. Ranghino, S. Romano, J.M. Lehn and G. Wipff

Montedison, G. Donegani, S.P.A., 28100 Novara, Italy
The Physics Dept, The University, I-27100 Pavia
Inst. Le Bel, 4, rue B. Pascal, F-67000 Strasbourg

INTRODUCTION

The 18-crown-6 ether, a simple model for macrocyclic ion receptors and carriers is flexible, and displays several conformations in crystals, depending on the nature of its environment, and on the presence and nature of the complexed cation (1,2): e.g., the uncomplexed crown has Ci symmetry, the sodium and potassium complexes are Cl and D3d, respectively. Molecular mechanics calculations on the isolated macrocycle (3) indicate that Ci is the most stable form compared to D3d and Cl (ΔE = 0, 1,9 kcal/mole).

We wanted to study the solvent effect on conformation, and therefore on complexation, with the following questions in mind: does the solvent stabilize the different conformers to a similar extent? If not, can it stabilize structures relevant to ion transport (e.g. Cl and D3d) more than others with no cavity suitable for complexation (e.g. Ci)? Does it stabilize the high energy structures with a non zero dipole moment (e.g. Cl) more than the centrosymmetric structures (e.g. Ci, D3d)?

We used the Monte Carlo approach to study the hydration of 18-crown-6 itself (without complexed cation) in the D3d, Ci and Cl rigid conformations. Three clusters made of 100 water molecules surrounding the 18-crown-6 "solute", have been considered, respectively and the "solute"-water and water-water interaction energies have been calculated for 600.000 moves at 300°K, using a 6-12-1 interaction potential. A more detailed report of this work will appear elsewhere (4).

RESULTS

We find that the solute-water interaction energy Esw is strongly conformation dependent and is significantly weaker for Ci than for Cl or D3d (see Table I).

Table I. Monte Carlo results: average water-water (Eww) and solute-water (Esw) interaction energies (in kcal/mole).

Conformation	Ci	D3d	Cl
Eww	-5.9	-6.0	-5.4
Esw	-29.4	-52.4	-54.3

Among the Ci and D3d conformers, with comparable intrinsic stabilities (3), D3d is much more favorably hydrated and should therefore become sigificantly more stable than Ci in water. The result is consistent with the experimental observation of a D3d conformation in molecular environments with polar C-H, N-H or O-H bonds (1,2). It is also important in view of ion complexation. Indeed, because of its shape and small cavity, Ci cannot bind a cation inside the ring, whereas D3d, which displays the largest cavity with inwards oriented oxygens can complex the alkali metal cations.

A second interesting energy result is that C1 is not found to be more solvated than D3d, despite the difference in dipole moments. Since both conformers have comparable hydration energies, with D3d intrinsically more stable than C1, we predict D3d to be more stable than C1 in an aqueous environment. It is thus likely that the observed dipole moment in solution comes from other conformers than C1 (e.g. C2, C1').

An analysis of the three clusters has been performed on the last 5000 configurations, considering three shells of water molecules : SH1OX is the first hydration shell of the oxygens, SH1S and SH2S are the first and first+second solvation shells of the solute, respectively. The average number N of water molecules in each shell and the corresponding solute-water interaction energies Esw in each shell (Table II) are conformation dependent, with Ci displaying the weakest Esw in each shell.

Table II. Contribution of hydration shells to the total hydration energy.

Conformer	Ci		D3d		C1	
	N	Esw	N	Esw	N	Esw
SH1OX	6.1	-18.5	5.5	-39.1	3.0	-33.5
SH1S	16.6	-23.8	20.0	-45.0	12.8	-40.3
SH2S	42.0	-25.7	46.4	-53.0	28.9	-44.6

We observe no proportionality between N and Esw, which indicates different patterns of hydrogen bonding, as seen graphically. Particularly in SH1OX, the oxygens are less hydrated in Ci than in D3d or C1 ($\Delta E \sim 15$ kcal/mole) despite the larger number of surrounding molecules! In fact, in Ci, each oxygen is H-bonded to about one water moecule in a linear arrangement; in C1, each of the three water molecules bridges two oxygen atoms. The interaction energy per water molecule is thus much weaker in Ci than in C1 (3.1 and 11.2 kcal/mole, respectively). In D3d, the situation is intermediate: among the three oxygens of each face of the ring, two are bridged by one H-O-H; another HOH is bound to the third oxygen, and to the H-O-H bridge. After the calculations were completed, the X-Ray structure of the 18-crown-6/1H$_2$O/paranitrophenol molecular complex has been published (5). It is gratifying to note that in this complex (i) the crown has the D3d structure, (ii) the water molecule has the same bridging positions as that obtained in our simulation, (iii) the phenolic OH is H bonded to HOH, like does the OH of one water molecule of the D3d-water cluster.

DISCUSSION

The cluster approximation with a solute kept rigid cannot give a full account of the hydration process: ideally, a great number of conformers should be sampled. Our calculations however clearly show the environmental effect on relative stabilities of the macrocyclic receptor; it is particularly interesting to find that **the intrinsically preferred Ci form, unefficient for ion binding, becomes less stable than D3d and Cl in water;** thus the macrocyclic receptor takes up in solution, a conformation which contains a preformed cavity suitable for cation binding.

The study makes also clear that the structure of the water in the first solvation shell depends markedly on the conformation of the solute, giving rise to different patterns of hydrogen bonds with the solute, and of water-water interactions. Particularly, in D3d a nice example of **cooperativity** between water molecules H-bonded one to the other has been found, whereas in Ci and Cl the water-water interactions around the solute are at least less favorable.

More generally, these results point out the role of the solvent in the conformational analysis of ionophores or receptors, which is crucial even in the event that they have identical binding sites. Better solvation is expected to occur if the solvation sites (e.g. O....O locations in 18-crown-6) fit into a water network, and allow for multiple bonding without disrupting the water network.

We have analyzed these results in terms of: (i) structure of the "solute"/Water "supermolecule", (ii) the conformation-dependent hydrophilic/total areas accessible to the solvent, (iii) the potential energy surface for one water molecule around the solute and (iv) the total density map for water (4). Studies on the conformation dependent electric field is actually in progress. No single approach accounts for the solvation patterns obtained by MC.

Acknowledgements

The authors thank R. Ripp and B. Chevrier for setting up programs to visualize the results and D. Moras for computer graphics facilities on the PS300 of the Laboratoire de Cristallographie Biologique, I.B.M.C. Strasbourg.

REFERENCES

(1) J. Dale, The Conformation of Free and Complexed Oligoethers, Isr. J. Chem. 20:3 (1980) and references cited therein.
(2) M. Dobler, "Ionophores and their Structures", Wiley-Interscience Pub. 1981.
(3) G. Wipff, P. Weiner, P. Kollman, A Molecular Mechanics Study of 18-Crown-6 and its Alkali Complexes: An Analysis of Structural Flexibility, Ligand Specificity, and the Macrocyclic Effect, J. Am. Chem. Soc. 104: 3249 (1982).
(4) G. Ranghino, S. Romano, J.M. Lehn, G. Wipff, Monte Carlo Study of the Conformation-Dependent Hydration of the 18-crown-6 Macrocycle, J. Am. Chem. Soc. 107:7873 (1985).
(5) M.R. Caira, W.H. Watson, F. Vögtle and W. Müller, A 1,4,7,10, 13, 16-Hexaoxacyclooctadecane (18-Crown-6)-2,4-Dinitrophenol-Water (1:2:2) Complex, Acta Cryst. C40:491 (1984).

SIMULATING PROTEIN DYNAMICS IN SOLUTION:

BOVINE PANCREATIC TRYPSIN INHIBITOR

Michael Levitt and Ruth Sharon

Department of Chemical Physics
Weizmann Institute of Science
Rehovot, Israel

INTRODUCTION

Aqueous solution provides the natural environment for protein and nucleic acid molecules. Water molecules interact with these biological macromolecules by attracting polar groups of atoms through hydrogen bonds and by repelling nonpolar groups through hydrophobic interactions. Both types of interaction have a clear effect on macromolecular stability in that intramolecular hydrogen bonding is weakened by the alternative hydrogen bonds to water and nonpolar group are forced to come together by the hydrophobic interaction.

In spite of its importance, the aqueous environment is often left out of theoretical studies of macromolecules. The reasons for this omission are three-fold. (a) Water is a complicated liquid and even the best pure water simulations do not reproduce all the experimentally measured properties. (b) A great many water molecules must be included to model a macromolecule surrounded by solution. (c) Water molecules interact strongly with other water molecules and the atoms of the macromolecule so that the energy of the system is dominated by the water. Our research has tried to overcome these problems by improving the model for the forces between water molecules, by using highly optimized programs, and by doing test calculations on a small protein which we have studied extensively in calculations without water.

The main results of our 42 picosecond simulation of BPTI protein surrounded by 1874 water molecules are:

(1) The protein behavior is very different when surrounded by water molecules in that the amplitude of atomic vibration is bigger, the stability of hydrogen bonds is generally lower and the rate of internal motions is lower.

(2) The interaction between water molecules and protein atoms depends very sensitively on the relative strength of hydrogen bond formation. There are twice as many hydrogen bonds to >C-O rather than >N-H groups; this has also been seen in high resolution x-ray and neutron diffraction studies of protein crystals (Edsall and McKenzie, 1983).

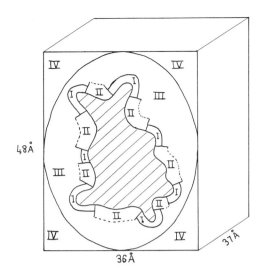

Fig. 1. Schematic drawing of BPTI protein in a box containing 1874 water molecules. The shells of water molecules used to analyse the results are indicated. Shell I, closer than 3.2 Å to polar protein atoms; shell II, closer than 4 Å to non-polar protein atoms; shell III, not in shells I, II or IV; shell IV, the fixed volume between the ellipsoid, which just touches the box, and the walls of the box itself. The periodic system is generated by placing copies of the box around the central box to give a three-dimensional P1 crystal lattice. When a water molecule leaves through one side of the box it automatically enters through the opposite side.

(3) The water behavior is less affected by the presence of the protein. For the 400 or so water molecules closer than 5 Å to the protein there is significantly slowed diffusion and rotational relaxation.

METHODS

The molecule chosen for this study of protein in solution is bovine pancreatic trypsin inhibitor, a small protein with 58 amino acids and 892 atoms (Deisenhofer and Steigemann, 1975; Walter and Huber, 1983).

Solution is reproduced by placing the protein in a box filled with water molecules and then treating the system as a periodic P1 crystal (see Fig. 1). This gives a total of 1874 water molecules; the entire system has 6514 atoms and 19542 degrees of freedom.

The bond, angle, torsional and nonbonded energy parameters are essentially the same as those used in previous molecular dynamics simulations (Levitt, 1983). Electrostatic interactions are included using atom centered point charges with protein parameters from amide crystals (Hagler et al. 1974) and water parameters from pure water simulations (Berendsen et al.1981).

The molecular dynamics trajectory of the entire system is calculated as before using the Beeman (1976) integration algorithm to move all 6514 atoms simultaneously. The time step of 0.002ps is found to work well in spite of the fact that over 60% of the atoms are light, fast moving hydrogen atoms.

RESULTS

Effect of Solvent on Protein

Table 1 compares properties of the protein calculated in vacuo and in solvent. Although both trajectories are the same length and all averages are calculated over the same 20ps period, there are many significant differences in the results. The r.m.s. deviation from the X-ray structure is significantly larger for the simulation in solvent. It is clear that merely including the surrounding solvent does not necessarily reduce the r.m.s. deviation from the X-ray conformation.

The amplitude of α-carbon atomic motion obtained in vacuo (0.13Å) is very much smaller than that obtained in solution (0.72Å). The value in solution is more similar to the values obtained in previous in vacuo simulations (Levitt, 1983), and also to the experimental value (0.68Å). The simulation in solution also shows larger values for the mean temperature factor and the fluctuation of the radius of gyration.

The rate of decay of auto-correlation of ring twisting motions is significantly slower in solution. This damping occurs for both the rings exposed to solution and those buried in the protein interior (e.g. tyr 21) indicating that the viscous damping effect of water is transmitted to the interior of the protein.

The greater amplitude of motion in solution relative to solution is seen in the temperature factors (B-values) for each class of atoms (Table 2). The B-values calculated in solution are in good agreement with those observed in the crystallographic analysis. This agreement is also seen for the individual values of each α-carbon atom (see Fig. 2).

Table 1. Overall Comparison of Properties of Vacuum and Solution Simulations

Property	In vacuo	In soln.	Expt.
All atom RMS deviation (Å)	1.61	2.32	0.9[a]
C^α rms fluctuation (Å)	0.13	0.72	0.68
Mean B for residue (Å)[b]	8.0	29.1	12.2
Radius of gyration (Å)	11.36	11.57	11.53
Rg rms fluctuation (Å)	0.04	0.06	?
Tyr 21 torsion decay rate(ps)[c]	0.1	0.3	?
Number of hydrogen bonds	36	27	25
Mean hydrogen bond free energy[d] (in kcal/mol)	-0.5	-0.1	?

[a] The rms deviation is calculated with respect to the old form x-ray coordinates. The experimental value refers to difference between the old (Deisenhofer and Steigemann, 1975) and new (Walter and Huber, 1983) forms of BPTI.

[b] The radius of gyration Rg is computed as
$Rg = \{1/n \sum |r_i - r|^2\}^{1/2}$
for all atoms except the non-polar hydrogen atoms.

[c] The decay time is calculated as $\int C(t)dt$, where $C(t)$ is the auto-correlation function for the relevant torsion angle.

[d] ΔG is the free energy of hydrogen bond formation defined by $\Delta G = -RT \ln(f/(1-f))$, where f is fraction of the time the hydrogen bond is formed.

Table 2. Comparison of Temperature Factors In Vacuo and in Solution

Atom Type	Number	Mean B-value (Å²)		
		In Vacuo[a]	In Soln.[b]	Experiment[c]
N	56	2.8	10.8	11.5
O	57	4.7	18.9	14.1
C	56	3.0	11.8	12.3
C	56	3.2	11.9	11.7
C	50	4.5	14.5	12.4
C	31	6.2	18.2	18.8
C	15	14.1	24.9	27.4

[a] Averaged over the period 20-40 ps of the trajectory.

[b] Averaged over the period 27.5-40 ps of the trajectory.

[c] Taken from the newly refined "old" form of BPTI (R-factor=17% at 1.2 Å resolution), as kindly provided by Huber and coworkers.

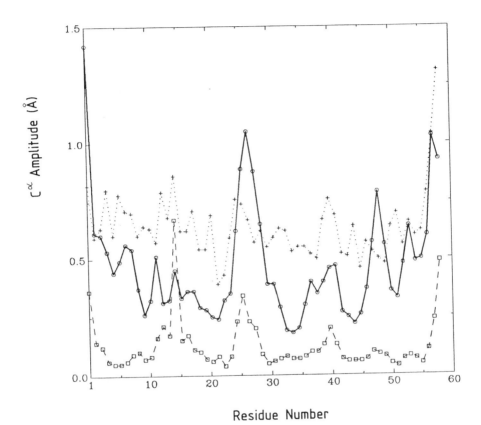

Fig. 2. Showing how the fluctuation of α-carbon atoms varies with residue number for BPTI in solution (solid line) and in vacuum (dashed line). The experimental fluctuation obtained from the crystallographic temperature factors is also shown (dotted line). Although the calculated fluctuations are much larger in solution than in vacuum, the peaks values seen in vacuum at residues 1, 14, 25, 40 and 58 are also peak values in solution. The solution values are in closer agreement with the experimental values (correlation coefficient of 0.65).

Table 3. Summary of Hydrogen Bonds Formed to Protein Groups

Group	In	Number Hydrogen Bonds Max.[a]	Number Hydrogen Bonds Actual	Average % Time Bonded	Water Interaction Energy[b] (kcal/mole)
>CO	peptide	58	44	68	-7.6
>NH	peptide	54	24	41	-5.4
-COO	asp,glu	6	0	-	-5.1
>CO	asn,gln	4	3	51	-7.6
-NH2	asn,gln	8	5	21	-6.2
>NH,-NH2	arg	30	17	18	-6.2
-NH3	lys	12	1	13	-2.1
-OH	ser,thr,tyr	8	8	100	-12.6

[a] Assuming one hydrogen bond for each O or H.

[b] Calculated as the minimum value of the interaction energy of the side chain group and a single water molecule. For two water molecules, the interaction energy is -7.1 kcal/mole.

The correlation coefficients between the X-ray B-values and the calculation are 0.65 in solution and 0.43 in vacuo. The in vacuo simulation shows very little motion for the main chain atoms.

Protein/Water Interaction

The strongest hydrogen bonds involve the hydroxyl (-OH) group, which has a water interaction energy of -12.6 kcal/mole (see Table 3). Next strongest are the >CO groups of the peptide group and the amide side chains of asn and gln. The hydrogen bonds to >NH and $-NH_2$ groups are less strong: they are formed for a short time and many of these groups do not form any stable hydrogen bond. There is a clear correlation between the observed tendency of the group to form a stable hydrogen bond and the strength of the group/water interaction energy. It is of interest that the uncharged -COO and $-NH_3$ groups make very poor solvent interactions. In reality, these groups would be ionized and make very strong hydrogen bonds with water molecules.

Effect of Protein on Solvent

The 1874 water molecules included in the simulation are divided into four classes as follows:

(I)   Closer than 3.2 Å to a polar atom of the protein (O or N).
(II) Closer than 4 Å to a non-polar atom of the protein (C or S).
(III) Not in classes (I), (II) or (IV).
(IV) Inside the volume defined by the box sides and an ellipsoid that just fits into the box.

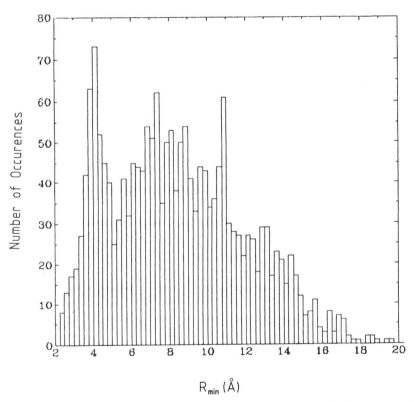

Fig. 3. Showing the distribution of r_{min} for all 1874 water molecules. The quantity r_{min} is the distance between a particular water molecule and the closest protein atom; it is averaged over the part of the trajectory analysed. The peak at 4 Å is unexpected and shows that the protein affects the structure of about 400 water molecules.

Table 4. Properties of Water molecules in Different Shells

Property	\multicolumn{5}{c}{Shell}				
	I	II	III	IV	All
Number	81	79	671	1041	1874
No H.bonds	54	0	10	1	66
Diffusion($Å^2$/ps)	0.17	0.25	0.29	0.41	0.35
Order param.	0.29	0.21	0.15	0.11	0.14
Rotation rate(ps)	19.6	16.0	13.6	11.3	12.7
Energies (Kcal/mol):					
Bond length	1.68	1.77	1.80	1.79	1.78
Bond angle	0.56	0.60	0.61	0.63	0.62
Nonbonded:					
Water/water	-19.39	-23.90	-24.69	-24.69	-24.43
Water/protein	-5.85	-1.14	-0.45	-0.04	-0.48
Total	-10.38	-10.15	-10.16	-9.95	-10.04
Relative to Shell IV	-0.38	-0.20	-0.21	0.00	-0.06

Diffusion constant calculated as $\langle dx^2 \rangle/6t$.

This classification (illustrated in Fig. 1) is used to calculate the average properties of water molecules in each shell (see Table 4). Shell I water molecules are least mobile as evidenced by a small diffusion constant and slow rotational relaxation time. These water molecules have lower bond length and bond angle strain energies than water molecules in bulk or in other shells. Although the water/water interaction energy in shell I is less favorable, the protein/water interaction more than compensates.

Shell II water molecules are also in contact with protein atoms but have very different properties from those in shell I. In many respects, the water molecules in shell II behave like those in shell III. Thus being in contact with non-polar atoms of the protein (shell II) is almost indistinguishable from being in the second hydration layer at a distance of more than 3.5 Å from the nearest protein atom. Water molecules in shell IV behave like bulk water, they diffuse and rotate most rapidly and have no interaction with the protein.

In classifying the water molecules, we have used r_{min} the distance from the particular water molecule to the nearest protein atom. The distribution of r_{min} for all 1874 water molecules (see Fig. 3) shows unexpected structure with peaks at 4.12 Å and 7.5 Å. There are 199 water molecules with r_{min} <4 Å and 399 water molecules in the first peak (r_{min} <5 Å). Water molecules in shells I and II are a subset of the peak 1 water molecule and number 160 in all. Thus, the peak does not arise from any special interaction between the protein and water molecules. Instead it seems to reflect the structure of water near any surface. It is not clear why the distribution should drop at r_{min}=5 Å after the first hydration shell has been filled.

DISCUSSION

This first simulation of the molecular dynamics of a protein in solution has revealed a number of unexpected features of this complicated system. While these findings are useful they are still preliminary. This work has developed computational machinery that can be used to simulate and analyse the motion of biological molecules in water. The analysis has been the most difficult aspect of the work. Without in-depth processing of the trajectory it would have been impossible to ensure that the work is free of errors, to understand the relationship between simulation and experiment and to discover new phenomena.

This research was supported by a grant from the United States-Israel Binational Science Foundation (BSF), Jerusalem, Israel.

REFERENCES

Beeman, D. 1976, J. Comput. Phys. 20 : 130-139.
Berendsen, H.J.C., Postma, J.P.M, van Gunsteren, W.F. & Hermans, J. 1981, in Intermolecular Forces ,Pullman, B., Ed., Reidel, Dordrecht, Holland 331-342
Deisenhofer, J. & Steigemann, W. 1975 , Acta Crystallogr. B31 : 238-250.
Edsall, J.T. & McKenzie, H.A. 1983, Adv. Biophys. 16 : 53-183.
Hagler, A.T., Huler, E. & Lifson, S. 1974, J.Amer.Chem.Soc. 96: 5319-5327.
Levitt, M. 1983, J.Mol.Biol. 168 : 595-620.
Walter, J. & Huber, R. 1983, J.Mol.Biol., 167 : 911-917.

5 DRUG DESIGN - SITE DIRECTED MUTAGENESIS

RATIONAL DESIGN OF DNA MINOR GROOVE-BINDING ANTI-TUMOR DRUGS

Richard E. Dickerson, Philip Pjura, Mary L. Kopka,
David Goodsell and Chun Yoon

Molecular Biology Institute
University of California at Los Angeles
Los Angeles, California 90024

Many of the most useful antitumor drugs act by binding directly to double-helical DNA, interfering with both replication and transcription. Some of these drugs intercalate between adjacent base pairs. A second important class consists of those that bind within the minor groove of B-DNA. These latter tend to show a sequence specificity, binding best to regions of several successive A.T base pairs. We have embarked on a planned course of study of the molecular structures of complexes of such groove-binding drugs with synthetic DNA oligomers, with two goals: to understand the basis for sequence specificity, and to design new drug analogues that are capable of binding specifically to any desired base sequence. Such sequence-reading molecules should be capable of being directed against key sequences typical of neoplastic rather than normal cells, or invader rather than host cells. If such molecules become a reality, they should have considerable importance in chemotherapy, directing their action against targeted cells and avoiding some of the more unpleasant side effects associated with chemotherapy.

NETROPSIN AND DISTAMYCIN

Netropsin and distamycin (Fig. 1) are among the most extensively studied antiviral, antitumor antibiotics. Their extensive literature has been analyzed elsewhere[1-3] and will not be repeated here. They bind only to double-helical B-DNA, and require a binding site consisting of A.T base pairs. Their base specificity is greater than for most DNA-binding drugs, making them intrinsically more interesting for study.

The repeating unit of both distamycin and netropsin is an amide plus a methylpyrrole. Naturally occurring forms as isolated from $\underline{Streptomyces}$ $\underline{distallicus}$ or $\underline{netropsis}$ appear in Fig. 1, but longer or shorter analogues can be prepared synthetically. A netropsin with $\underline{x}$ pyrrole rings, termed netropsin-$\underline{x}$ or Nt $\underline{x}$, will have $\underline{x}$ + 1 amides, and will require a binding site of at least $\underline{x}$ + 2 consecutive A.T base pairs, for reasons that become obvious from the x-ray crystal structure analysis to be described.

We have solved the crystal structure of a complex of netropsin-2 with a double-helical B-DNA dodecamer of sequence: C-G-C-G-A-A-T-T-^{Br}C-G-C-G, as reported in references 1-3. Figure 2 shows how the drug molecule binds within the minor groove of the double helix, and the small adjustments made to the DNA helix in the process of binding.

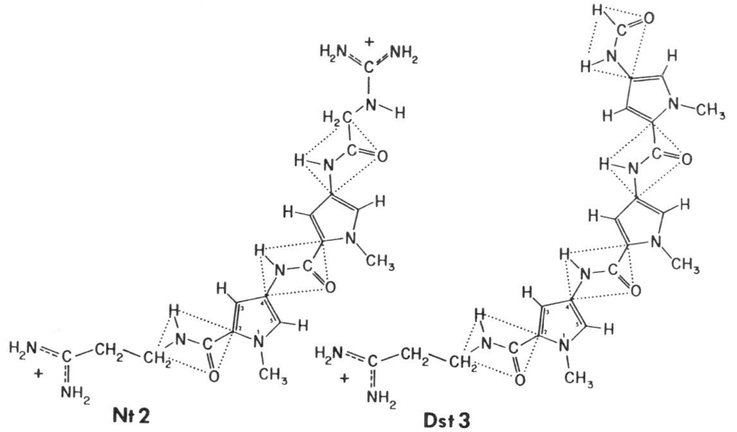

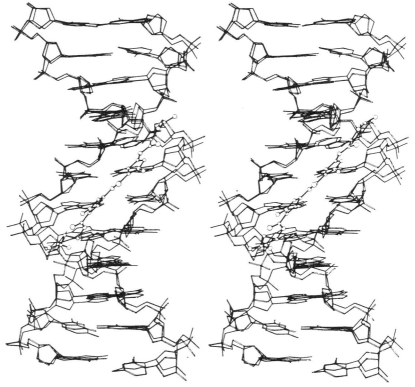

Fig. 1. Structures of netropsin-2 (left) and distamycin-3 (right), the naturally-occurring forms. Netropsin has two cationic ends; in distamycin one of these is replaced by formamide.

Fig. 2. Stereo pair drawing of the binding of netropsin (ball-and-stick) within the minor groove of the B-DNA oligomer (black stick bonds). The structure of the DNA prior to netropsin bonding is indicated by open stick bonds. Netropsin fits snugly in the center of the minor groove, even opening it slightly to make room. It also bends the helix axis back by 8° across the binding site.

The way in which netropsin binds to DNA is shown in Figure 3. Each of the three amide NH makes a bifurcated hydrogen bond to adenine N3 or thymine O2 atoms on opposite strands of the helix, on adjacent base pairs. Fig. 3 makes it immediately obvious why the number of A.T base pairs should be one greater than the number of amide groups (or two greater than the number of pyrroles). Stabilization of the DNA/drug complex has three components: the hydrogen bonds just mentioned, hydrophobic interactions between organic rings on the drug and the walls of the DNA minor groove, and electrostatic attraction between the two cationic ends of netropsin and the negative potential generated by the DNA phosphates.

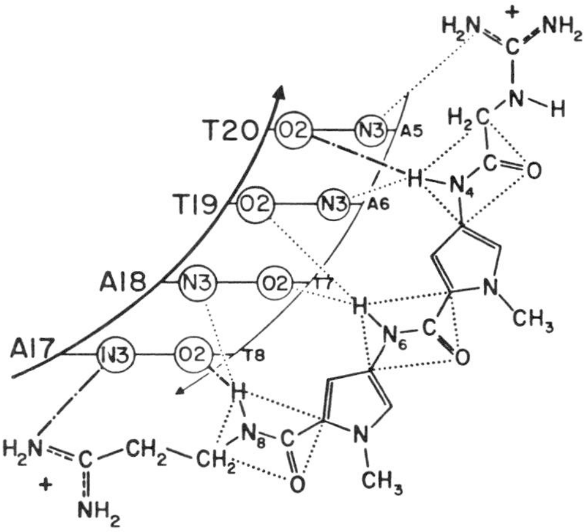

Fig. 3. Schematization of hydrogen bonds between netropsin and adenine N3 or thymine O2 atoms on the floor of the minor groove. N-N or N-O separations of 3.2 Å or less are -----; longer bonds are dotted. Bases are numbered 1 - 12 in one strand, and 13 - 24 in the other.

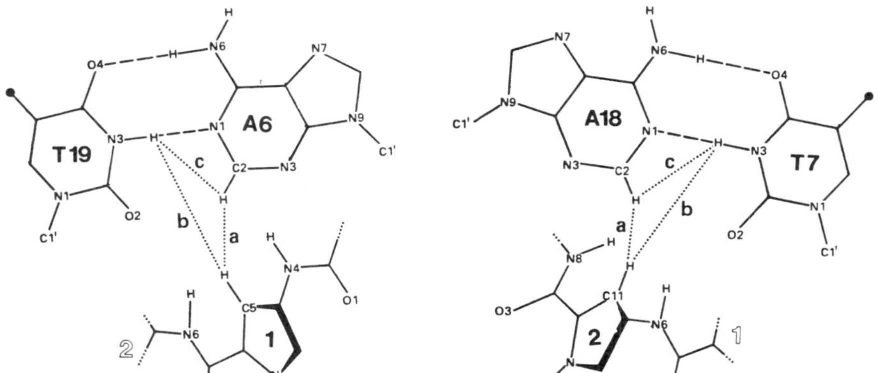

Fig. 4. The central two base pairs in the dodecamer, showing their relation to pyrrole rings 1 and 2 of netropsin. Proton-proton distances a, b and c, and distances between C and N atoms to which they are attached, are given in Table I.

211

Neither the hydrogen bonds shown in Fig. 3 nor electrostatic attractions are responsible for the observed A.T base preference of netropsin. The bridging of adenine N3 and thymine O2 atoms is identical to that in the spine of hydration that runs down the minor groove of drug-free DNA and helps stabilize the B form of the double helix.[4,5] But the bridging amide NH are not close to the C2 protons of the adenines, which provide the only real differentiation between an A.T and a G.C base pair in the minor groove. The "reading" of base sequence is accomplished by close van der Waals packing contacts between the two pyrrole rings and the edges of the central adenines. There is no room for a -NH_2 group on the C2 atom, as would be present with guanine. As Table I shows, the C2-H of adenine and C5-H or C11-H of netropsin are around 2 Å apart, and the carbon atoms to which they are attached are 4 Å apart. If a -NH_2 group were to be attached to the adenine C2, then its N would be only 3 Å away from the netropsin ring carbon, unacceptably close. Farther out from the central two A.T base pairs, the methylene -CH_2- "knees" of the netropsin molecule exert the same kind of steric selectivity as with the central pyrroles. In essence, the drug molecule uses braille to read the DNA sequence.

Table I. Critical Distances in DNA/Netropsin Contacts

Type of Separation	Length	Interactions: Nt1 with A6.T19	Nt2 with T7.A18
H-H Distances	a	2.35 Å	2.00 Å
	b	4.57	4.31
	c	2.64	2.73
N-C or C-C Distances	a'	4.06	3.87
	b'	5.97	5.71
	c'	3.56	3.68

The geometry shown in Fig. 4 has been confirmed by nuclear magnetic resonance measurements[6,7] that observe a NOE from one proton to another across distances marked a and c in Fig. 4, but not across distance b. The latter, indeed, is longer than the 4 Å limit conventionally associated with observation of a NOE effect. All of the nmr data are fully in agreement with our x-ray structure, and the conclusion in reference 7 that the netropsin molecule lies asymmetrically along one side of the minor groove is erroneous[8].

The binding of netropsin to DNA is summarized in Figure 5, with heavy arrows indicating bifurcated hydrogen bonds to the floor of the minor groove, and the four hatched interactions being tight van der Waals contacts that read A.T base pairs and reject G.C. Distamycin has been found to be slightly more tolerant of G.C "mistakes" at one end of its binding site[9]. It may be that the uncharged end can be displaced from the floor of the groove to make room for an intrusive guanine -NH_2 as schematized in Fig. 6.

Now that we know why netropsin and distamycin require a run of A.T base pairs as their binding site, can one conceive of synthetic netropsin analogues that could be tailored to read any desired sequence of A.T and G.C base pairs? How can we train a netropsin molecule to read G.C?

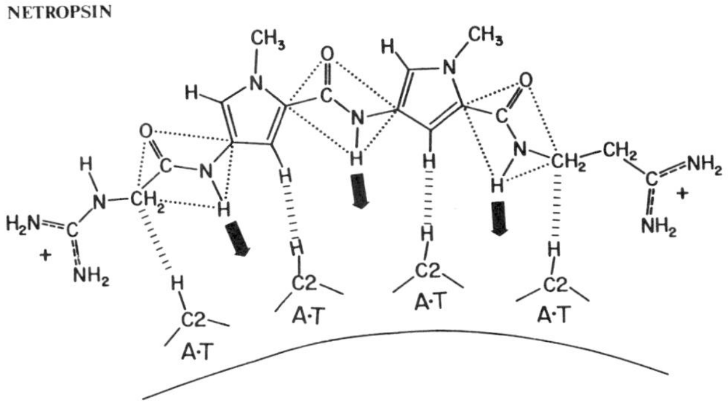

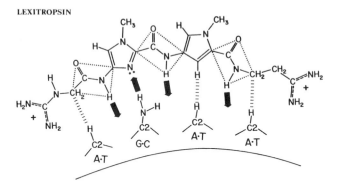

Fig. 5. Schematic of binding of netropsin to DNA

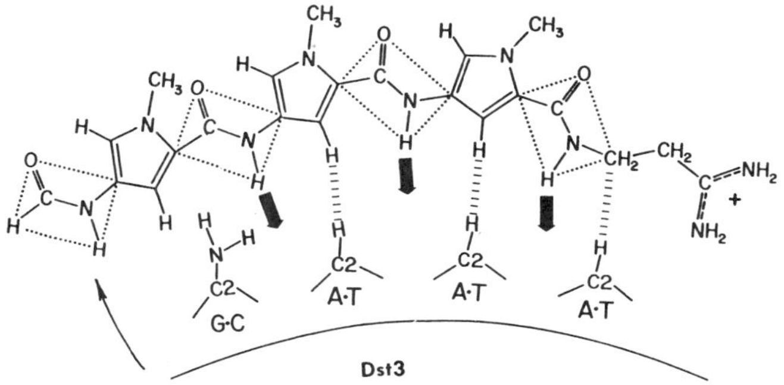

Fig. 6. Possible explanation of G·C tolerance of distamycin

Fig. 7. A lexitropsin, or a base sequence-reading analogue of netropsin. Substitution of imidazole for pyrrole permits formation of a new hydrogen bond to guanine.

LEXITROPSINS

One way not only to allow but to require the presence of a G·C base pair in the central binding region would be to replace the pyrrole by an imidazole or furan, as shown in Figure 7. Not only would space then be created for the guanine -NH$_2$ group; a new hydrogen bond would be formed to help stabilize the complex. Since one can synthesize longer analogues of distamycin or netropsin, it should be possible in principle to design a netropsin analogue that was capable of recognizing, and binding selectively to, any desired sequence of ten or twelve A.T or G.C base pairs. We have embarked on a collaborative program with Prof. J. William Lown of Alberta, in which he will synthesize a series of lexitropsins, we will prepare DNA oligomers with complementary sequences, and he will study the complexes in solution with nmr while we study them in the crystal by x-ray diffraction. Eighteen different lexitropsins have been synthesized, including that shown in Figure 7, and a DNA oligomer of sequence C-G-T-G-T-T-A-A-C-A-C-G has been prepared to go with this particular lexitropsin. If the molecule shown in Figure 7 only tolerates G.C base pairs, then it probably will bind once more to the T-T-A-A center of the oligomer. But if it actively prefers G.C because of the new hydrogen bond, then it probably will bind to T-G-T-T or the symmetry-related A-A-C-A.

One possible difficulty may arise with this approach, involving a fundamental structural mismatch between drug and DNA. This problem will be considered again after discussion of a second and quite new structure analysis of a groove-binding DNA-drug complex.

HOECHST 33258

Hoechst 33258, shown in Figure 8, is a potent carcinogen and a DNA stain for microscopy. Both aspects arise because it binds tightly to double-stranded DNA. Footprinting studies[10] indicate that, although it also prefers A.T base pairs in its binding site, it occasionally extends across into the adjacent G.C base pair region. We have recently solved the structure of the complex of Hoechst 33258 with C-G-C-G-A-A-T-T-C-G-C-G, and find that it, like netropsin, binds within the minor groove.

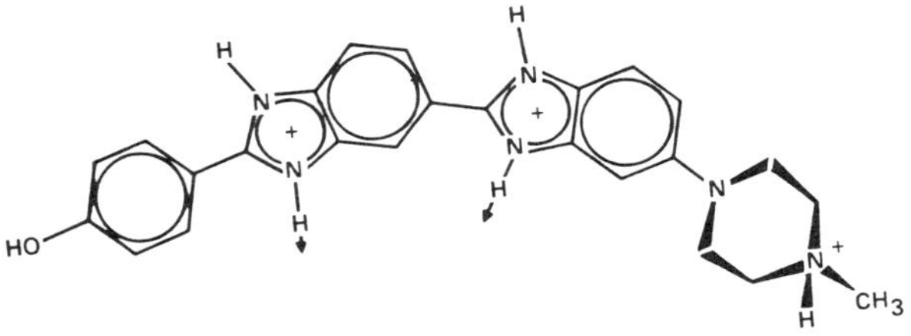

Fig. 8. Molecular structure of Hoechst 33258. Its repeating unit is a benzimidazole, and the ends are capped by a phenol and a piperazine ring. The compound is normally encountered as the tri-HCl salt.

Structure Analysis of the Hoechst/DNA Complex

The structure analysis will be outlined briefly since it has not appeared elsewhere, although a full report will be published later. Cocrystals of the DNA/drug complex were grown at 5°C under the same conditions as for the native dodecamer[11]. Crystals were grown by vapor diffusion against a reservoir of higher MPD (2-methyl-2,4-pentanediol) concentration. They appeared after 1-2 week equilibration against 30% MPD, and the final MPD concentration was raised to 35% to prevent re-dissolving. Diffractometer data were collected to 1.9 Å resolution, but the limit of reliable data was judged to be 2.2 Å. Unit cell and space group suggested that the crystals of the DNA/drug complex probably were isomorphous with those of the native C-G-C-G-A-A-T-T-C-G-C-G dodecamer. Use of these atomic coordinates as a trial structure gave a residual error or R factor of 46.4%. A test with the unbent MPD7 helix structure gave poorer agreement (53.0%), as did the

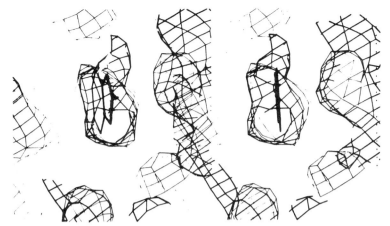

Fig. 9. Stereo sections through the $(2F_o - F_c)$ electron density map, showing an edge view of the first benzimidazole ring, prior to introduction of information about it into the structure analysis. Note walls of minor groove to left and right, and below.

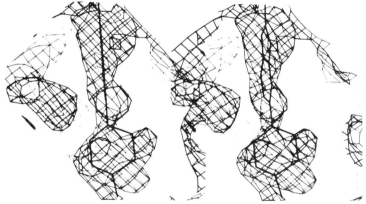

Fig. 10. Top view of the second benzimidazole and the piperazine ring in the "down" position. The opening in the cage density at center leads up to the "up" piperazine image, clipped from this stereo view.

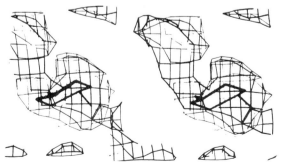

Fig. 11. View from the second benzimidazole (not shown) toward the piperazine ring in its "down" position. Density leading to upper left is the "up" piperazine image.

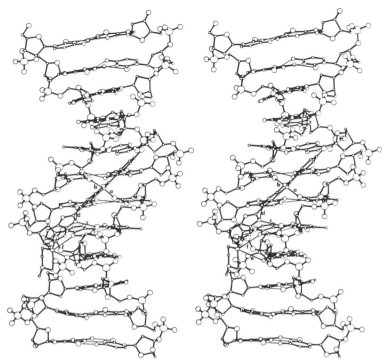

Fig. 12. Front view of the Hoechst/DNA complex with the piperazine ring in the "down" position. Distances a-f are probable hydrogen bonds.

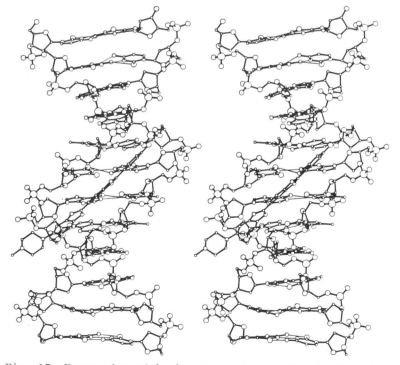

Fig. 13. Front view with the piperazine in the "up" position.

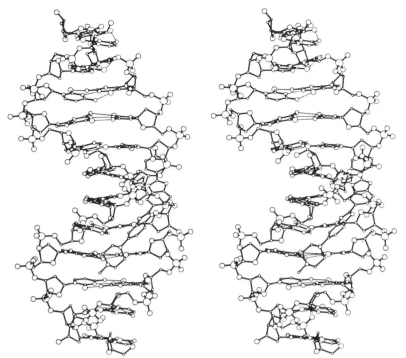

Fig. 14. Side view with piperazine ring down.

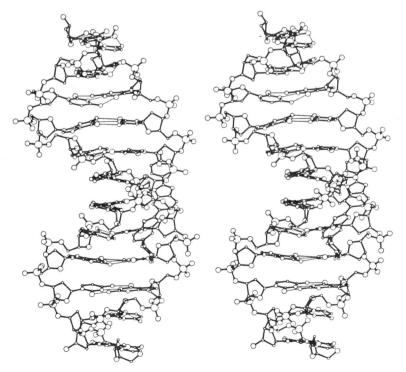

Fig. 15. Side view with piperazine ring up.

refined netropsin/DNA coordinates (51.7%). Using the native DNA coordinates as a starting point, 28 cycles of Jack-Levitt restrained energy refinement dropped the residual error to R = 31.7%. Gradual addition of 40 water molecules where indicated by ($F_O - F_C$) difference maps led to R = 28.0%. At this point, before any information about Hoechst was incorporated into the analysis a ($2F_O - F_C$) map was displayed on an Evans and Sutherland Picture System II, using FRODO. Three views of the Hoechst image and the neighboring DNA density appear in Figs. 9 - 11. With only 2.2 Å resolution, and with no contribution to phasing from the drug molecule, the Hoechst image of necessity is rough. But it could be fitted unambiguously with the ideal Hoechst framework with one exception: the puckered piperazine ring clearly was statistically distributed between two positions. In the "down" position it lay on the floor of the minor groove, but in the "up" position it extended out of the groove. Four more cycles of refinement with both choices have led to tentative R values of 28.1% for the up position and 27.9% for the down. Refinement is not complete as of the moment of writing, and will continue. The piperazine ring probably will ultimately have to be refined with partial occupancy in each of its two loci.

Results: The Hoechst/DNA Complex

Stereo pair drawings of the complex of Hoechst 33258 with the dodecamer C-G-C-G-A-A-T-T-C-G-C-G are shown in Figs. 12 - 15. In each case, bases in the first strand are numbered C1 to G12, and those in the second strand are C13 to G24. Base pair C1·G24 is at the top. The drug molecule associates with DNA in much the same way as netropsin, but with interesting differences. The hydrogen bonding is depicted in Fig. 16, which should be compared with Fig. 3. The NH on the first benzimidazole makes a bifurcated hydrogen bond with the O2 of thymines T7 and T19 across the center of the AATT region, and the NH on the second benzimidazole bridges A18 and T8. Hence Hoechst 33258 has "slipped" down by one position to occupy the lower two of the three bridge sites filled by netropsin and shown in Fig. 2. The piperazine ring laps over into the G.C region of the minor groove, which is intrinsically wider than the A.T zone and hence better able to accommodate it. Indeed, the inner nitrogen atom of the piperazine may form a weak hydrogen bond with the NH_2^- group on guanine G16, as sketched in Fig. 16. Distances between C and/or N atoms that are potentially hydrogen-bonded, shown as a - f in Fig. 12 are: a = 3.76 Å, b = 3.34 Å, c = 2.76 Å, d = 3.34 Å, e = 3.89 Å, and f = 3.76 Å. (These distances must be regarded as only provisional, until refinement is completed.) Some separations are long, for conventional single hydrogen bonds, but are not out of line with what is encountered in cases of bifurcated hydrogen bonds under constrained circumstances[13].

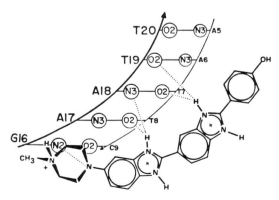

Fig. 16. Binding of Hoechst 33258 to C-G-C-G-A-A-T-T-C-G-C-G.

The dual positions that we observed for the piperazine ring indicate how weakly it is held within the minor groove. Electrostatic interactions between the piperazine cation and the negative potential generated by the phosphate backbones may be equally strong in either position of the ring. Note from Figs. 12 and 13, however, that movement from the down to the up position is not simply a matter of motion of the piperazine ring alone. Benzimidazole 2, to which it is attached, must be rotated by 180° about its bond to the first benzimidazole. Hence the distribution between the two positions of piperazine within the minor groove of DNA must represent the population of two conformations about the bond between benzimidazoles in the drug molecule in solution, prior to complex formation with DNA. There can be no equilibration between Fig. 12 and Fig. 13 once the drug is bound to the DNA. But the lack of a close interaction between piperazine and the floor of the minor groove is clearly not sufficiently serious to prevent drug binding when the bond joining benzimidazoles is turned as in Fig. 13.

As with netropsin, the "reading" of A.T base pairs by Hoechst 33258 appears to involve packing of aromatic ring edges--in this case the phenol and the six-membered rings of benzimidazoles--so tightly against the C2-H of adenines that there would be no room to permit the $-NH_2$ of guanine. The binding site in this complex involves four base pairs: A6·T19, T7·A18, T8·A17, and C9·G16. But we can see no reason, from this structure, why the uppermost AT base pair A5·T20 should be protected by binding of Hoechst. This seems to present a quandary when compared with Dervan's finding from footprinting studies[10] that Hoechst protects a five-base-pair A.T site, with occasional protection of an adjacent G.C base pair. But then these same experiments with netropsin indicate a five A.T base pair protection site, whereas the x-ray analysis can only account for four pairs. This may only represent a difference in the kind of information about binding sites returned by x-ray crystallography and solution footprinting protection studies.

The dual occupancy of the piperazine ring at a G.C base pair is suggestive of the tolerance of distamycin for a G.C base pair at the boundary of its binding site[10]. It supports the hypothesis, presented in Fig. 6, that distamycin accommodates G.C by lifting its uncharged formamide end up from the floor of the minor groove, whereas electrostatic interactions prevent this from occurring as readily with netropsin.

In summary, this not yet completely refined structure analysis of the Hoechst 33258/DNA complex indicates that Hoechst joins netropsin and distamycin as base sequence-specific minor groove-binding drugs, with Hoechst being somewhat less specific because it is shorter. One might imagine synthesizing longer Hoechst analogues with more benzimidazole units, capable of recognizing longer A.T regions, except for one serious potential difficulty that must now be addressed.

ISOLEXINS

Both the amide-pyrrole unit of netropsin and the benzimidazole unit of Hoechst 33258 share a common failing: they are 25% too long for the natural DNA repeat[14]. When an extended alternating amide-pyrrole chain is hydrogen-bonded to the floor of an infinite B helix as observed in netropsin, its helical twist angle is 46°, whereas that for the DNA is only 36°. The comparable twist angle for Hoechst is 43°. Hence one would anticipate that longer polymers using these subunits would gradually get out of phase with the DNA helix, and would bind poorly if at all. Yet the development of the lexitropsin concept described earlier depends on a continued proper registration between the sequence-reading drug polymer and the DNA that it is reading.

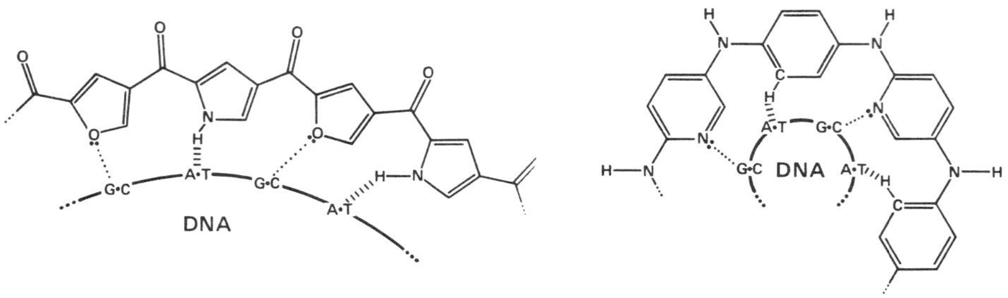

Fig. 17. Two potential isolexins, or isohelical sequence-reading drug polymers. Left: alternation of carbonyl with furan (for G.C) or pyrrole (for A.T). Right: alternation of amine with pyridyl (for G.C) or benzyl (for A.T).

We recently have carried out a systematic investigation of many potential repeating polymers capable of binding within the minor groove of B-DNA. A polymer whose main chain torsion angles have been twisted so that its chain follows along the path of the minor groove, at the proper distance from the floor of the groove for hydrogen bonding, is termed "isohelical" with the DNA helix. Polymers that could be made isohelical then were examined for two criteria: whether their intrinsic helical repeat matched that of the DNA, and whether variants of the polymer subunits offered the possibility of discriminating between G.C and A.T base pairs[14]. Two very promising candidates turned up, as shown in Fig. 17. That with five-membered rings can be regarded as a netropsin-like polymer, systematically shortened by deletion of all main chain amine groups, -NH-. (Deletion of the carbonyls instead leads to a very similar polymer.) A new basis must then be found for binding of the polymer to the DNA and for recognition of sequence. But as can be seen at left in Fig. 17, a pyrrole ring can recognize A.T via close van der Waals contacts between NH and adenine C2-H, whereas a furan can both recognize G.C and help hold the polymer in place by virtue of a hydrogen bond from the guanine $-NH_2$ to the furan O. Such extended polymers no longer are lexitropsins properly speaking, since they are not analogues of netropsin. But as sequence-reading polymers, chosen to be isohelical with the DNA in a way that both netropsin and Hoechst 33258 are not, they may be termed "isolexins". The synthesis of the isolexin shown at left in Fig. 17 is now under way by Tammy Smith in this laboratory. When synthesized, it will be tested for tightness of binding to a properly chosen DNA sequence, and for its specificity for that sequence over any other.

CHEMOTHERAPUTIC IMPLICATIONS

If the approaches discussed above are successful, we should be able to design and synthesize an isolexin chain that is capable of recognizing and binding preferentially to any desired succession of ten or so G.C vs. A.T base sequences. It may be more difficult to find ways of discriminating between A.T and T.A base pairs, but this may not be essential. Even with discrimination only between types of base pairs, without regard to orientation, a ten base pair reading isolexin would read only one random sequence out of $2^{10} = 1024$. An isolexin designed to target a particular DNA sequence in the control region of an oncogene, or in a genome of a parasitic cell, would still bind to normal or host DNA 0.1% of the time. But this side binding would be fully acceptable since most antineoplastic drugs in use today are totally non-discriminating, binding to all DNA. Reduction of the unpleasant side effects of chemotherapy to one-tenth of one percent of their present levels would be regarded as a major success, and would fully repay the efforts that have gone into this research.

ACKNOWLEDGEMENTS

We would like to acknowledge with thanks the help and active collaboration provided by Prof. J. William Lown of the University of Alberta in Edmonton, Alberta. This work was supported by National Institutes of Health grant GM-31299 and American Cancer Society grant #NP-504.

REFERENCES

1. M. L. Kopka, P. Pjura, C. Yoon, D. Goodsell and R. E. Dickerson, The binding of netropsin to double-helical B-DNA, in "Structure and Motion: Membranes, Nucleic Acids and Proteins" E. Clementi, G. Corongiu, M. H. Sarma and R. H. Sarma, Eds., Adenine Press, New York (1985), p. 461.
2. M. L. Kopka, C. Yoon, D. Goodsell, P. Pjura and R. E. Dickerson, The molecular origin of DNA-drug specificity in netropsin and distamycin, Proc. Natl. Acad. Sci. USA 82:1376 (1985).
3. M. L. Kopka, C. Yoon, D. Goodsell, P. Pjura and R. E. Dickerson, Binding of an antitumor drug to DNA: Netropsin and C-G-C-G-A-A-T-T-^{Br}C-G-C-G. J. Mol. Biol. 183:553 (1985).
4. H. R. Drew and R. E. Dickerson, Structure of a B-DNA dodecamer. III. Geometry of hydration. J. Mol. Biol. 151:535 (1981).
5. M. L. Kopka, A. V. Fratini, H. R. Drew and R. E. Dickerson. Ordered water structure around a B-DNA dodecamer. J. Mol. Biol. 163:129 (1983).
6. D. J. Patel, Antibiotic-DNA interactions: Intermolecular nuclear Overhauser effects in the netropsin-d(C-G-C-G-A-A-T-T-C-G-C-G) complex in solution. Proc. Natl. Acad. Sci. USA 17:6424 (1982).
7. M. H. Sarma, G. Gupta and R. H. Sarma, Netropsin specifically recognizes one of the two conformationally equivalent strands of poly(dA)·poly(dT). One dimensional nmr study at 500 MHz involving NOE transfer between netropsin and DNA protons. J. Biomol. Str. and Dyn. 2:1085 (1985).
8. R. E. Dickerson and M. L. Kopka, Nuclear Overhauser data and stereochemical considerations suggest that netropsin binds symmetrically within the minor groove of poly(dA)·poly(dT), forming hydrogen bonds with both strands of the double helix. J. Biomol. Str. and Dyn. 2:submitted (1985).
9. R. S. Youngquist and P. B. Dervan, Sequence-specific recognition of B-DNA by oligo(N-methylpyrrolecarboxamide)s. Proc. Natl. Acad. Sci. USA 82:2565 (1985).
10. K. D. Harshman and P. B. Dervan, Molecular recognition of B-DNA by Hoechst 33258. Nucl. Acids Res. 13:4825 (1985).
11. R. Wing, H. Drew, T. Takano, C. Broka, S. Tanaka, K. Itakura and R. E. Dickerson, Crystal structure analysis of a complete turn of B-DNA. Nature 287:755 (1980).
12. A. V. Fratini, M. L. Kopka, H. R. Drew and R. E. Dickerson. Reversible bending and helix geometry in a B-DNA dodecamer: CGCGAATT^{Br}CGCG. J. Biol. Chem. 257:14686.
13. G. A. Jeffrey and J. Mitra, Three-center (bifurcated) hydrogen bonding in the crystal structures of amino acids. J. Am. Chem. Soc. 106:5546 (1984).
14. D. Goodsell and R. E. Dickerson, Isohelical analysis of DNA groove-binding drugs. J. Med. Chem., submitted (1985).

PROTEIN CRYSTALLOGRAPHY AND DRUG DESIGN

Wim G.J. Hol

Laboratory of Chemical Physics
University of Groningen
Nijenborgh 16, 9747 AG Groningen, The Netherlands

INTRODUCTION

Safe and reliable drugs belong to the most impressive scientific achievements of mankind. One would only wish that there were more drugs such as penicillin which allows recovery from serious and painful bacterial infections in a matter of days, if not hours. Unfortunately, the discovery of such compounds with almost miraculous curing properties is a very rare event indeed.

The design of new drugs is, however, undergoing a revolutionary change at present. The reason for this is the tremendous progress being made in a number of disciplines: protein purification methods, gene sequencing procedures, recombinant DNA techniques, protein crystallography, computer graphics, and theoretical biochemistry. This allows in an increasing number of cases the elucidation of the three-dimensional structure of potential target proteins at the atomic level. Inspection of such protein structures provides medicinal chemists with detailed suggestions as to which kind of molecules may be selective and efficient inhibitors. Perhaps just as important, three-dimensional structures of target proteins also show which substituents do *not* have to be tried out experimentally as it is obvious that the resulting derivatives of the initial compound cannot possibly fit into the active site. It may therefore be concluded that laboratories which are able to bring about a close and mutually stimulating collaboration between the most modern biochemical techniques and medicinal chemistry have fascinating possibilities for developing new drugs.

GENERALLY USEFUL PROTEIN STRUCTURES

Before going to consider the "proper" potential target proteins with known three-dimensional structure, it may be worthwhile to point out that there are a number of protein structures known which have great potential medical relevance (Table 1). The hormone insulin was one of the first proteins with known three-dimensional structure, and is used presently in making various insulin analogues with hopefully improved properties. The immunoglobulins have received considerable attention from protein crystallographers, which yielded numerous structures and culminated in the recent determination of the first structure of an antibody-protein complex (1). Neocarzinostatin and actinoxanthin may become important in the development of new anti-cancer drugs, but like all other proteins in Table 1 they are not potential target enzymes to be blocked by pharmaceutical compounds and solely for that reason they will receive no further attention here.

Table 1. Protein structures with potential medical applications

Immunoglobulins
Insulin
Glucagon
Prealbumin
Uteroglobin
Actinoxanthin
Neocarzinostatin
Retinol binding protein

THREE CASES

One can consider three different situations for the development of new drugs on the basis of target protein structures:
(i) a human protein is not functioning properly and, hence, only that protein has to be considered; no proteins from infective agents need to be studied;
(ii) the potential target is a unique protein from an infectious organism and, consequently, no analogous human protein needs to be taken into account;
(iii) the potential target is a protein from a harmful organism which has to be inhibited in such a manner, however, that the human isozyme is not impaired.

Of course, the boundaries between these cases are by no means strict but in the next three sections examples will be described which belong largely to the classes outlined above.

DRUGS WHICH HAVE TO "CORRECT" HUMAN PROTEINS

The three-dimensional structure of human hemoglobin (2,3) is at present a source of inspiration for the design of new drugs against sickle cell anemia (see e.g. 4,5,6). The general aim is often to alter the three-dimensional structure of deoxy hemoglobin in such a manner that (i) no deoxy Hb fibers are formed which prevent passage of erythrocytes through capillaries; and, (ii) the oxygen transport capacity of the blood is not diminished too severely. Any effective drug must be able to strike a delicate balance between these two somewhat conflicting requirements.

Other examples of drugs which affect only human enzymes are given in Table 2. Angiotensin converting enzyme (ACE) - with unknown structure - is nevertheless listed because knowledge of the three-dimensional structures of Zn-protrases has been very helpful in the development of the

Table 2. Protein crystallography and drug design
Human proteins

Hemoglobin	Sickle cell anemia
Phospholipase A_2	Inflammation
Dihydrofolate reductase	Cancer
Calmodulin	
α_1-Proteinase inhibitor	Thrombosis
Angiotensin Converting Enzyme (ACE)	Hypertension
Renin	Hypertension
Oncoproteins	Cancer

anti-hypertension drug captopril (7). Renin has been included because succesful crystallization has been reported (8) and, moreover, computer graphics has allowed the construction of a hypothetical model of this enzyme (9) which is being used in the design of selective inhibitors one of which may one day also become a useful anti-hypertensive drug.

Cancer falls squarely into the category of diseases for which succesful drugs have to interact specifically with human proteins. The structure of the widely-used anti-cancer drug methotrexate bound to a bacterial dihydrofolate reductase is known in great detail (10), but a complex with the human enzyme would be more helpful.

The so-called "onco proteins" are receiving immense interest for obvious reasons. In particular the observation that point mutations (11,12,13) of the *ras* gene give rise to carcinogenic properties of its product, p21 protein, is worth mentioning. Such mutations result in crucial differences in enzymes in normal and cancer cells which could be exploited in the design of selective inhibitors of cancer cells. This has been stressed by Wierenga & Hol when they derived their hypothetical model for part of the p21 protein (14,15).

Phospholipase A_2 is included in Table 1 because it is a crucial enzyme in the initial steps of the synthesis of prostaglandins, leukotrienes and thromboxanes (16). The pancreatic protein has been studied in great detail (17,18,19). The recent observation that the pancreatic and cellular phospholipases A_2 are immunologically related (20), suggests that the known structures may form good starting points for the design of new drugs against inflammation.

INHIBITING UNIQUE PROTEINS FROM INFECTIOUS ORGANISMS

Penicillin is the prime example of a compound which inhibits a vital metabolic step in an infectious agent which step has no counterpart in humans. Even though penicillin is tremendously succesful, resistance occurs and it is therefore of great potential importance that the structure of penicillin-binding proteins (21,22) and of β-lactamases (23) have been solved recently, or are being solved. The structures of these proteins may lead to suggestions for new compounds which surpass penicillin in activity. Also combinations of compounds, such as penicillin plus β-lactamase inhibitors, may be developed and might become clinically useful in cases of penicillin resistance.

Table 3. Protein crystallography and drug design
Proteins unique for parasites.

Penicillin target enzymes	Bacterial infections
β-Lactamases	Bacterial infections
Choleratoxin (CT)	Intestinal disease
Enterotoxin (LT)	Intestinal disease
Diphteria toxin (DT)	Diphteria
Rhinovirus	Common cold
Polio virus	Polio
Haemagglutinin	Influenza
Neuraminidase	Influenza
Variable Surface Coat Protein	Sleeping sickness
Glycosomal Enzymes	Sleeping sickness

Structures of viruses and of viral coat proteins are obviously of tremendous importance for the development of new vaccines against several common human diseases. But also from the point of view of rational drug design, the crystallographic studies on rhino- and poliovirus (24,25) and also on the influenza virus coat proteins haemagglutinin (26) and neuraminidase (27), are of potential use. By careful inspection of these structures compounds might be found which interfere with the assembly of the coats of these viruses in such a manner for instance that the immune system can attack the constant parts of these viral coats in an efficient manner.

Bacterial toxins are clearly potential targets for the design of new drugs. In particular choleratoxin (CT) excreted by *Vibrio cholerae*, and the closely related heat-labile enterotoxin (LT), excreted by enterotoxic *E. coli* strains, seem to be important (28). For both toxins, crystallographic studies are under way (29,30). The B-subunits bind to gangliosides in the membranes of gut epithelial cells whereafter the A-subunit is transported through the membrane into the cell where it catalyses the ADP-ribosylation of the G_s protein of the adenylate cyclase system. As a consequence, the cAMP concentration in the cell raises and water and salt are excreted with severe diarrhoea as a result. Accurate structures of choleratoxin and *E. Coli* enterotoxin provide many starting points for the design of drugs. The development of new vaccines may also benefit once this structural knowledge becomes available.

SELECTIVE INHIBITION OF PROTEINS FROM INFECTIOUS ORGANISMS

Dihydrofolate reductase (DHFR) is a key enzyme in numerous organisms which has been the target of several widely employed drugs. Cancer is cured in many instances by the DHFR-inhibitor methotrexate - as mentioned in section 2. Selective inhibition of malarial DHFR by pyremethamine has lead to a great reduction in the occurrance of the malaria several years ago, but resistance and other factors have made malaria again a formidable problem in many tropical countries. No three-dimensional structure of malaria DHFR is known yet, but the group of D.A. Matthews and J. Kraut has succeeded in unravelling the structures of chicken (31-33) and several bacterial DHFR's with great accuracy. Very recently, this work had lead to a fascinating publication where not less than ten pharmaceutically interesting compounds were bound to its target enzyme (34). The detailed study of these complexes at the atomic level must be a marvellous stimulus to the entire field of developing new bactericides.

Another example of a project aimed at developing selective inhibitors is a collaboration between the groups of Opperdoes in Brussels, Borst in Amsterdam and ours in Groningen. We focus on the features of the "glycosomes": unique organelles of the trypanosomes - the causative agent of sleeping sickness in man. In these microbody-like glycosomes nine glycolytic enzymes form a multi-enzyme complex whose proper functioning is vital for this parasite. We have succeeded in crystallizing triosephosphate isomerase from *Trypanosoma brucei*. A 3 Å X-ray structure has been obtained inspite of the fact that only 600 µg of protein was available (35). Crystals of hexokinase and aldolase have also been obtained but were of too small dimensions. Promising-looking crystals of glyceraldehyde phosphate dehydrogenase have still to be investigated (R.K. Wierenga, personal communication). This project clearly has still some serious hurdles to pass before its goal has been reached.

CONCLUSION

Although this contribution has become a somewhat dry summary of promising projects, without real succes stories, with a narrow focus on proteins and on X-ray crystallography, and suffers also from a total lack of attention for the many difficulties in developing new drugs which cannot be resolved by accurate three-dimensional structures of target proteins, I am convinced that we are witnessing the beginning of a new and exciting era of drug design where protein crystallography plays a central role. An attempt is being made to give a somewhat more lively account of this field in a forthcoming review paper (36).

ACKNOWLEDGEMENT

I like to thank Dr. Rik K. Wierenga for numerous stimulating discussions.

REFERENCES

1. A.G. Amit, R.A. Mariuzza, S.E.V. Phillips and R.J. Poljak, Nature 313: 156-158 (1985).
2. B. Shaanan, J. Mol. Biol. 171: 31-59 (1983).
3. G. Fermi, M.F. Perutz, B. Shaanan and R. Fourme, J. Mol. Biol. 175: 159-174 (1984).
4. P. Goodford in: "Drug Action at the molecular level" (G.K.C. Roberts, ed.) pp. 109-126 (1977).
5. J.A. Walder, R.Y. Walder and A. Arnone, J. Mol. Biol. 141: 195-216 (1980).
6. D.J. Abraham, M.F. Perutz and S.E.V. Phillips, Proc. Natl. Acad. Sci. 80: 324-328 (1983).
7. D.W. Cushman, H.S. Cheung, E.F. Sabo and M.A. Ondetti, Biochemistry 16: 5484-5491 (1977).
8. M.A. Navia, J.P. Springer, M. Poe, J. Boger and K. Hoogsteen, J. Biol. Chem. 259: 12714-12717 (1984).
9. B.L. Sibanda, T. Blundell, P.M. Hobart, M. Fogliano, J.S. Bindra, B.W. Dominy and J.M. Chirgwin, FEBS Lett. 174: 102-111 (1984).
10. J.T. Bolin, D.J. Filman, D.A. Matthews, R.C. Hamlin and J. Kraut, J. Biol. Chem. 257: 13650-13662 (1982).
11. C.J. Tabin, S.M. Bradley, C.I. Bargmann, R.A. Weinberg, A.G. Papageorge, E.M. Scolnick, R. Dhar, D.R. Lowy and E.H. Chang, Nature 300: 143-149 (1982).
12. E.P. Reddy, R.K. Reynolds, E. Santos and M. Barbacid, Nature 300: 149-152 (1982).
13. E. Taparowsky, Y. Suard, O. Fasano, K. Shimizu, M. Goldfarb and M. Wigler, Nature 300: 762-765 (1982).
14. R.K. Wierenga and W.G.J. Hol, Nature 302: 842-844 (1983).
15. W.G.J. Hol and R.K. Wierenga in "X-ray crystallography and drug action", Clarendon Press, Oxford (Eds. A.S. Horn & C.J. de Ranter), pp. 151-168 (1984).
16. B. Samuelsson, Science 220: 568-575 (1983).
17. B.W. Dijkstra, K.H. Kalk, W.G.J. Hol and J. Drenth, J. Mol. Biol. 147: 97-123 (1981).
18. B.W. Dijkstra, G.J.H. van Nes, K.H. Kalk, N.P. Brandenburg, W.G.J. Hol and J. Drenth, Acta Cryst. B38: 793-799 (1982).
19. B.W. Dijkstra, R. Renetseder, K.H. Kalk, W.G.J. Hol and J. Drenth, J. Mol. Biol. 168: 163-179 (1983).
20. M. Okamoto, T. Ono, H. Tojo and T. Yamano, Biochem. Biophys. Res. Comm. 128: 788-794 (1985).

21. O. Dideberg, P. Charlier, G. Dive, B. Joris, J.M. Frère and J.M. Ghuysen, Nature 299: 469-470 (1982).
22. J.A. Kelly, J.R. Knox, P.C. Moews, G.J. Hite, J.B. Bartolone, H. Zhao, B. Joris, J.M. Frère and J.M. Ghuysen, J. Biol. Chem. 260: 6449-6458 (1985).
23. J. Moult, L. Sawyer, O. Herzberg, C.L. Jones, A.F.W. Coulson, D.W. Green, M.M. Harding and R.P. Ambler, Biochem. J. 225: 167-176 (1985).
24. E. Arnold, J.W. Erickson, G.S. Fout, E.A. Frankenberger, H.J. Hecht, M. Luo, M.G. Rossmann and R.R. Rueckert, J. Mol. Biol. 177: 417-430 (1984).
25. J.M. Hogle, J. Mol. Biol. 160: 663-668 (1982).
26. I.A. Wilson, J.J. Shekel and D.C. Wiley, Nature 289: 366-373 (1981).
27. J.N. Varghese, W.G. Laver and P.M. Colman, Nature 303: 35-40 (1983).
28. J.L. Middlebrook and R.B. Dorland, Microbiol. Rev. 48: 199-221 (1984).
29. P.B. Sigler, M.E. Druyan, H.C. Kiefer and R.A. Finkelstein, Science 197: 1277-1279 (1977).
30. S.E. Pronk, H. Hofstra, H. Groendijk, J. Kingma, M.B.A. Swarte, F. Dorner, J. Drenth, W.G.J. Hol and B. Witholt, J. Biol. Chem. (1985) in press.
31. K.W. Volz, D.A. Matthews, R.A. Alden, S.T. Freer, C. Hansch, B.T. Kaufman and J. Kraut, J. Biol. Chem. 257: 2528-2536 (1982).
32. D.J. Filman, J.T. Bolin, D.A. Matthews and J. Kraut, J. Biol. Chem. 257: 13663-13672 (1982).
33. D.A. Matthews, J.T. Bolin, J.M. Burridge, D.J. Filman, K.W. Volz, B.T. Kaufman, C.R. Beddell, J.N. Champness, D.K. Stammers and J. Kraut, J. Biol. Chem. 260: 381-391 (1985).
34. D.A. Matthews, J.T. Bolin, J.M. Burridge, D.J. Filman, K.W. Volz and J. Kraut, J. Biol. Chem. 260: 392-399 (1985).
35. R.K. Wierenga, W.G.J. Hol, O. Misset and F.R. Opperdoes, J. Mol. Biol. 178: 487-490 (1984).
36. W.G.J. Hol, Angew. Chemie, in preparation.

ENZYME MECHANISM: WHAT X-RAY CRYSTALLOGRAPHY CAN(NOT) TELL US

Johan N. Jansonius

Department of Structural Biology, Biozentrum
University of Basel, CH-4056 Basel, Switzerland

INTRODUCTION

Any claim that X-ray crystallography can make a contribution towards the elucidation of enzyme mechanisms implies that the spatial structure of a protein in a crystal lattice corresponds to that in solution. Is that the case? This question, which regularly has been asked (and is still being asked), has been considered by e.g. Drenth (1972), Matthews (1977), and Fersht (1977). All gave an affirmative answer. Some of the strongest arguments in favor are:
- The structures of many proteins have been determined repeatedly and under different conditions. Identical results have been obtained.
- Related proteins are invariably found to have the same polypeptide chain folding.
- Crystal lattice contacts are in general too weak and too limited in area to disturb the protein folding to any significant extent. The molecules retain considerable motility.
- Properties of the molecule in solution can usually be explained very well by the crystal structure.
- NMR data confirm crystallographically determined structures of proteins (this is not always true for the much more flexible peptides).
- Enzymes often retain nearly full catalytic activity in the crystal, especially if no large conformational changes occur during catalysis.

In spite of the above arguments we should not close our eyes for the limitations of the crystal structure. If there is a population of conformers of a given protein in solution, as might be the case for very small proteins and especially for flexibly hinged multidomain proteins, only one of these will usually be selected in the crystal. Furthermore, flexible loops in the polypeptide chain of a protein may be frozen into a particular conformation in the crystal, due to stabilization by a neighboring molecule. Finally, it should be realized that rotation around single bonds in sidechains of surface residues is essentially unrestricted. Various experiments document the considerable motility of protein molecules, not only in solution, but also in the crystal. As far as that is concerned, we are in a way deceived by the time and space average picture of our electron density maps. The dynamics of protein structure and its relation to function are discussed in a review by Bennett and Huber (1984).

Having established the scope and limitations of the crystal structure of proteins, we may now ask what it can do (and not do) for us in the study of enzyme mechanism. Important contributions that X-ray crystallography can, in favorable cases, make are the following:
- Accurate models of their spatial structures are obtained.
- If the crystal packing and the solvent are favorable, the structures of inhibitor complexes that mimic catalytic intermediates can be determined.
- Extrapolating from the structures of the free enzyme and inhibitor derivatives, model building using computer graphics produces models of catalytic intermediates.
- Combination of these models with all relevant information from different sources may lead to a proposal for the stereochemical pathway of the catalyzed reaction.

It should be realized, however, that the mechanistic proposal in which the X-ray studies often culminate is by no means the final answer. It should rather be regarded as a working hypothesis, to be tested by whatever techniques that suggest themselves to reach this goal.

Obviously, the energetic aspects and, related to this, the time course of an enzyme catalyzed reaction are not addressed by X-ray crystallography, other than, to some extent, in cryocrystallographic experiments. The scope of these, to date, has however been exceedingly limited. The kinetic experiments related to such questions will in most cases have already been carried out before the spatial structure of an enzyme has been elucidated.

Conformational changes can be revealed by X-ray crystallography, but their effect on the catalytic rate or on specificity again must be studied by other means.

Finally, the movement of water molecules, of protons and hydroxide ions remains uncertain, although the X-ray data may give hints about probable events of this kind in catalysis.

In the following, crystallographic studies on the mechanism of action of three enzymes will be discussed, as more or less representative examples, chosen subjectively on the basis of personal interest of the author. The enzymes selected are the plant thiol protease papain (work carried out by J. Drenth et al., University of Groningen), the thermostable bacterial neutral protease thermolysin (B.W. Matthews and coworkers, University of Oregon) and chicken mitochondrial aspartate aminotransferase studied in the author's own laboratory at the University of Basel.

PROTEOLYTIC ENZYMES

The first two enzymes to be discussed are proteases, which hydrolyze peptide bonds. A nucleophilic attack is carried out on the carbonyl carbon of the scissile peptide bond in one of two possible ways.
a) A reactive nucleophilic group of the enzyme itself, e.g. a serine γ-oxygen in serine proteases (the chymotrypsin and subtilisin families) or a cysteine γ-sulfur in thiol proteases (the papain family), carries out the attack directly. In this case a so-called acylenzyme intermediate is formed (an ester or thiolester in the examples given) with release of the first product, upon protonation of its α-amino nitrogen atom. Subsequently, an activated water molecule in turn attacks the ester carbonyl carbon, giving rise to the second product.
b) Alternatively, the first nucleophilic attack can directly be carried out by an activated water molecule. In this case no covalent intermediate is produced and both products are formed simultaneously. Such a mechanism

probably applies to the metalloproteases thermolysin and (perhaps) carboxypeptidase.

We will discuss below the X-ray crystallographic evidence for the catalytic mechanisms of papain and thermolysin.

Papain

This enzyme can be isolated from the dried latex of the tropical tree Carica papaya. It consists of a single polypeptide chain of 212 amino acid residues of known sequence. Amongst these are seven cysteines, six of which are paired in three disulfide bridges, the seventh, Cys 25, being essential for catalysis (Glazer and Smith, 1971). The enzyme was crystallized from 70% methanol/water, a medium in which it is inactive (Drenth et al., 1962; 1971). Its spatial structure was elucidated originally at 2.8 Å resolution (Drenth et al., 1968; 1971) and recently refined at 1.65 Å resolution (Kamphuis et al., 1984). An independent structure determination and refinement at 2.0 Å resolution in a different crystal form (type D, Drenth et al., 1971) was carried out in the author's laboratory (Priestle et al., 1984).

Papain is a two-domain protein. The active site is situated in a pronounced cleft between the two domains (Wolthers, 1971; Drenth et al., 1976). The thiol group of Cys 25 is positioned near the N-terminus of a 5-turn helix (helix 1, residues 24-42) and thus "feels" the positive electric field of its dipole (Hol et al., 1978). The imidazole of His 159 (in the second domain) lies exactly opposite Cys 25 (in the first domain) at less than 3.5 Å distance. These two residues are believed to be essential for catalysis. Schechter and Berger (1967) have established that papain specifically recognizes seven residues of the substrate, four (P_4-P_1) on the N-terminal and three (P_1'-P_3') on the C-terminal side of the scissile peptide bond. The enzyme prefers phenylalanine at the P_2 position (Schechter and Berger, 1968). The subsites S_4-S_1 (that bind residues P_4-P_1) have been defined crystallographically in the following way (Drenth et al., 1976). Chloromethylketone derivatives of three peptides with high affinity for these subsites were reacted in solution with Cys 25 to give the corresponding alkylated derivatives R-CO-CH_2-S-papain. These were crystallized and X-ray data were collected to 2.8 Å resolution. Difference electron density maps demonstrated that these peptides occupied the cleft between the domains. Antiparallel pleated sheet hydrogen bonding was indicated between the NH and CO of Gly 66 and the CO and NH of the P_2 residue of the inhibitor. The carbonyl oxygen of residue P_1 pointed away from His 159 towards the NH of Cys 25 and the NH_2 of the sidechain amidogroup of Gln 19. The analogy with the "oxyanion hole" (Robertus et al., 1972) in serine proteases was recognized and used in building models of putative acylenzymes (a known intermediate in catalysis by thiol proteases; Glazer and Smith, 1971) and of the tetrahedral intermediate, the first, transient, intermediate in the acylation step. Interestingly, Drenth et al. (1976) could not only show that the resulting models (nearly) fit the papain active site (a widening of $\sim$ 1 Å of the active site cleft in the derivative structures was indicated by the difference map), but also that the tetrahedral intermediate (the putative transition state) makes stronger H-bonds in the "oxyanion hole" than the acylenzyme. This finding strengthens the proposal, since enzymes are believed to have active site structures which are complementary to the transition state of the catalyzed reaction (Fersht, 1977). Another observation was that a $\sim 30°$ rotation of the imidazole group of His 159 around $C\beta$-$C\gamma$ (which experimentally had been shown to be possible) would bring it in the right position and orientation to donate a proton to the leaving group in the tetrahedral intermediate. Since the pH-activity profile of papain corresponds to a bell-shaped curve with a low pK of ~ 4 and a high pK ~ 8 (Glazer and Smith, 1971), such a proposal required that the high pK be attributed to His 159, and the pK of 4 to Cys 25, the two acting as an ion pair between pH 4 and 8. This, at the time, controversial assignment has recently been confirmed by NMR studies (Lewis et al., 1981). The abnormal pK's of Cys 25 and His 159 can at least qualitatively be explained by the effect of the electric dipole of helix 1 (Van Duijnen et al., 1979). The proposal by Drenth et al. (1975; 1976), in which the strongly nucleophilic thiolate anion of Cys 25 attacks the carbonyl carbon of the scissile peptide bond and His 159 protonates the leaving group in the resulting tetrahedral intermediate, producing the first product and an acylenzyme intermediate which is subsequently hydrolyzed, again with the assistance of His 159, seems very reasonable. So far, only one family of thiol proteases has been discovered, and all key residues in the proposal of Drenth et al. (1976) are invariant (Kamphuis et al., 1985). Nevertheless, further crystallographic evidence for substrate binding is ultimately needed to confirm the proposed mechanism.

Thermolysin

This thermostable enzyme from B. thermoproteolyticus is a single-chain protein consisting of 316 amino acid residues of known sequence. Its spatial structure has been determined by Matthews and coworkers (Colman et al., 1972) and highly refined (Holmes and Matthews, 1982). Thermolysin (TLN) is a two-domain protein, the structure of which is stabilized by four calcium ions, one in the mainly pleated sheet containing N-terminal domain and three in the largely helical C-terminal domain. The active site lies in a deep groove. At its base the essential zinc ion is bound by His 142, His 146 and Glu 166, its fourth ligand being a water molecule. Glu 143 is properly positioned for catalyzing a direct or (more likely) indirect attack on the carbonyl carbon of the scissile peptide bond in analogy to Glu 270 in carboxypeptidase A (Quiocho and Lipscomb, 1971). The catalytic sites of these two enzymes show strong similarity (Kester and Matthews, 1977). The Zn ion in both has the function to coordinate and thereby polarize the carbonyl oxygen atom. His 231 in TLN was originally believed to be a candidate for protonation of the scissile amido nitrogen (Kester and Matthews, 1977), a function attributed to Tyr 248 in carboxypeptidase A (Quiocho and Lipscomb, 1971). In the terminology of Schechter and Berger (1967), TLN recognizes six residues of the substrate: $P_3-P_2-P_1-P'_1-P'_2-P'_3$. The main specificity is for a large hydrophobic residue (e.g. Leu, Phe) at position P'_1 (Morihara and Tsuzuki, 1970).

Numerous inhibitor complexes of TLN have been studied crystallographically by Matthews and coworkers. Some of the key structures have been highly refined (Monzingo and Matthews, 1984; Hangauer et al. 1984; and references therein). The results show that the P'_1 sidechain is buried in a hydrophobic pocket, while the main chain CO and NH groups of residues P_2, P_1, P'_1 and P'_2 of the substrate can make hydrogen bonds to the enzyme. The most revealing results were obtained with phosphoramidon (N-(α-L-rhamnopyranosyl-oxyhydroxyphosphinyl)-L-leucyl-L-tryptophan; Weaver et al., 1977) with an inhibition constant of 2.8×10^{-8} M and N-(1-carboxy-3-phenylpropyl-L-leucyl-L-tryptophan) (CLT, $K_i = 5 \times 10^{-8}$ M; Monzingo and Matthews, 1984). These compounds were shown to straddle the zinc ion with their free phosphoramidate oxygens (phosphoramidon) and N-carboxyphenylpropyl carboxylate oxygens (CLT). The analogy of these two structures to tetrahedral intermediates in peptide hydrolysis $R-C(OH)(O^-)-N(H)R'$, coupled to their strong inhibiting power strongly suggested five-coordination of the zinc ion in a non-covalently bound tetrahedral intermediate in cataly-

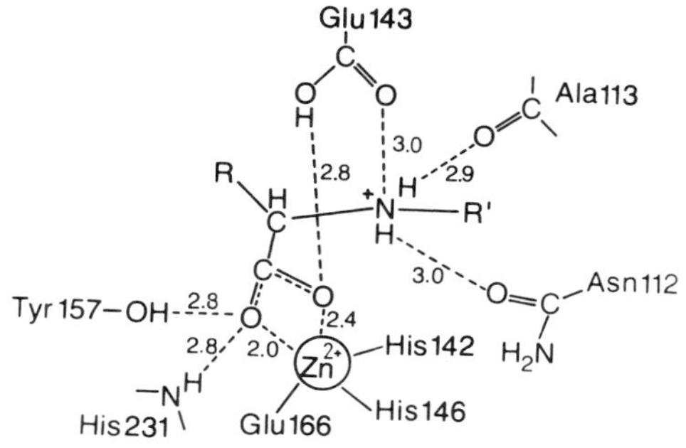

Fig. 1. Schematic drawing of the interactions of CLT with residues in the active site of thermolysin. Only the region including the catalytic groups is shown. The dashed lines indicate presumed hydrogen bonds, with the exception of the 3.0 Å distance between the Glu 143 oxygen and the CLT nitrogen atom (after Fig. 5 of Monzingo and Matthews, 1984)

sis by TLN. Such a possibility had already been suggested by Holmes and Matthews (1981) on the basis of the binding properties of hydroxamate inhibitors. Figure 1 shows schematically the binding of the crucial part of CLT in the TLN active site. The two carboxylate oxygens are at distances of 2.0 Å (proximal oxygen) and 2.4 Å (distal oxygen) from the zinc ion. Interestingly, the proximal oxygen accepts hydrogen bonds from both His 231 and Tyr 157, strongly suggesting stabilization of a putative tetrahedral intermediate by oxyanion binding in an analog of the oxyanion hole (Robertus et al., 1972) in serine proteases. Obviously, this is incompatible with a role for His 231 as proton donor. The uncharged (while covered by the inhibitor) Glu 143 has one of its carboxylate oxygens at 2.8 Å from the distal N-carboxymethyl oxygen, presumably hydrogen bonding to it, and the other at 3.0 Å from the leucyl nitrogen atom of the inhibitor. This suggested a role for Glu 143 as proton shuttle from the attacking water molecule (corresponding to the distal oxygen in the inhibitor) to the nitrogen of the scissile peptide bond of the substrate. As already remarked by Kester and Matthews (1977), Glu 143 is too far from the zinc ion to be able to attack the substrate carbonyl carbon directly, when the carbonyl oxygen approaches the zinc ion. Subsequent extensive model building studies (Hangauer et al., 1984) led to the following proposal. In the Michaelis complex the substrate (scissile) carbonyl oxygen probably does not yet bind to the zinc ion. As the transition state is approached it will insert itself, thereby pushing the zinc liganding water molecule towards Glu 143. This residue, becoming more and more covered by the substrate, grabs a proton from the water molecule which is thus activated toward nucleophilic attack on the neighboring carbonyl carbon. The resulting tetrahedral intermediate is stabilized by stronger binding to the zinc ion of both oxygens (one of which has a negative charge) and by hydrogen bonding of the proximal oxygen (in the view of Fig. 1) in the oxyanion hole provided by His 231 and Tyr 157. Protonation of the amido nitrogen of the scissile bond by Glu 143 precedes the breaking of this bond and release of the two products.

Monzingo and Matthews (1984) suggest that a similar mechanism applies to carboxypeptidase A, including five-coordination of the zinc ion and a double role for Glu 270. This would imply that Tyr 248 would have a different function (if any) than the proton-donating role suggested by Lipscomb (Quiocho and Lipscomb, 1971; Lipscomb, 1980). Studies of the esterase activity of carboxypeptidase A have indicated five-coordination of the zinc ion in a tetrahedral intermediate. However, the latter seems to be covalently bound to the enzyme, presumably as a mixed anhydride to Glu 270 (Kuo and Makinen, 1982; Kuo et al., 1983).

COVALENT CATALYSIS: ASPARTATE AMINOTRANSFERASE

Aspartate aminotransferase (AAT, EC 2.6.1.1) is an α_2 dimeric enzyme of $M_r \sim 90000$. Two different nuclear genes in higher animals code for a cytosolic form (cAAT) of 412 amino acids per chain and the precursor of a mitochondrial enzyme (mAAT) which is processed to a length of 401 amino acids per chain upon import into this organelle (Christen and Metzler, 1985). These enzymes catalyze reversible amino group transfer from aspartate to 2-oxoglutarate and from glutamate to oxaloacetate (Braunstein, 1973). The coenzyme pyridoxal phosphate (PLP) binds the amino acid substrate covalently as an aldimine in the first of a series of known covalent intermediates (Figure 2), which can be recognized by their electronic absorption spectra. The slowest steps which are catalyzed very effectively by the enzyme are deprotonation of the substrate α-carbon in the "external" aldimine II to produce the transient quinonoid intermediate III which is subsequently reprotonated at C4' (the original aldehyde carbonyl carbon), resulting in the ketimine intermediate IV. Hydrolysis of this intermediate yields the pyridoxamine (PMP) form of the enzyme (V) and the oxoacid product.

X-ray crystallographic studies, underway in Moscow and Iowa City on chicken and pig cAAT and in Basel on chicken mAAT, have established identical folding patterns and highly similar behavior of the two enzymes with respect to inhibitor binding (Christen and Metzler, 1985). The most extensive studies to date have been carried out on mAAT. A partially refined model (R=23%) exists of the unliganded PLP-enzyme (space group P1) and a highly refined model (R=19%) of another form (space group $C222_1$) in which the enzyme is inhibited by maleate. The current resolution of both structures is 2.3 Å. Difference electron density maps at 2.8 Å of the apoenzyme and the PMP-enzyme helped to define the coenzyme conformations and revealed that these structures are essentially identical to the PLP-enzyme. A 2-methylaspartate derivative, an external aldimine model, differed (at 2.3 Å resolution) from the maleate derivative only at the coenzyme and inhibitor positions and their immediate environment (Jansonius and Vincent, 1986). Information from the above structures and several further derivatives of the P1 form at 2.8 Å resolution (Kirsch et al., 1984) gave the following picture of the structure and dynamics of this enzyme. mAAT is an elongated molecule (105 Å x 60 Å x 50 Å). The two subunits are related by a twofold axis. Each consists of a large PLP-binding domain (residues 48-325) and a small domain (residues 16-47 and 326-410), in the numbering scheme corresponding to the pig cAAT amino acid sequence, to which chicken mAAT is 48% homologous. The N-terminal arm, residues 3-15, detaches itself from the small domain and makes strong van der Waals contacts to a pocket underneath the large domain of the other subunit, notably through the indoles of Trp 5 and Trp 6. The active sites are pockets at the end of a cleft between the small domain of the subunit to which they belong and the large domain of the other. They lie on opposite sides of

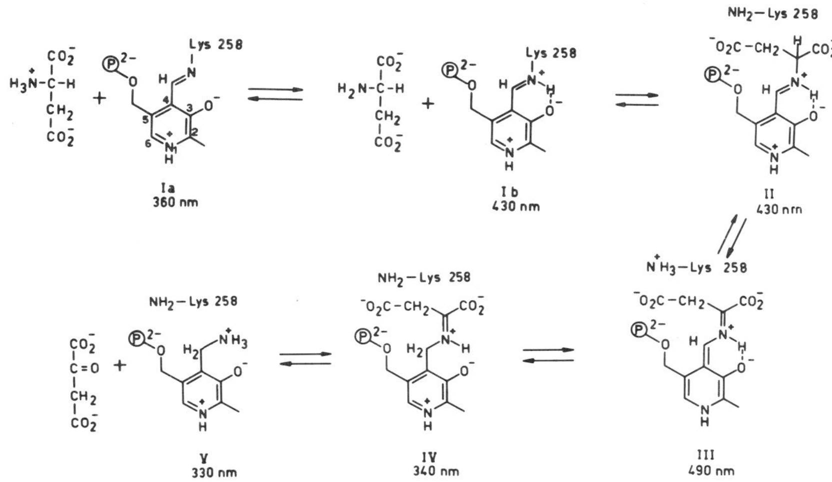

Fig. 2. The main steps in the catalytic conversion of aspartate into oxaloacetate by aspartate aminotransferase. The approximate absorption maxima of the various species are indicated, as is the presumed role of Lys 258 in proton transfer. Ia, PLP-enzyme with unprotonated "internal" aldimine; Ib, PLP-enzyme with protonated "internal" aldimine; II, "external" aldimine intermediate; III, quinonoid intermediate; IV, ketimine intermediate; V, pyridoxamine enzyme. In Ia, the numbering of the atoms in the pyridine ring is shown. Side chain atoms would be indicated by a prime, e.g. C4', C5', O5'.

the dimer. The back wall of each active site is formed by the coenzyme, which is engaged in an aldimine linkage with Lys 258. The protonated state of its pyridine N1 is stabilized by a hydrogen bond/salt bridge to the carboxylate group of Asp 222, the ionized state of the 3'-OH by a short hydrogen bond provided by the phenolic hydroxyl of Tyr 225. Thus a highly reactive form of PLP is stabilized, which in solution is only a minor conformer. The phosphate group receives no less than eight hydrogen bonds to its three non-ester oxygens. Two of these are provided by Arg 266, which together with the positive field at the N-terminus of an α-helix (residues 108-122) largely compensates the double negative charge of the phosphate group. The bridging oxygen lies in front of (i.e. on the substrate side of) the (extended) plane of the pyridine ring, the torsion angle C4-C5-C5'-O5' being ~ -37°. The aldimine double bond lies behind the pyridine ring, rather than being coplanar. The substrate binding pocket is surrounded by Trp 140 (below), Ser 296 and Asn 297 (at the left), Tyr 70* (other subunit) and Gly 38 (above) and Asn 194 (at the right). At its entrance one finds Arg 292* at the left and Arg 386 at the right. The helix formed by residues 16-25 lies in front of the entrance to the pocket.

A comparison of the unliganded PLP-enzyme with the maleate derivative shows first of all that in the latter the small domain has reoriented, so as to close the door to the active site pocket behind the inhibitor maleate. This molecule makes hydrogen bonds/salt bridges with Arg 292* and Arg 386, the latter having approached with the small domain. The sidechains of Ile 17 and Leu 18 are in van der Waals contact with maleate. Above maleate, a reorientation of residues 37-39 causes the sidechain of

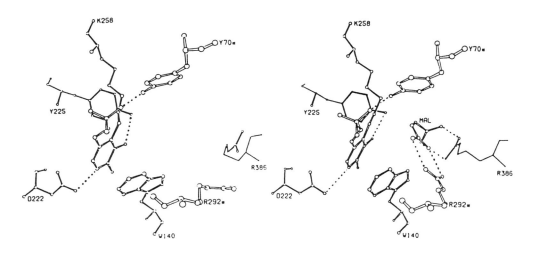

Fig. 3. The coenzyme and selected active site groups in the unliganded PLP-enzyme (left) and the maleate derivative (right) in the same view direction in which the molecular twofold axis is vertical and the active site entrance is at the right. Hydrogen bonds are indicated by dotted lines. The standard one-letter code for aminoacids is used, MAL stands for maleate. Note the rearrangement of both Arg 292* and Arg 386 and the approach of the latter (with the small domain) to make double hydrogen bonds and salt bridges with the two carboxylate groups of maleate. The coenzyme rearrangement upon maleate binding, described in the text, is also clearly visible.

Val 37 to be in contact with the inhibitor directly as well. The result of all this is complete occlusion of bulk water.

The maleate derivative shows an absorption maximum at 440 nm, due to protonation of the coenzyme aldimine (Figure 2, structure Ib). This protonation creates the possibility of a strong "conjugated" (Braunstein, 1973) hydrogen bond to the negatively charged 3'-oxygen. This explains the rotation of the coenzyme pyridine ring around N1-C4. This rotation not only causes near-coplanarity of the pyridine ring with the aldimine double bond, but also brings about a reduction of the torsion angle C4-C5-C5'-O5' to $\sim -16°$. Protonation of the "internal" aldimine is obligatory prior to nucleophilic attack of the amino acid substrate on C4' (the transaldimination step). We believe therefore that the reduction of the torsion angle around C5-C5' corresponds to an approach towards the transition state in this step which presumably is the eclipsed conformation, with a minimum distance between O5' and C4' (Jansonius and Vincent, 1986). Figure 3 shows, side by side, pictures of the coenzyme and key residues in the active sites of the unliganded PLP-enzyme and the maleate derivative.

In the 2-methylaspartate derivative the inhibitor binds through an "external" aldimine with the coenzyme. The latter has tilted forward by $\sim 30°$, pivoting aroung N1, which remains hydrogen bonding to Asp 222. The torsion angle around C5-C5' is now $\sim +90°$. The movement has essentially been carried out by rotations around the three bonds between C5 and the phosphorus atom. The observed reduction of the torsion angle around C5-C5' upon protonation seems to exclude the "oscillatory rotor" mechanism (Arnone et al., 1984) in which the O5'-C5-C5 arm with the pyridine ring attached swings around the P-O5' bond "over the top" (a $\sim 270°$ rotation) ra-

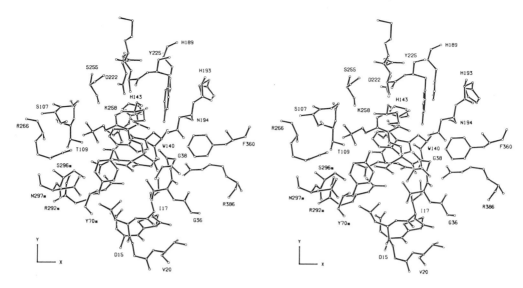

Fig. 4. Stereo view from "above" (along the molecular two-fold axis) of the active site of chicken mAAT in the covalent "external aldimine" complex with 2-methylaspartate. The carboxylate groups of the inhibitor coincide with those of maleate in the corresponding derivative (Fig. 3), but the coenzyme pyridine ring has rotated forward by $\sim 30°$. The pyridine N1, which remains hydrogen bonded to Asp 222, is essentially the pivot of this motion, which is carried out by rotations around C5-C5' and O5'-P and, to a lesser extent, C5'-O5'. The sidechain of Lys 258 "hangs" above the coenzyme in an ideal position for proton transfer from the substrate Cα to the coenzyme C4'.

ther than directly forward as suggested for mAAT (Kirsch et al., 1984; Jansonius and Vincent, 1986).

The 2-methylaspartate moiety fits perfectly in the "closed" active site. The 2-methyl group expels a water molecule which is present in the maleate derivative. This structure illustrates nicely that the Lys 258 ϵ-amino group, released uncharged behind the coenzyme upon transaldimination, is boxed in completely. It is the only group which is sterically and electronically capable to deprotonate the α-carbon of the substrate. It cannot obtain an additional proton in any other way. Figure 4 shows, in stereo, the tight fit of the covalently bound inhibitor 2-methylaspartate in the "closed" active site of mAAT and the strategic position of Lys 258.

Thus, the described crystallographic structures give answers to many questions concerning the catalytic mechanism of AAT. Computer graphics allowed very reasonable models to be built for the various intermediates of catalysis. It turns out that in the closed active site, there is room above the substrate for only a single, isolated, water molecule, capable of carrying out hydrolysis of the ketimine intermediate after Lys 258 has removed the α-proton and subsequently protonated C4'. The again uncharged ϵ-amino group of Lys 258 presumably takes over a proton of this water molecule and thus activates it for attack on the ketimine carbon. Via a tetrahedral intermediate the oxoacid product is produced and released upon the conformational change towards the open structure.

CONCLUSION AND OUTLOOK

The three examples above illustrate the amount of information on enzyme mechanism obtainable from X-ray crystallographic studies at different resolutions for systems of varying complexity. The work on papain shows the kind of problems one has to cope with in case of instability of the active enzyme and a solvent which drastically decreases inhibitor binding constants. Here the only solution was the use of covalently binding inhibitors which simultaneously solved the problem of the hyper-reactivity of the thiol group and the weak binding of competitive Michaelis-type inhibitors. The studies on aspartate aminotransferase illustrate the necessity of cocrystallization of enzyme and inhibitors when conformational changes occur. This situation made the solution of another crystal structure unavoidable.

Very reasonable mechanistic proposals could be made in all three described cases. These can now be tested further. The stereochemical pathway of these catalytic processes is quite well understood due to the crystallographic studies. It remains difficult, however, to explain rate constants on this basis. Our understanding of intermolecular interactions and of the hydrophobic effect is still too limited to compute energy levels of enzyme-ligand complexes, even if their structures are known very accurately.

However, a considerable improvement in our understanding of enzyme mechanism and of the forces that stabilize macromolecular structure can soon be expected from extensive studies of enzymes and other proteins modified in a systematic way by site directed mutagenesis of their genes. Such studies are already underway in many laboratories. Alternatively, the binding properties of modified ligands of a macromolecule provide similar and perhaps even more accurate information (see the contribution by F.A. Quiocho in this volume). Thus, the combined efforts of geneticists, protein chemists, enzyme kineticists and crystallographers should lead to much more insight into enzyme catalysis in the next five to ten years.

ACKNOWLEDGMENTS

I thank Dr. M.G. Vincent for help in preparing some of the figures and critical reading of the manuscript, and Ms. U. Abrecht for expert typing. The crystallographic study of mitochondrial aspartate aminotransferase was supported in part by grant 3.224.82 of the Swiss National Science Foundation.

REFERENCES

Arnone, A., Rogers, P.H., Hyde, C.C., Makinen, M.W., Feldhaus, R., Metzler, C.M., and Metzler, D.E., 1984, Crystallographic and chemical studies on cytosolic aspartate aminotransferase, in: "Chemical and Biological Aspects of Vitamin B_6 Catalysis," A.E. Evangelopoulos, ed., Part B, Alan R. Liss, New York.

Bennett, W.S., and Huber, R., 1984, Structural and functional aspects of domain motions in proteins, CRC Critical Reviews in Biochemistry, 15:291.

Braunstein, A.E., 1973, Amino group transfer, in: "The Enzymes," P.D. Boyer, ed., 3rd ed., Vol. 9, Academic Press, New York.

Christen, P., and Metzler, D.E., eds., 1985, "Transaminases," J. Wiley and Sons, New York.

Colman, P.M., Jansonius, J.N., and Matthews, B.W., 1972, The structure of thermolysin: an electron density map at 2.3 Å resolution, J.Mol. Biol., 70:701.

Drenth, J., 1972, The comparison of protein structures in the crystalline state and in solution, in: "Enzymes, Structure and Function," FEBS Proceedings, Vol. 29, Elsevier, Amsterdam.

Drenth, J., Jansonius, J.N., Koekoek, R., Marrink, J., Munnik, J., and Wolthers, B.G., 1962, The crystal structure of papain-C. I. Two-dimensional Fourier synthesis, J.Mol.Biol., 5:398.

Drenth, J., Jansonius, J.N., Koekoek, R., Swen, H.M., and Wolthers, B.G., 1968, Structure of papain, Nature (London), 218:929.

Drenth, J., Jansonius, J.N., Koekoek, R., and Wolthers, B.G., 1971, The structure of papain, Advan.Protein Chem., 25:79.

Drenth, J., Kalk, K.H., and Swen, H.M., 1976, Binding of chloromethyl ketone substrate analogues to crystalline papain, Biochemistry, 15:3731.

Drenth, J., Swen, H.M., Hoogenstraaten, W., and Sluyterman, L.A.Ae., 1975, A mechanism of papain action, Proc.Kon.Ned.Acad.Wetensch. Amsterdam, C78:104.

Fersht, A., 1977, "Enzyme Structure and Function," W.H. Freeman and Company, San Francisco.

Glazer, A.N., and Smith, E.L., 1971, Papain and other plant sulfhydryl proteolytic enzymes, in: "The Enzymes," P.D. Boyer, ed., Vol. 3, 3rd ed., Academic Press, New York.

Hangauer, D.G., Monzingo, A.F., and Matthews, B.W., 1984, An interactive study of thermolysin-catalyzed peptide cleavage and inhibition by N-carboxymethyl dipeptides, Biochemistry, 23:5730.

Hol, W.G.J., Van Duijnen, P.Th., and Berendsen, H.J.C., 1978, The α-helix dipole and the properties of proteins, Nature (London), 273:443.

Holmes, M.A., and Matthews, B.W., 1981, Binding of hydroxamic acid inhibitors to crystalline thermolysin suggests a pentacoordinate zinc intermediate in catalysis, Biochemistry, 20:6912.

Holmes, M.A., and Matthews, B.W., 1982, Structure of thermolysin refined at 1.6 Å resolution, J.Mol.Biol., 160:623.

Jansonius, J.N., and Vincent, M.G., 1986, Structural basis for catalysis by aspartate aminotransferase, in: "Biological Macromolecules and Assemblies," F. Jurnak and A. Mc Pherson, eds., Vol. 3, J. Wiley and Sons, New York, in press.

Kamphuis, I.G., Drenth, J., and Baker, E.N., 1985, Thiol proteases. Comparative studies based on the high-resolution structures of papain and actinidin, and on amino acid sequence information for cathepsins B and H, and stem bromelain, J.Mol.Biol., 182:317.

Kamphuis, I.G., Kalk, K.H., Swarte, M.B.A., and Drenth, J., 1984, Structure of papain refined at 1.65 Å resolution, J.Mol.Biol., 179:233.

Kester, W.R., and Matthews, B.W., 1977, Comparison of the structures of carboxypeptidase A and thermolysin, J.Biol.Chem., 252:7704.

Kirsch, J.F., Eichele, G., Ford, G.C., Vincent, M.G., Jansonius, J.N., Gehring, H., and Christen, P., 1984, Mechanism of action of aspartate aminotransferase proposed on the basis of its spatial structure, J.Mol.Biol., 174:497.

Kuo, L.C., Fukuyama, J.M., and Makinen, M.W., 1983, Catalytic conformation of carboxypeptidase A, J.Mol.Biol., 163:63.

Kuo, L.C., and Makinen, M.W., 1982, Hydrolysis of esters by carboxypeptidase A requires a pentacoordinate metal ion, J.Biol.Chem., 257:24.

Lewis, S.D., Johnson, F.A., and Shafer, J.A., 1981, Effect of cysteine-25 on the ionization of histidine-159 in papain as determined by proton NMR spectroscopy. Evidence for a His-159—Cys-25 ion pair and its possible role in catalysis, Biochemistry, 20:48.

Lipscomb, W.N., 1980, Carboxypeptidase A mechanisms, Proc.Natl.Acad.Sci USA, 77:3875.

Matthews, B.W., 1977, X-ray structure of proteins, in: "The Proteins," H. Neurath and R.L. Hill, eds., 3rd ed., Vol 3, Academic Press, New York.

Monzingo, A.F., and Matthews, B.W., 1984, Binding of N-carboxymethyl dipeptide inhibitors to thermolysin determined by X-ray crystallography: a novel class of transition-state analogues for zinc peptidases, Biochemistry, 23:5724.

Morihara, K., and Tsuzuki, H., 1970, Thermolysin: Kinetic study with oligopeptides. Eur.J.Biochem., 15:374.

Priestle, J.P., Ford, G.C., Glor, M., Mehler, E.L., Smit, J.D.G., Thaller, C., and Jansonius, J.N., 1984, Restrained least-squares refinement of the sulphydryl protease papain to 2.0 Å, Acta Cryst. A40 Suppl., C-17.

Quiocho, F.A., and Lipscomb, W.N., 1971, Carboxypeptidase A: a protein and an enzyme, Advan.Protein.Chem., 25:1.

Robertus, J.D., Kraut, J., Alden, R.A., and Birktoft, J.J., 1972, Subtilisin; a stereochemical mechanism involving transition state stabilization, Biochemistry, 11:4293.

Schechter, I., and Berger, A., 1967, On the size of the active site in proteases. I. Papain, Biochem.Biophys.Res.Commun., 27:157.

Schechter, I., and Berger, A., 1968, On the active site of proteases. III. Mapping the active site of papain; specific peptide inhibitors of papain, Biochem.Biophys.Res.Commun., 32:898.

Van Duijnen, P.Th., Thole, B.Th., and Hol, W.G.J., 1979, On the role of the active site helix in papain, Biophys.Chem., 9:273.

Weaver, L.H., Kester, W.R., and Matthews, B.W., 1977, A crystallographic study of the complex of phosphoramidon with thermolysin. A model for the presumed catalytic transition state and for the binding of extended substrates, J.Mol.Biol., 114:119.

Wolthers, B.G., 1971, Explorations of the active site of papain, Doctoral thesis, University of Groningen, The Netherlands.

THE ROLE OF SEPARATE DOMAINS AND OF INDIVIDUAL AMINO-ACIDS IN ENZYME CATALYSIS, STUDIED BY SITE-DIRECTED MUTAGENESIS

D.M.Blow*, P.Brick*, K.A.Brown*, A.R.Fersht+ and G.Winter@

*Blackett Laboratory and +Department of Chemistry
Imperial College, London SW7 2BZ and @Laboratory of
Molecular Biology, MRC Centre, Cambridge CB2 2QH, England

Protein engineering is the production of a novel or altered protein - a 'factitious' protein - by the creation or manipulation of a gene which controls its synthesis, for a specific purpose. To achieve the intended purpose, an ability to predict the properties of the factitious protein is required.

At the current state of the art, we have to admit that our powers of prediction are limited, and our methods crude. So, assessment of the actual properties of the factitious protein is needed to validate the success of the engineering work, and forms an essential part of protein engineering.

The theme of this paper is the interplay between protein engineering and protein crystallography. Effective protein engineering obviously depends on an appreciation of the structure-function relationships of the factitious protein, and so depends largely on protein crystallography for structural data. In the first place, this will come from the structure of the wild-type protein. Without this knowledge, redesign of a protein is a matter of guesswork, and could often be done better by evolution under a controlled selective pressure, as in a chemostat. Even poor structural knowledge is vastly more effective than none, especially because, in practice, protein engineering proceeds far more rapidly than crystallography.

Crystallography is also important to make a structural check on a newly engineered protein, to see whether the design has worked out as planned. Because of the relative speeds of different kinds of work, it will not be possible to work out the structure of every protein which has been engineered. But an example will be presented which will show the usefulness of a structure analysis. In this case, the plan was a speculative one, and understanding of the change in enzymatic properties is still incomplete.

It is less obvious that protein engineering will also become a handmaid of protein crystallography, an addition to the battery of techniques which the crystallographer may deploy, to advance the determination of structure directly. In the example to be given, structure analysis of a factitious mutant has greatly improved the wild-type structure.

The examples will be taken from our work on tyrosyl-tRNA synthetase, an enzyme whose detailed structure determination has proved difficult (Bhat et al.,1982; Blow & Brick, 1985). The main cause of difficulty has been the high level of disorder encountered in various parts of the structure. Of the 419 amino-acids comprising the polypeptide chain of the monomer, the C-terminal 100 are so disordered that no density can be identified for most of this part of the molecule. In a number of other parts of the molecule the B-factor rises above 40 $Å^2$, making the continuity of the peptide chain questionable, the assignment of secondary structure doubtful, and the positioning of amino-acid side chains impossible. But the β-structure which forms the framework of the α/β domain formed by the 200 amino-terminal amino-acids is very well-ordered. Many parts of the structure adjacent to the six-stranded β-sheet can be refined to reasonably accurate positions with little difficulty.

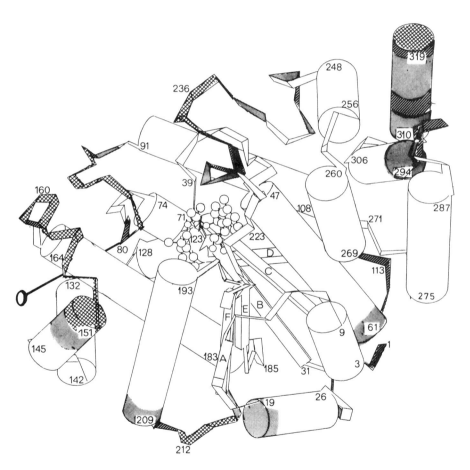

Fig.1 Schematic diagram of the main chain conformation of one monomer unit, highlighting disordered regions of chain. The thick line at the left represents the dimer axis. Stippled: B > 20 A, main chain conformation questionable; shaded: B > 27 A, main chain conformation doubtful; cross-hatched: B > 35 A, main chain conformation unknown. The inhibitor tyrosinyl adenylate is shown as a ball-and-stick model.

Figure 1 shows the structure. It has been refined from the measurable diffraction data, which extend to 2.1A. The intensities beyond 2.3Å are so near background that they contribute little to the precision of the structure determination, while significantly increasing the overall R-factor. At such resolution, the parts of the structure with $B > 30$ Å^2 are making negligible contribution to the Bragg reflections, and their interpretation could not in any case be improved by an extension of the resolution.

Figure 1 also shows a tyrosyl adenylate molecule, the activated amino-acid which is formed by the enzyme before the tyrosine moiety is transferred to a tRNA molecule (Rubin & Blow, 1981). It will be noted that most parts of the protein adjacent to the tyrosyl adenylate intermediate are well ordered. This may not surprise those with an entropic view of enzyme action, who believe that an enzyme whose structure is already ordered so as to accept a substrate will require less additional free energy to reach the transition state. But it allows many of the amino-acids closely involved in interaction with the intermediate to be identified, and to define their interactions with reasonable precision.

Figure 2 is a schematic diagram showing the polar interactions which the intermediate probably makes with the enzyme. We have been criticised by some purists for showing this diagram (and indeed some details of it have had to be changed as refinement progressed). They say that even if the approximate conformation of the enzyme and intermediate are known, there is insufficient information to assign hydrogen bonds. The strength of a hydrogen bond depends on its length and linearity. If the length is wrong by 0.3A, or the angle by 30°, incorrect conclusions will be drawn. This criticism would have been justified 20 years ago, when there were no other techniques to check the results of protein crystallography, and when every word of a structural paper had to be taken literally by an enzyme chemist. But in a protein engineering environment, the situation is different. Hypotheses can be put forward from structural data, and they can be checked by other techniques.

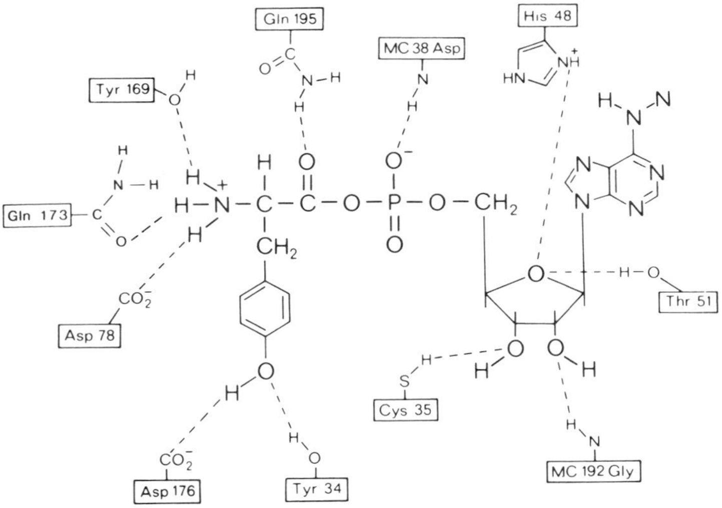

Fig.2 Close interactions of polar groups of the enzyme with polar groups of the tyrosyl adenylate intermediate, indicating possible hydrogen bonds.

Table 1   Sequence homologies between aminoacyl-tRNA synthetases

synthetase	species														
tyrosyl-	(B.stear.)[1]	34Y	C	G	F	D	P	T	A	D	S	L	H	I	G	H	L	A	T	I52
tyrosyl-	(E.coli)[2]	36Y	C	G	F	D	P	T	A	D	S	L	H	L	G	H	L	V	P	L54
methionyl-	(yeast)[3]	201T	S	A	L	P	Y	V	N	V	P	H	L	G	A	I	I	G	S219
methionyl-	(E.coli)[4]	11T	C	A	L	P	Y	N	N	A	S	I	H	L	G	H	M	L	E	H29
isoleucyl-	(E.coli)[5]	53H	D	G	P	Y	A	N	G	S	I	H	I	G	H	S	V	N	K71
glutaminyl-	(E.coli)[6]	30T	R	F	P	E	P	N	G	Y	L	H	I	G	H	A	K	S	I48

[1] Winter et al. (1983)
[2] Barker et al. (1982a)
[3] Walter et al. (1983)
[4] Barker et al. (1982b)
[5] Webster et al. (1984)
[6] Hoben et al. (1982).

ACAABP

By now all the side-chain interactions indicated in Fig. 2 have been the subject of protein engineering experiments. In many cases the energy attributable to a particular interaction has been measured by its effect on enzyme kinetic constants (Fersht et al., 1985). Mutants of Asp 38 and Asp 78 have so far failed to produce useful quantities of protein. Studies of Thr 51 mutants have suggested that in the wild-type the interaction of Thr 51 with the adenylate is very weak, and produces no binding energy (Wells & Fersht, 1985). In fact our latest results suggest lengths between 3.2 and 3.8 A for the distance O4' to O$^\gamma$(51), in maps for complexes of the enzyme with intermediate and inhibitor. The Thr 51 interaction is probably not a hydrogen bond.

A number of these interactions are with a part of the structure (residues 34-61) found to have a high structural homology with methionyl-tRNA synthetase (Blow et al., 1983). There is a significant sequence homology between the two enzymes, which becomes more convincing when extended to include the sequences of eukaryotic methionyl-tRNA synthetase and the isoleucyl- and glutaminyl- enzymes (Table 1).

MUTANT TP51

Although now discounted, the possible Thr 51 interaction highlighted an interesting point, which led to discovery of an enzyme with dramatically improved properties. The Ramachandran conformational angles of residues 46-60 all lie in the α-helical region : in loose terms, residues 46-60 form an α-helix (though $\phi(48)$ is low). But it was noted that in the E.coli enzyme, residue 51 is proline (Fig. 3), which of course cannot lie in the middle of a regular α-helix. While the crystallographers were pointing out that the amino-terminus of the helix is irregular, and the NH of Thr 51 is not aligned to make an α-helical hydrogen bond, the genetic engineers went ahead and made the TP51 mutant, which has proline at this position. In this mutant, $k_{cat}/K_M(ATP)$ is increased by 200 000, largely because of a decrease of $K_M(ATP)$ (Wilkinson et al., 1984). This corresponds to a reduction of the transition state free energy for adenylation of tyrosine by about 2 kcal/mol.

The geneticists also invented an interesting method for detecting interaction between mutation sites (Carter et al., 1984). The idea is that if two mutants each alter k_{cat}/K_M (that is, they each alter the transition state free energy of the catalysed reaction), then in the simplest case, the double mutant will have a value of k_{cat}/K_M corresponding to the sum of the two effects. If there is a substantial departure from additivity, this is interpreted as evidence of interaction between the effects of the two mutations. This would be expected, for example, if one mutation caused a structural change at the site of the second mutation.

In the case of mutation of Cys 35 and Thr 51, Carter et al. (1984) found the effects additive (within 0.4 kcal/mol). On the other hand, there was gross interaction between mutation at His 48 and at Thr 51. The mutant HG48 enzyme has a lower k_{cat}/K_M than the wild-type enzyme, and this k_{cat}/K_M is unchanged in the double mutant HG48TP51. In this case, the TP51 mutation does not cause a big increase in activity. This was taken to suggest a structural change affecting His 48 by the mutation TP51, such as would be expected if the α-helix linking them were distorted.

The structure of the TP51 mutant has now been determined at 2.5 A

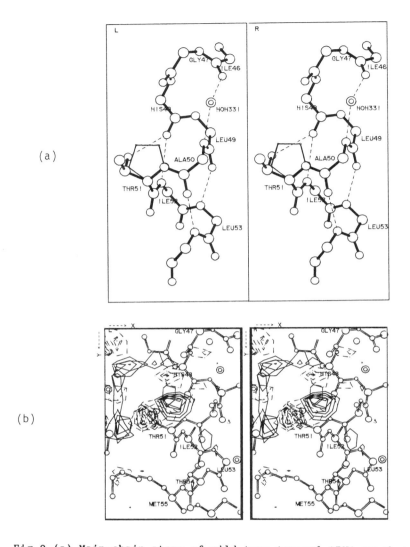

Fig 3 (a) Main-chain atoms of wild-type tyrosyl-tRNA synthetase, residues 46-54, with Thr51 side chain. The Pro51 side chain observed from the electron density of the TP51 mutant is shown in thinner bonds. Beginning with CO48...NH52, regular helical hydrogen bonds are made, but NH51 makes no hydrogen bond, and the CO46...NH50 interaction is through a well ordered water molecule. (b) Stereo view of the difference electron density for TP51, close to Thr51. Near Thr 51, positive density is observed for the C_γ and C_δ atoms of proline, with negative density close to the threonine side chain, on the side away from the C_δ and C_γ positions. The positive density at the left is part of the density for adenine of the tyrosinyl adenylate inhibitor, which was present in the TP51 crystals at low occupancy.

resolution. The crystals were isomorphous with the wild-type enzyme. The difference map shows the threonine being directly replaced by proline, without any significant structural change. As already noted, the threonine NH is not in any case able to form an α-helix hydrogen bond. The slightly distorted helix can accommodate the proline side chain without any other structural change.

THE TRUNCATED MUTANT

Serious difficulties were encountered in the refinement of the structure of tyrosyl-tRNA synthetase. Considerable improvements to the m.i.r. map by density modification techniques (Bhat & Blow, 1982, 1983) brought the interpretation to the point where structure-factor least-squares refinement can usefully begin. But there were still severe difficulties in making a satisfactory fit between a model and the electron density, mostly but not exclusively in regions of high protein temperature factor. The conventional R-factor was still above 30%.

Part of the difficulty was clearly due to the absence from the map (at a nominal 2.1Å resolution) of almost all density for residues 320-419. These residues may be disordered, with B factors of 60 $Å^2$ or more; or they may have a number of alternative conformations. In either case they barely contribute to the density (apart from a small 'island' of 1500$Å^3$ of positive density with no particular features).

Waye et al. (1983) used a technique called 'M13 splint deletion', exploiting two restriction enzyme sites at codons 317 and 417 to delete 100 codons from the gene at the carboxyl terminus. Expression of this gene in E.coli produced a 'truncated' enzyme of 319 amino acids, corresponding to 1-317,418,419 of the wild-type enzyme.

The truncated enzyme was found to crystallise more readily than the wild-type enzyme, and P.Brick undertook its structure analysis. It was hoped that the deletion of 100 disordered amino-acids would yield less noisy diffraction data, leading to a more accurate molecular model.

Crystals were obtained from the truncated enzyme at neutral pH in the presence of tyrosine using polyethylene glycol as precipitant. The space group is $P2_1$, with cell dimensions a = 95.4Å, b = 67.1Å, c = 61.4Å, β = 90.78°. The crystallographic asymmetric unit contains both subunits of the dimeric molecule. Data were collected to 2.5Å resolution.

The orientation and position of the truncated molecule in its unit cell were determined by molecular replacement techniques. The search object comprised the 270 better ordered amino-acids from the refined wild-type structure. Restrained structure-factor least-squares refinement (Konnert & Hendrickson, 1980) with alternating cycles of model building lowered the R-factor to 21%.

The new molecular model has greatly improved stereochemistry and it has been possible to identify about 50 water molecules in each subunit. Most of these are situated in and around the active site cleft. The temperature factors are generally higher than those of the wild-type enzyme, and residues that had high temperature factors in the wild-type model generally retain high temperature factors in the model of the truncated enzyme. In the truncated enzyme there is clear side-chain density for some less ordered residues, providing a more reliable model for these.

The truncated enzyme structure, providing two independent new views of the monomer, has allowed significant improvements to be made in the wild-type model. The coordinates for the one subunit of the truncated enzyme were then transformed back into the unit cell of the wild-type crystals and refined. The resulting coordinates gave a new set of phases, providing electron-density and difference maps for further model-building. The rebuilt structure has now refined much better to an R-factor of 23%.

It had already been demonstrated that a fragment of a protein may provide a molecule more amenable to crystallization and structure determination (Waller et al., 1971; Zelwer et al., 1982; Joyce & Grindley, 1983; Ollis et al., 1985). In this case it has been used to improve the accuracy of structural information about the intact molecule, which will be valuable in the design and structure analysis of other factitious enzymes.

FUTURE PROSPECTS

It is difficult to overestimate the potential of protein engineering. Materials technology is entering a new era. Since life started, until the 1980's, the development of protein molecules has proceeded by evolution. The results have been phenomenal : examples from this meeting are the development of the light-harvesting proteins and the photoreaction centre with their exquisitely-tuned spectral sensitivities and rate constants; or the continuing battle between the immune system and the bacterial and viral infective agents.

Now our ability to create new proteins is constrained only by our limited imagination and the crudity of our techniques of design. In design, at least, we shall rapidly gain experience.

The role of protein crystallography is central, but is limited by the slow rate of producing results. Ideally, the structure of every factitious protein will be checked by experiment. In practice this means expressing the protein on the milligram scale, purifying it, crystallising it, and solving its structure. Neither the expression, the purification, nor the crystallisation can be totally routine. Perhaps no more than a small proportion of factitious proteins will ever be studied at high resolution, but it is certain that more manpower and more equipment are required for the crystallographic work.

REFERENCES

Barker, D.G., Bruton, C.J. & Winter, G. 1982a, FEBS J. 150 : 419-423.
Barker, D.G., Ebel, J-P., Jakes, R. & Bruton, C.J. 1982b . Eur. J. Biochem. 127 : 449-457.
Bhat, T.N. & Blow, D.M. 1982, Acta Cryst. A38 : 21-29.
Bhat, T.N., Blow, D.M., Brick, P. & Nyborg, J. 1982, J. Mol. Biol. 158: 699-709.
Bhat, T.N. & Blow, D.M. 1983, Acta Cryst. A39 : 166-170.
Blow, D.M., Bhat, T.N., Metcalfe, A., Risler, J.L., Brunie, S. & Zelwer, C. 1983, J. Mol. Biol. 171 : 571-576.
Blow, D.M. & Brick, P. 1985, Aminoacyl-tRNA Synthetases, in: "Biological Macromolecules and Assemblies. Vol. II", F.A. Jurnak & A. McPherson, ed., Wiley, New York.
Carter, P.J., Winter, G., Wilkinson, A.J. & Fersht, A.R. 1984, Cell 38 : 835-840.

Fersht, A.R., Shi, J-P., Knill-Jones, J., Lowe, D.M., Wilkinson, A.J., Blow, D.M., Brick, P., Carter, P., Waye, M.M.Y. & Winter, G. 1985, Nature 314 :235-238.
Hoben, P., Royal, N., Cheung, A., Yamao, F., Biemann, K. & Soll, D. 1982, J. Biol. Chem. 257 :11644-11650.
Joyce, C.M. & Grindley, N.D.F. 1983, Proc. Natl. Acad. Sci. 80 : 1830-1834.
Konnert, J.H. & Hendrickson, W.A. (1980), Acta Cryst. A36 : 344-350.
Ollis, D.L., Brick, P., Hamlin, R., Xuons, N.G. & Steitz, T.A. 1985 Nature 313 :762-766.
Rubin, J. & Blow, D.M. 1981, J. Mol. Biol. 145 : 489-500.
Waller, J.P., Risler, J.L., Monteilhet, C. & Zelwer, C. 1971, FEBS Letters, 16 :186-188.
Walter, P., Gangloff, J., Bonnet, J., Boulanger, Y., Ebel, J-P. & Fasiolo, F. 1983, Proc. Nat. Acad. Sci., 80 : 2437-2441.
Waye, M.M.Y., Winter, G., Wilkinson, A.J. & Fersht, A.R. 1983, EMBO J. 2 :1827-1829.
Webster, T., Tsai, H., Kula, M., Mackie, G.A. & Schimmel, P. 1984, Science 226 :1315-1317.
Wells, T.N.C. & Fersht, A.R. 1985, Nature 316 :656-657.
Wilkinson, A.J., Fersht, A.R., Blow, D.M., Carter, P. & Winter, G. 1984, Nature 307 :187-188.
Winter, G., Koch, G.L.E., Hartley, B.S. & Barker, D.G. 1983, Eur. J. Biochem. 132 : 383-387.
Zelwer, C., Risler, J.L. & Brunie, S. 1982, J. Mol. Biol. 155 : 63-81.

SELECTED AND DIRECTED MUTANTS OF T4 PHAGE LYSOZYME

Tom Alber, Terry M. Gray, Larry H. Weaver, Jeffrey A. Bell,
Joan A. Wozniak, Sun Daopin, Keith Wilson, Sean P. Cook,
Edward N. Baker and Brian W. Matthews

Institute of Molecular Biology and Department of Physics
University of Oregon
Eugene, Oregon 97403

The lysozyme from bacteriophage T4 is being used as a model system to determine the factors that influence the folding and stability of proteins. The three-dimensional structure of the protein is known and lysozymes with modified properties arising from single amino acid substitutions have been obtained by classical selection techniques as well as by site-directed mutagenesis. By rationalizing the stabilities of the mutant proteins in terms of their observed three-dimensional structures we are attempting to quantitate the contributions that single amino acids make to protein stability. Our studies to date lead to the following conclusions. (1) Thermal stability is global. (2) Different types of interaction contribute to stability. (3) Changes in stability need not be associated with large structural changes. (4) An increase in the stability of an α-helix may enhance the stability of the protein as a whole. (5) Hydrogen bonds contribute to stability. (6) Protein structures can readily adjust to changes in amino acid sequence.

INTRODUCTION

Experimental studies of protein stability do not, in general, provide a method of distinguishing the individual contribution of each amino acid in its unique environment in the native structure. We are attempting to tackle this problem by comparing the structures and stabilities of mutants of phage T4 lysozyme that differ in sequence by a single amino acid. The mutants are of two distinct classes: selected and directed. Selected mutants are identified after "random" chemical mutagenesis using genetic screens to detect lysozymes that are temperature-sensitive[1] or have enhanced thermal stability[2]. Directed mutants are obtained by oligonucleotide-directed mutagenesis, which allows the planned substitutions of selected residues with any of the naturally-occurring amino acids.

This is a report of work in progress and, as such, follows closely the report given in Reference 3.

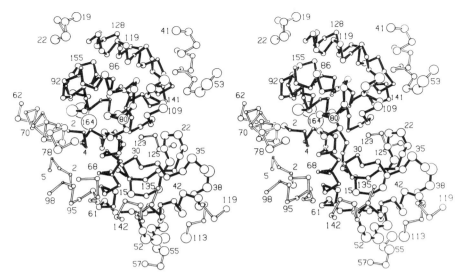

Figure 1. Stereo drawing of the backbone of T4 phage lysozyme in its crystallographic environment. The circle at each alpha carbon position has a diameter proportional to the refined crystallographic "thermal factor". One complete lysozyme molecule is drawn with solid bonds. Segments of neighboring molecules that surround the reference molecule and make contact with it are drawn with open bonds.

STRUCTURE AND DYNAMICS OF WILD-TYPE LYSOZYME

The structure of the lysozyme from bacteriophage T4 has been determined crystallographically and refined at 1.7Å resolution[3-5].

The crystallographic refinement gives information concerning the molecular dynamics of the structure in the crystal, and, by inference, in solution as well. Figure 1 shows the apparent "thermal displacements" of the backbone of the molecule. Analysis of these individual motions indicates that the lower lobe in the figure undergoes a pronounced "hinge-bending" motion that opens and closes the active site cleft[5]. X-ray analysis of the binding of mono-, di- and trisaccharides to T4 lysozyme originally suggested the necessity of such motion for access of substrate to the "closed" active site cleft observed in the crystal structure[6].

TEMPERATURE-SENSITIVE LYSOZYME MUTANTS

Using classical genetic methods developed by Streisinger and coworkers[1] we have obtained a number of temperature-sensitive (ts) lysozymes[7-10]. Such ts mutants reveal amino acid substitutions that alter the stability of the lysozyme molecule yet still allow the protein to fold and be active at the permissive temperature. The mutant proteins might lack interactions that stabilize the wild-type protein or, alternatively, the altered amino acid might introduce new interactions that destabilize the folded form. As discussed below, the sites of these ts mutations are good candidates for further analysis by site-directed mutagenesis. The temperature-sensitive lysozymes that we have studied in most detail are listed in Table 1. The thermodynamics of unfolding of some of these lysozymes have been determined by J. Schellman and his coworkers[11].

Table 1. Some temperature-sensitive mutant lysozymes studied crystallographically

Lysozyme	Origin	Activity(a)	ΔT_m(b)	Resolution(c)	R(d)
Wild type	--	100	--	1.7	19.6
Arg 96 → His	ts	100	-14	1.9	19.5
Met 102 → Thr	ts	60	-13	2.1	18.1
Ala 146 → Thr	ts	55	-9	2.1	19.0
Gly 156 → Asp	ts	50	-5	1.7	17.8
Thr 157 → Ala	ts	90	-13	1.9	18.9

(a) Rate of hydrolysis of cell walls relative to wild-type.
(b) Melting temperature (°C) relative to wild-type, measured at pH 3. [From Refs. (10), (11) and personal communication from W.A. Baase and W.J. Becktel.]
(c) Resolution of crystallographic data (Å).
(d) Crystallographic residual at the present stage of refinement.

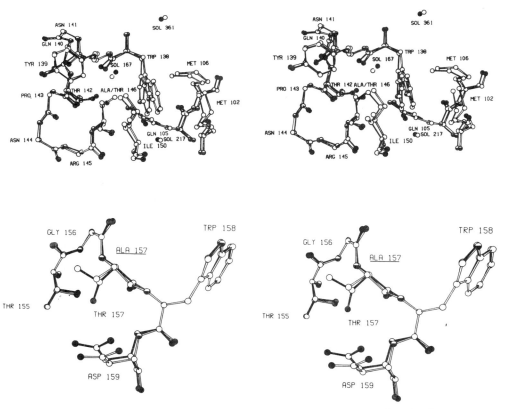

Figure 2. Stereo drawings showing the structures of mutant lysozymes (open bonds) superimposed on the wild-type structure (solid bonds) in the vicinity of the amino acid substitution. (a) Temperature sensitive mutant (Ala 157) and wild-type (Thr 157). (b) Temperature sensitive mutant (Thr 146) and wild-type (Ala 146).

Structural changes in two of the mutant lysozyme structures that are associated with temperature-sensitive mutations are shown in Figures 2(a) and 2(b). These figures are based on refined crystallographic coordinates. For the surface substitution Thr 157 → Ala there is very little change in the protein structure associated with the mutation. The largest movement is in the adjacent side-chain of Asp 159 which, in the mutant structure, moves approximately 0.6Å away from the altered amino acid (Fig. 2(a)). In the case of the internal substitution Ala 146 → Thr, the larger side-chain in the mutant structure causes neighboring groups to move outwards to make room for the additional atoms. The indole of Trp 138 moves about 0.7Å. This shift causes adjustments in the side-chains of Met 102 and Met 106 which in turn largely fill a cavity that exists in the wild-type structure.

In general, our structural studies of ts mutant lysozymes are consistent with the view that different amino acids can contribute to stability via a variety of interactions including hydrophobic contacts, hydrogen bonding and ionic interactions. As discussed in the following section, site-directed mutagenesis can be used to help differentiate between these different contributions.

SITE-DIRECTED MUTAGENESIS

Because of the structures of the natural amino acids, the ts substitution Thr 157 → Ala is especially susceptible to analysis using site-directed mutagenesis. Thr 157 is located in a bend between two α-helices and is partly buried and partly exposed to solvent. The substitution Thr 157 → Ala was identified as a temperature-sensitive lysozyme with melting temperature (at pH 2) 13° lower than that of wild-type. The consequences of this substitution are shown in detail in Figure 2(a) and, in schematic form in Figure 3.

Several factors appear to underly the decrease in stability. These include loss of van der Waals interactions of both the γ-carbon and the γ-hydroxyl, loss of hydrogen bonds (to Thr 155 and from the peptide NH of Asp 159) and, possibly, changes in the structure of the surrounding solvent. By making a series of substitutions at position 157 (and at position 155 as well), we are attempting to discriminate between these contributing factors. For example, substitution with serine at position 157 should reveal the contribution of the hydrogen bonds to stability. Substitution with valine (or isoleucine) should show the importance of hydrophobic interactions, and so on. Of course one must also consider that these substitutions may change the free energy of the unfolded form of the protein.

By using a degenerate mixture of DNA primers it is possible to generate a number of different amino acid substitutions in a single experiment[10,12]. Production of large quantities of the mutant lysozymes was facilitated by subcloning the genes into an E. coli expression vector based on the tac and lac UV5 promoters[13]. The substitutions at position 157 obtained to date include Ser, Leu, Ile, Asn, Asp, His, Gly, Glu and Arg (as well as Ala 157, the original ts mutant). Additional substitutions are in progress. Oscillation photography[14,15] was used to measure diffraction data to 1.7Å resolution for the mutant lysozymes. Sections of the difference density maps at 2.0Å resolution for the two substitutions Thr 157 → Arg and Thr 157 → Asn are shown in Figure 4.

For the Thr 157 → Arg substitution the loss of the γ-hydroxyl of the threonine is clearly shown by the negative peak at this position.

Table 2. Relative stabilities of lysozymes with different substitutions at position 157

Amino acid at position 157	Melting temperature (°C ± 1°, measured at pH 2) relative to wild-type
Threonine (wild-type)	0
Asparagine	-2
Aspartic Acid	-2
Serine	-3
Glycine	-3
Leucine	-4
Glutamic Acid	-4.5
Arginine	-5
Histidine	-6
Isoleucine	-9
Alanine (Ts mutant)	-13

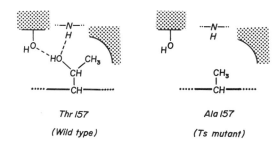

Figure 3. Schematic drawing showing some of the interactions of threonine 157 in wild-type lysozyme and the consequences of replacing the threonine by alanine in the temperature-sensitive mutant lysozyme.

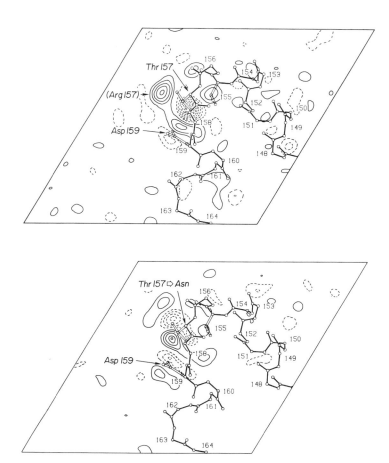

Figure 4. The figure shows a single section taken from 2.0Å resolution difference electron density maps calculated for two site-directed amino acid substitutions. Superimposed on the difference electron density is the trace of the polypeptide backbone that lies near to the section. Except for Thr 155, Thr 157 and Asp 159 (open bonds), all protein side-chains have been omitted. (Residues 148-155 form part of one α-helix and residues 159-164 are in another helix.) The maps are calculated with amplitudes ($F_{mutant} - F_{wild-type}$) and phases from the refined wild-type structure. Positive contours (solid) and negative contours (broken) are drawn at equal increments of 2σ where σ is the standard deviation of the difference density throughout the unit cell. (a) Arg 157 minus Thr 157. (b) Asn 157 minus Thr 157.

The γ-carbon of the arginine side-chain occupies essentially the same position as the γ-carbon of the threonine and the distal end of the arginine side-chain then curves around so as to form a salt linkage with Asp 159. The movement of this acid group toward the guanidinium of the arginine can be seen in Figure 4(a).

When Thr 157 is replaced by Asn, the oxygen of the Asn side chain occupies a position similar to that of the γ-hydroxyl of the threonine, and appears to form essentially the same hydrogen bond network as present in the wild-type structure. This is achieved in spite of the difference in geometry of the threonine and asparagine side chains. In contrast to the Thr 157 → Arg substitution, the Thr 157 → Asn replacement causes Asp 159 to move away from the altered amino acid (Fig. 4(b)).

The relative stabilities of ten mutant lysozymes with substitutions at position 157 are given in Table 2. All substitutions at position 157 isolated to date decrease the stability of the protein. The original ts mutant, Thr 157 → Ala, has significantly lower stability than that resulting from any other substitution at position 157.

CONCLUSIONS

The analysis of the mutant lysozymes described here has a number of implications for protein stability in general.

1. <u>Thermal stability is global</u> The temperature-sensitive mutations that we have characterized to date occur in separated locations. There is not one highly localized region with which temperature sensitivity is associated. The observations are consistent with the hypothesis that protein stability is global, with individual amino acids contributing additively to the overall stability of the protein.

2. <u>Local environment is important to stability</u> In comparing the amino acid substitutions that occur in the different temperature-sensitive mutants, no simple pattern emerges (Table 1). In different instances one sees changes in charge, hydrophobicity, solvent structure, packing and hydrogen bonding. Perhaps the most striking example is provided by the mutant lysozymes Thr 157 → Ala and Ala 146 → Thr. These are inverse substitutions that would be expected to have opposite effects on the free energy of the unfolded state, yet both reduce the thermostability of the folded protein. Clearly the consequence of these mutations depends on the environment and interactions of the substituted amino acid within the folded protein and is not solely due to an effect on the stability of the unfolded state.

3. <u>Different types of interaction contribute to protein stability</u>
The variety of amino acid substitutions in different temperature-sensitive mutants shows that different effects including hydrophobic interactions, electrostatic interactions, packing defects, solvent structure and hydrogen bonding all influence stability. Some substitutions no doubt destabilize the protein by differentially stabilizing the unfolded form.

The substitution Thr 157 → Ala seems to destabilize through loss of hydrogen bonds involving O^γ and also by the loss of van der Waals interactions with O^γ and C^γ of the threonine. The alteration Ala 146 → Thr disrupts the packing within the molecule and introduces a new hydrogen bonding group that may not be easily accommodated by the wild-type structure. Replacement of Arg 96 by histidine causes a net loss of intramolecular hydrogen bonds. We do not observe that temperature sensitivity is necessarily associated with the loss of ion pairs (<u>c.f.</u> Ref. 16).

In most cases the ts lysozymes have either less hydrophobic surface buried or more hydrophobic surface exposed to solvent, or a combination of both.

4. **Changes in stability need not be associated with large structural changes** One of the striking results of this study is the close correspondence between the three-dimensional structures of the different temperature-sensitive mutants and the wild-type structure. For mutants Arg 96 → His and Thr 157 → Ala the amino acid substitutions are on the surface of the protein and the structural changes are quite localized. In the case of the internal substitution Ala 146 → Thr there are changes that propagate through neighboring residues, but such changes are, in general, not more than 0.5-0.8Å (Fig. 2(b)). There is no evidence that temperature sensitivity is necessarily associated with an obvious, dramatic, change in the structure of the protein.

However it should be noted that the crystallographic analyses (to date) are at room temperature, well below the melting temperature of wild-type lysozyme (54°C at pH 3). At room temperature we do not see any obvious changes in thermal fluctuations or "looseness" of the mutant structures relative to the wild-type structure, as has been inferred from the compensating changes that are observed in the entropies and enthalpies of unfolding[11].

5. **Stability of secondary structure** One way of stabilizing proteins could be to stabilize the folded conformations of individual α-helices and β-sheets, and perhaps turns as well (e.g. see Ref. 17). For all temperature-sensitive mutations within helices the helical propensity[17-20] either decreases or is ambiguous, whereas for the one example of a substitution not in a helix the helical propensity increases in the mutant. This supports the idea that the stabilization of individual helices tends to increase the stability of the protein as a whole.

6. **Hydrogen bonds contribute to protein stability** The finding that the γ-OH of Ser 157 is sufficient to raise the T_m of T4 lysozyme by 10°C relative to Ala at this position (Table 2) suggests that the hydrogen bond network shown in Figure 3(a) contributes more to the stability of the protein than does the van der Waals interactions of the γ-methyl group. In addition, crystallographic analysis of the Thr 157 → Asn mutant shows that the oxygen of the asparagine side-chain occupies a position close to that of the threonine hydroxyl in wild type lysozyme and forms similar hydrogen bonds. As with the serine substitution, preservation of the hydrogen bond network correlates with the retention of stability. The original temperature sensitive mutant, with an alanine at position 157, has two unsatisfied hydrogen-bonding groups in the folded protein (Fig. 3) and is the least stable of all the lysozymes with substitutions at residue 157.

In the case of Thr 157 → Arg (Fig. 4(a)), the arginine side-chain loses van der Waals and/or hydrogen bonding interactions because it has no counterpart to the γ-oxygen of the wild-type threonine. However it appears to compensate, in part, for this loss by the formation of a new salt bridge with Asp 159 (Fig. 4(a)). Although this salt bridge contains two hydrogen bonds between the side-chain carboxyl and guanidinium moieties, it does not stabilize the protein to the same extent as the interactions of the wild-type threonine 157 side chain. Thus, different hydrogen bonds make different contributions to protein stability.

An interesting situation occurs for the substitution of Thr 157 with glycine. Since the alanine substitution at position 157 is destabilizing, it might be expected that substitution with glycine would have an even larger effect. In fact, the observed effect is just the opposite (Table 2). We have suggested[21] that the absence of a side chain, in the case of glycine may allow a bound water molecule to stabilize the protein by restoring the hydrogen bond network shown in Figure 3. Very recent crystallographic studies show that this is indeed the case. Thus, protein structures can also be stabilized by bound water molecules.

7. <u>Protein evolution</u> All of the substitutions made to date at position 157 have stabilities between that of the wild-type protein and the original temperature-sensitive mutant. The observation that no substitution is more stable than threonine suggests that in this case the wild-type protein has achieved optimum stability at this site. However we believe it unlikely that proteins have evolved to have maximum stability. Indeed, it is possible to select lysozyme mutants that are more thermostable than the wild-type protein[2].

One of the striking results of the site-directed mutagenesis is that most of the mutant lysozymes have stabilities closer to the wild-type (Thr 157) protein than to the temperature-sensitive mutant (Ala 157). In addition, amino acids that give rise to the most stable proteins need not be chemically or structurally similar. The protein must compensate in different ways for these chemically disparate substitutions. This suggests that proteins may have evolved to accommodate amino acid substitutions that may occur during evolution. Protein structures may be more resistant to destabilization by amino acid substitution than has been imagined, and mutations that lead to temperature sensitivity could be the exception rather than the rule.

ACKNOWLEDGEMENTS

This work was supported in part by grants from the National Institutes of Health (GM 21967; GM 20066), the National Science Foundation (PCM 8312151) and the Murdock Charitable Trust.

REFERENCES

1. G. Streisinger, F. Mukai, W. J. Dreyer, B. Miller, and S. Horiuchi, Cold Spring Harbor Symp. Quant. Biol. 26 : 25 (1961).
2. T. Alber and J. A. Wozniak, <u>Proc. Natl. Acad. Sci.</u> U.S.A. 82 : 747 (1985).
3. T. Alber, T. M. Gray, L. H. Weaver, J. Bell, J. A. Wozniak, K. Wilson, S. Daopin, and B. W. Matthews, P&S Biomedical Sciences Symposia. "Biological Organization: Macromolecular Interactions at High Resolution". R.M. Burnett and H.J. Vogel, eds. In press (1985).
4. S. J. Remington, W. F. Anderson, J. Owen, L. F. Ten Eyck, C. T. Grainger, and B. W. Matthews, J. Mol. Biol. 118 : 81 (1978).
5. L. H. Weaver and B. W. Matthews, <u>J. Mol. Biol.</u>, submitted.
6. W. F. Anderson, M. G. Grütter, S. J. Remington, L. H. Weaver, and B. W. Matthews, <u>J. Mol. Biol.</u> 147 : 523 (1981).
7. M. G. Grütter, R. B. Hawkes, and B.W. Matthews, <u>Nature</u> 277 : 667 (1979).

8. M. G. Grütter, L. H. Weaver, T. Gray, and B. W. Matthews, In "Bacteriophage T4" (C. K. Matthews, E. Kutter, G. Mosig, P. Berget, eds.), pp. 356-360. American Society for Microbiology, Washington, D.C. (1983).
9. M. G. Grütter, L. H. Weaver, T. Alber, T. M. Gray, and B. W. Matthews, Manuscript in preparation.
10. T. Alber, M. G. Grütter, T. M. Gray, J. A. Wozniak, L. H. Weaver, B-L. Chen, E. N. Baker, and B. W. Matthews, UCLA Symposium on Molecular and Cellular Biology, Protein Structure, Folding and Design. Alan R. Liss, in press (1985).
11. R. Hawkes, R. Grütter, and J. Schellman, J. Mol. Biol. 175 : 195 (1984).
12. T. Alber, D. Muchmore, and B. W. Matthews, Manuscript in preparation.
13. D. C. Muchmore, C. B. Russell, and F. W. Dahlquist, Manuscript in preparation.
14. M. G. Rossmann, J. Appl. Crystallogr. 12 : 225 (1979).
15. M. F. Schmid, L. H. Weaver, M. A. Holmes, M. G. Grütter, D. H. Ohlendorf, R. A. Reynolds, S. J. Remington, and B. W. Matthews, Acta Crystallogr. A37 : 701 (1981).
16. M. G. Perutz and H. Raidt, Nature 255, 256 (1975).
17. P. Argos, M. G. Rossmann, U. M. Grau, H. Zuber, G. Frank, J. D. Tratschin, Biochemistry 18 : 5698 (1979).
18. P. Y. Chou and G. D. Fasman, Ann. Rev. Biochem. 47 : 251 (1978).
19. M. Levitt, Biochemistry 17 : 4277 (1978).
20. M. Sueki, S. Lee, S. P. Powers, J. B. Denton, Y. Konishi, and H. A. Scheraga, Macromolecules 17 : 148 (1984).
21. T. Alber and B. W. Matthews, In Protein Modification and Design, Alan R. Liss, Inc. In press.

6 VIRUSES

THE STRUCTURE OF A HUMAN COMMON COLD VIRUS (RHINOVIRUS 14) AND ITS FUNCTIONAL RELATIONS TO OTHER PICORNAVIRUSES

Michael G. Rossmann[1], Edward Arnold[1], John W. Erickson[1a], Elizabeth A. Frankenberger[1b], James P. Griffith[1], Hans-Jürgen Hecht[1c], John E. Johnson[1], Greg Kamer[1], Ming Luo[1], Anne G. Mosser[2], Roland R. Rueckert[2], Barbara Sherry[2] and Gerrit Vriend[1]

[1]Department of Biological Sciences, Purdue University, W. Lafayette, Indiana 47907; [2]Biophysics Lab, University of Wisconsin, 1525 Linden Drive, Madison, Wisconsin 53706

INTRODUCTION

Picornaviruses are associated with serious diseases in humans and other animals, and they comprise one of the largest families of viral pathogens. For example, the common cold, poliomyelitis, foot-and-mouth disease and hepatitis can be caused by these viruses. They are among the smallest RNA-containing animal viruses (1-3). Their molecular weight is around 8.5×10^6 and they contain about 30% by weight RNA. Their external diameter is roughly 300 Å and they form icosahedral shells. Picornaviridae have been subdivided into four genera on the basis of their buoyant density, pH stability and sedimentation coefficients: enterovirus (e.g. polio, hepatitis A and coxsackie viruses), cardiovirus (e.g. encephalomyocarditis and Mengo viruses), aphthovirus (e.g. foot-and-mouth disease virus) and rhinovirus. They differ also in the number of known serotypes. For instance there are three known serotypes for polioviruses, seven for foot-and-mouth disease viruses (FMDV) and at least 89 for human rhinoviruses (HRV). Accordingly, it has been possible to produce effective vaccines for poliomyelitis and, with greater difficulty, for foot-and-mouth disease, but not for the common cold.

Picornavirions contain 60 protomers (4), each composed of four structural proteins VP1, VP2, VP3 and VP4 corresponding to genes 1D, 1B, 1C and 1A, respectively (for nomenclature see 5). Their molecular weights in HRV14 are 32,000, 29,000, 26,000 and 7,000. The capsid protein VP0, corresponding to gene 1AB, is cleaved into its components VP4 and VP2 only in the final stages of assembly (4,6,7). The cleavage occurs at an Asn-Ser peptide in HRV14, and hence is not effected by the viral protease 3C whose specificity is for Gln-Gly peptides. The assembly of the capsid occurs in a series of steps culminating in the insertion of RNA into capsids to produce

---

Present Addresses: [a]Department of Physical Biochemistry, AP-9A D-47E, Abbott Laboratories, Abbott Park, North Chicago, Illinois 60064; [b]Department of Agronomy, Purdue University, W. Lafayette, Indiana 47907; [c]F. G. Roentgenstrukturanalyse, Universitaet Wuerzburg, Zentralbau Chemie, Am Hubland, D-8700 Wuerzburg, West Germany.

mature virions with the concomitant cleavage of VP0 (3,8-11).

The virions contain a single, positive strand of RNA with a protein, VPg, covalently attached to the 5' end (12-17). The RNA, containing one long open reading frame, is translated into a polyprotein which is processed into its component proteins in a series of steps (18). The gene order is essentially the same for HRV, FMDV, poliovirus and encephalomyocarditis virus (EMCV). The initial cleavage of capsid precursor protein from the polyprotein is probably due to a host cell protease, but subsequent cleavages are mostly dependent on the release of viral protease excised from the polypeptide (19).

Considerable effort has been devoted to mapping topological relationships among VP1, VP2, VP3 and VP4 within the capsid (cf. 1) using chemical labelling of the surface of intact particles (20-22), treatment with cross-linking reagents (23-25), reaction with specific antibodies and cross-linking with UV light (26). The consensus is that VP1 is the most external and immuno-dominant protein, while VP4 is inaccessible from the outside but can be cross-linked with RNA on the interior. Heat treatment or mild denaturing agents cause a conformational change to the capsid, thus altering the response to antisera. Surprisingly, the internal capsid protein VP4 can dissociate and escape from the capsid during antigenic conversion (3).

The RNA, and hence by inference the polyprotein, has been sequenced for all three strains of poliovirus (27-30), for various FMDV strains (31-34), for two strains of rhinovirus (35-37), for EMCV (38) and for hepatitis A virus (39,40). Protein-to-protein comparisons between HRV, poliovirus, EMCV, FMDV and hepatitis A virus sequences show that HRV and poliovirus are closely related, whereas sequence homology of HRV to EMCV, FMDV or hepatitis A virus is not immediately obvious, particularly for the structural proteins. Homologies for non-structural proteins between HRV and poliovirus range from 44% in the protease gene 3C to 65% in the RNA polymerase gene 3D (35-37). Comparison of the structural proteins of HRV14 with poliovirus shows a 60% homology for VP4 and VP2 while VP3 and VP1 show only 47% and 44% conservation of amino acids. The polymerase, which is among the most conserved picornavirus proteins, has been shown to be homologous to the polymerase of the cowpea mosaic plant virus (41,42) as well as those of other plant and animal viruses (43-46).

An animal's immunological response to a virus is one of the major defenses against disease. Antibodies can bind to viruses but they do not necessarily neutralize infectivity (cf. 47). In spite of the 60-fold equivalence of each potential binding site on the virus, as few as four neutralizing antibodies per virion can be sufficient to inhibit infectivity of poliovirus (48). Neutralizing antibodies usually change the isoelectric point of the picornavirions (49,50), indicating that a conformational change frequently accompanies neutralization. Antibodies may neutralize by interfering with cell attachment, membrane penetration or virus uncoating (51,52). Antibodies that bind to poliovirus may require bivalent attachment for neutralization of the virus (53,48). Extensive studies have been reported (see below) on mapping the antigenic surfaces of HRV14 (54,55), polio and FDMV. Four major immunogenic sites have been identified for HRV14. One of the HRV14 immunogenic sites on VP1 coincides with the dominant immunogen of polio.

In spite of the sequence and surface similarities of picornaviruses, they have different host and tissue specificity. Abraham and Colonno (56,57) have shown, using 24 rhinovirus serotypes in competition binding assays, that, while the majority recognize one receptor, a second smaller group recognizes a different receptor. HRV14 (sequenced by Callahan et al.

(36) and by Stanway et al. (35)) belongs to the larger receptor group while HRV2 (sequenced by Skern et al. (37)) belongs to the smaller one. Minor et al. (58) have been able to produce monoclonal antibodies that block cellular receptors of poliovirus as have Colonno and co-workers for the large rhinovirus receptor group (57). Krah and Crowell (59) have characterized some properties of HeLa cell receptors for group B coxsackieviruses. They found that concanavalin A and other lectins adsorbed to receptors and inhibited virus attachment, a finding similar to that of Lonberg-Holm (60).

X-ray diffraction studies of crystalline picornaviruses have been limited. Coxsackievirus crystals (61) and poliovirus crystals (62) were reported a long time ago and rhinovirus strain 1A crystals a little later (63). Preliminary examination of the polio crystals (64) showed the particles to possess icosahedral symmetry. However, the technical problems involved in a complete structure determination were not solved until the elucidation of the small plant RNA viruses tomato bushy stunt virus (TBSV, 65), southern bean mosaic virus (SBMV, 66) and satellite tobacco necrosis virus (STNV, 67). This encouraged the renewal of the crystallographic study of poliovirus (68), and stimulated work on rhinovirus (69,70) as well as Mengo virus (71). It was shown that rhino and poliovirus crystals can be roughly isomorphous, suggesting close similarities of the viral capsids, a result later supported by sequence homologies. Comparison of the X-ray diffraction patterns of rhino and Mengo virus crystals also indicated significant structural homologies (71).

THE STRUCTURE DETERMINATION

HRV14 crystals were prepared as described by Arnold et al. (70). The crystals are cubic with $a = 445.1$ Å belonging to space group $P2_13$. There are four particles per crystal cell with each virion situated on a crystallographic threefold axis along the body diagonal. Thus, one-third of the virus, or 20 icosahedral asymmetric units, is in the crystallographic asymmetric unit. Packing considerations showed that the particle had to be near (0,0,0) or (1/4,1/4,1/4) given the choice of origin as defined in International Tables A (72).

Initial 5 Å resolution data were collected on rotating anode X-ray generators and these were used to compute a rotation function (73) which determined the particle orientation in the cell (70). However, the crystals suffered appreciable radiation damage and high resolution data were not practicably attainable. This was solved by using X-ray radiation from a synchrotron source where $0.3°$ oscillation photographs extended to 2.6 Å resolution. Some experiments were performed at the DESY synchrotron in Hamburg, but the data used for the results given here were collected at CHESS (Cornell High Energy Synchrotron Source). A new crystal was used for every exposure. A total of 83 film packs were eventually included in the native data with an R-factor on intensities of 11.0% for $F^2 > 1\sigma$. Surveys for suitable isomorphous heavy atom derivatives eventually produced two related compounds, namely 1 mM $KAu(CN)_2$ and 5 mM $KAu(CN)_2$. Full high resolution data were collected of the former, but only partial low resolution ($0.6°$ to $0.8°$ oscillation angles) of the latter.

The mean differences between heavy atom derivative and native data sets were not encouraging. Not only were the differences rather small in proportion to the size of the native amplitudes (8% for 1 mM $KAu(CN)_2$ and 12% for 5 mM $KAu(CN)_2$) but also the size of the differences increased at a resolution greater than 7 Å. Thus, there appeared to be little substitution and some concern about lack of isomorphism, particularly at higher resolution.

In spite of the lack of confidence in the available heavy atom derivative data, systematic search procedures (74,75) were applied to 6 Å difference Pattersons. The first three-dimensional searches were based on locating the self-Patterson vectors within an icosahedral virion given its known orientation. Trial heavy atom sites were used systematically to explore all possibilities in an icosahedral asymmetric unit between 80 and 160 Å radius. Plausible sites were then explored further using partial four-dimensional searches in which the virion position was slightly varied and cross-Patterson vectors between particles were also considered. The results showed clearly a single site, A, on the icosahedral threefold axes at a radius of 111.0 Å for the 1 mM $KAu(CN)_2$ compound and a possible substitution at the same site in the 5 mM $KAu(CN)_2$ derivative. The best particle position was found to be at $\underline{x} = \underline{y} = \underline{z} = 0.0006$. Three-dimensional searches were then conducted for a second site assuming the particle position as well as the A site, yielding site B in the 5 mM but not in the 1 mM $KAu(CN)_2$ derivative. These sites were checked with a reciprocal space "feedback" technique (76). The heavy atom parameters for both heavy atom data sets were used for determining phases to 6 Å resolution. The resultant electron density map, averaged over the 20 different non-crystallographic asymmetric units, showed clearly the viral protein envelope between about 100 and 150 Å radius.

The 6 Å map was improved with two cycles of molecular replacement averaging and back-transformation (77-79). A difference map of the two Au heavy atom data sets with respect to the native data, based on the improved phases, showed an unambiguous third, C, site in the 5 mM $KAu(CN)_2$ derivative. In contrast, difference maps of two other heavy atom data sets (KUO_2F_5 and mercurochrome) showed no interpretable sites. These were used to compute phases from 25 to 5 Å resolution. The resultant map was the starting point for five cycles of molecular replacement. The electron density map was exceptionally clear showing that there were probably three β-barrels per icosahedral asymmetric unit.

Phase extension beyond 5 Å, to eventually 3.0 Å resolution, was performed in 20 steps. In each step, the resolution was extended by three reciprocal lattice points. Two to three molecular replacement real-space averaging cycles were usually sufficient to improve the R-factor to less than 30% and correlation coefficient to greater than 0.5 in the current outermost resolution shell. Both $\underline{R}$ and $\underline{C}$ measure the agreement of the observed structure amplitudes and those calculated by Fourier back-transformation of the averaged cell. Attempts at increasing the resolution in larger steps led to erroneous phases, probably the result of satisfying the non-crystallographic symmetry constraints in isolated outer regions of reciprocal space independently of the established phases at lesser resolution. The particle envelope was defined by an external and internal sphere of 163 Å and 104 Å radius, respectively. Density outside the envelope was set to zero. The external sphere was truncated by planes tangential to the interparticle contacts, at a radial distance of 157 Å to avoid particle overlap, while allowing all of the protein density to be included within the envelope. Electron densities used for averaging were computed on grids with spacing less than one-fifth of the resolution of the data.

The final electron density map was averaged and displayed in sections at 1 Å intervals perpendicular to an icosahedral twofold axis. A mini-map of the 3.5 Å resolution electron density was used for the original chain tracing and amino acid identification. An atomic model was then built into the final 3.0 Å resolution map using an Evans and Sutherland PS300 computer graphics system and the FRODO program written by Alwyn Jones (80), modified for the PS300 at Rice and Purdue Universities.

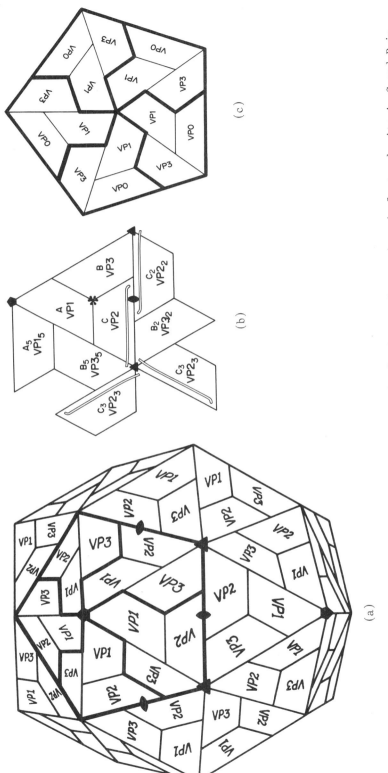

Figure 1. Relation of the pseudo-equivalent VP1, VP2 and VP3 subunits to the quasi-equivalent subunits A, C and B in TBSV and SBMV. (a) Icosahedral capsid. (b) shows the ordered amino-terminal arm βA present only in the C subunit of the plant viruses and VP2 of HRV14. The amino end of the arm interacts with two other VP2 arms across the threefold axis, while the carboxy end of the arm interacts with a VP2 across the twofold axis. An * indicates the position of the quasi-threefold axis in SBMV and TBSV analogous to the pseudo-threefold axis in HRV14. Subscripts designate the symmetry operation required to obtain the given subunit from the basic triangle. (c) The thickly outlined VP1, VP3, VP0 unit corresponds to the 6S protomer, and the 15-mer cap to the 14S pentamer observed in assembly experiments.

267

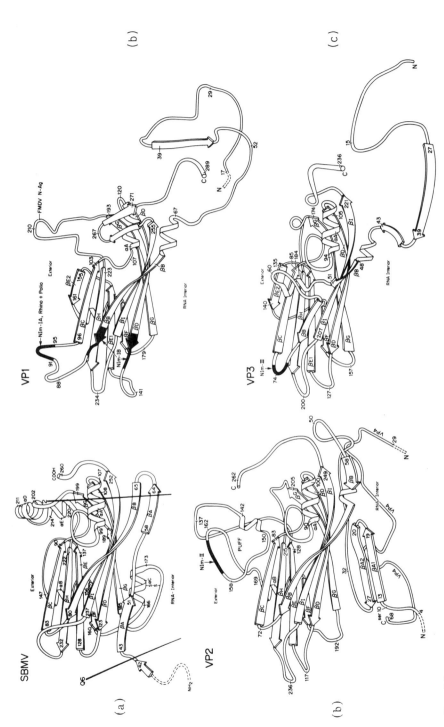

Figure 2. Diagrammatic drawings showing the polypeptide fold of SBMV and of each of the three larger capsid proteins of HRV14. The nomenclature of the secondary structural elements is derived from that of SBMV (99). Amino acid sequence numbers, appropriate for each protein, are also shown. (a) SBMV, (b) VP1 of HRV14, (c) VP2 and VP4 of HRV14, (d) VP3 of HRV14. (Adapted from a drawing of SBMV by Jane Richardson.)

Phase determination based on non-crystallographic symmetry was first suggested by Rossmann and Blow (73,81). It was used to compute very low resolution maps of SBMV (82) and polyoma virus (83) and also was used for phase extension in the determination of STNV (83a) and hemocyanin (84). However, there was considerable trepidation that the molecular replacement method might fail in extending the phases all the way from low resolution (5 Å) to high resolution (3 Å). This is the first time that such phase extension has been used successfully.

The high quality of the map reflects the power and exactness of the non-crystallographic constraints as opposed to the approximations required in an isomorphous replacement phase determination. The non-crystallographic constraints, however, are only meaningful in the presence of precise data and the availability of a sufficiently powerful computer to perform the large number of iterations during phase extension. Hence, our results were possible only due to the use of the CHESS synchrotron and the Cyber 205 supercomputer. The success of these techniques holds promise for a rapid series of structural determinations of viruses and other biological assemblies of high symmetries, by reducing the dependence upon finding satisfactory isomorphous heavy atom derivatives.

THE STRUCTURE

The particle consists of an icosahedral protein shell (Fig. 1a) surrounding an RNA core. The lack of visible structure in the central cavity results from the random orientation of the asymmetric RNA molecule. Both the tertiary fold of the VP1, VP2 and VP3 polypeptide chains and their quaternary organization within the HRV14 capsid are closely similar to the two published high resolution structures of $T = 3$ (180 identical subunits per capsid) RNA plant viruses, TBSV (65) and SBMV (66). Although the subunit organization within the icosahedron is somewhat different (85), a similar tertiary structure has also been found for $T = 1$ (60 subunits per capsid) STNV (67). The radial position and orientation of structurally equivalent atoms of HRV14 and SBMV generally agree to better than 3 Å relative to the icosahedral symmetry axes. In the plant viruses, the three quasi-equivalent subunits A, B and C have the same amino acid sequence but cannot have identical geometrical environments (86). For SBMV, there are 260 amino acids per subunit but the first 63 residues are associated with the RNA and, therefore, do not have icosahedral symmetry. These are said to be in the "random" domain. In the C subunit, unlike the A and B subunits, residues 38 to 63 are ordered with their ends forming a "β-annulus" about the icosahedral threefold axes (Fig. 1b). Residues 64 to 260 of SBMV are referred to as the "shell" domain and form an eight-stranded anti-parallel β-barrel (Fig. 2a). The β-barrels in HRV14, like those in SBMV, are wedge-shaped with the thin end (left in Fig. 2) pointing toward the five- or threefold (quasi-sixfold) axes. There are four excursions of the polypeptide chain toward the wedge-shaped end. Each excursion makes a sharp bend or "corner". The most exterior (top left in Fig. 2) corner is formed between the β-sheets βB and βC, the second corner down is formed between βH and βI, the third corner is between βD and βE1 while the most internal corner connects βF and βG. The βF-βG corner is the site of a 25-residue insertion in SBMV, including the α-helix αC, that is not present in any of the viral proteins of picornaviruses, nor in TBSV or STNV.

The three larger capsid proteins VP1, VP2 and VP3 in HRV14 are oriented and situated at essentially the same radius and position as the A, C and B subunits in SBMV, respectively. The capsid proteins of HRV14 are related by a pseudo-threefold axis, analogous to the quasi-threefold axis in SBMV and TBSV, at the * in Figure 1b. However, VP1 and VP3 have additions at their amino and carboxy termini. The amino ends, as in plant viruses, are in contact with the RNA, but unlike plant viruses, they are more acidic than basic and have only a very few amino-terminal residues "disordered" (lacking icosahedral symmetry). The first 64 of the 73 amino-terminal residues of VP1 reside under VP3, while the first 42 of the 71 amino-terminal residues of VP3 are under VP1. Thus, the predominant positions of VP1 and VP3 at the RNA-protein interface are exchanged relative to their positions at the exterior surface.

The first 25 of the 69 residues of the internal structural protein VP4 are not seen in the electron density map, implying that they lack icosahedral symmetry. VP4 is positioned in part below VP1 and VP2 with its visible amino end surrounding the fivefold axis. The carboxy ends of VP1 and VP3 are external and function in part to associate proteins within a protomer (Fig. 1c).

Large sequence insertions relative to the typical shell domain form protrusions on VP1, VP2 and VP3 and create a deep cleft or "canyon wall" on the viral surface. The canyon separates the major part of five VP1 subunits (in the "North") clustered about a pentamer axis from the surrounding VP2 and VP3 subunits (in the "South"), thus forming a moat around the VP1 protrusions on the fivefold axis. The South canyon walls are lined with the carboxy-terminal ends of VP1 and a large sequence insertion in VP1 corresponding to helices αD and αE in the equivalent SBMV capsid protein. The North canyon wall is partially lined with the carboxy terminus of VP3. VP2 is hardly associated at all with the canyon, while VP1 is the major contributor to the residues lining the canyon. The canyon is ~25 Å deep and 12 to 30 Å wide.

Due to the additional elaborations which VP1 has on the surface relative to VP2 and VP3, its overall shape is that of a kidney, with the depression forming a large part of the canyon. The first 16 residues of VP1 (Fig. 2b) are not seen in the electron density map. The shell domain of VP1 in HRV14 starts at residue 74. The small sequence insertion between βB and βC in rhinovirus and poliovirus is not found in FMDV. This loop forms a major immunogen in HRV14 (NIm-IA to be discussed below) and poliovirus. Five carbonyl oxygens of fivefold-related Tyr 142 residues converge on the vertex to chelate a presumably cationic ligand on the fivefold axis. The residues in VP1 of HRV14, that are analogous to αD and αE helices in SBMV, protrude to the surface and form part of the South rim of the canyon, but do not form helices in HRV14. There is an 8-residue insertion in polio and FMDV relative to HRV14 at the most external portion of this segment. These additional residues contain the major antigenic site of FMDV and have been predicted to form an α-helix (87). The carboxy-terminal 23 residues of VP1 line the South rim of the canyon in HRV14. There is little conservation of these residues among picornaviruses.

The first three residues of VP2 (Fig. 2c) are not seen in the electron density. The proximity of serine 10 in VP2 to the carboxy terminus of VP4 suggests that the VP0 cleavage, which occurs during the virus maturation step, may be autoproteolytic. There is no histidine in the immediate vicinity of Ser 10 and the carboxy end of VP4. However, nucleotide bases of the RNA might act as proton acceptors in the autocatalysis. Thus the insertion of RNA into the growing capsids might trigger the change of VP0 into VP2 and VP4 (cf. serine proteases (88,89)). The first 9 residues of VP2 are likely to have been rearranged after cleavage. Portions of the region between residues 10 and 72 are involved in contacts between twofold-related VP2 polypeptide chains (Fig. 1b). The αD and αE helices are absent in VP2 (Fig. 2c). There is a large 43-residue insertion, in the VP2 position corresponding to βE2 of VP1 and VP3, forming an external mushroom-shaped "puff". This is positioned adjacent to the VP1 elaborations, associated with the major antigenic site in FMDV, which line the South canyon wall. The most external residues of this puff correspond to NIm-II of HRV14. In contrast to VP1 and VP3, the carboxy terminus of VP2 has no extension beyond the shell domain.

All residues of VP3 (Fig. 2d) can be seen in the electron density. The 26 amino-terminal residues form a fivefold β-barrel about the pentamer axis analogous to the β-annulus (65) about trimer axes in SBMV and TBSV. This fivefold annulus extends down into the RNA to a radius of 111 Å. The polypeptide emerges from the β-annulus, circles around the base of the VP1 shell domain while making extensive contact with the RNA, emerges on the viral surface near residue 61 and then enters the shell domain at residue 72. The top corner of the VP3 shell domain, between βB and βC, is the NIm-III site of HRV14, structurally equivalent to NIm-IA in VP1. Helix αB of SBMV is replaced by an extended chain in VP3 and the external helices αD and αE of SBMV are absent as in VP2. When the shell domains of VP3 and VP1 are superimposed, much of the amino- and carobxy-terminal axes are about structurally equivalent.

IMMUNOGENIC SITES ON HRV14

Amino acid residues within the major neutralization immunogens of HRV14 have been identified by Sherry and Rueckert (54) and by Sherry et al. (55). They isolated mouse hybridoma lines which secrete monoclonal antibodies that neutralize HRV14. Each was then used to select several viral mutants resistant to neutralization by that antibody. Finally, every monoclonal antibody was assayed for its ability to neutralize the mutants. The results revealed four major immunogenic neutralization sites. Each immunogenic site was composed of overlapping epitopes where a given mutant was resistant to many or all of the antibodies directed against that site. These results were obtained without knowledge of the three-dimensional structure. When, however, the electron density map became available, it was immediately clear that the substitutions that could confer resistance to neutralization, regardless of their location in the amino acid sequences, were localized into four distinct areas corresponding exactly to the proposed immunogens. Moreover, these residues invariably faced outward toward the viral exterior. The possible effects which a given mutation might have on the capacity of the antibody to neutralize the virus have not yet been fully scrutinized in light of the structure. For a fuller understanding of these phenomena, it will be necessary to differentiate between mutants that resist neutralization while still maintaining their ability to bind antibodies and those mutants which entirely fail to bind antibodies.

The immunogenic site NIm-IA is an insertion between βB and βC at the top-most corner of the VP1 wedge. It is the most external portion of the complete virion being 160 Å from the center. Residues 91 and 95, associated with NIm-IA, are on the extreme external portion of the loop.

Residues associated with site NIm-IB on VP1 are on the carboxy ends of βB and βD, which are situated on either side of the amino end of strand βI. No mutations conferring resistance have yet been found on the βI strand. The NIm-IB site is close to the fivefold axis, a little below the canyon rim.

NIm-II on VP2 is at the extreme outside of the puff, 155 Å from the viral center. This immunogenic puff is adjacent to the external loop formed by the sequence insertion in VP1 corresponding to αD and αE of SBMV. One of the residues (210E on VP1) that is associated with NIm-II is, indeed, on VP1.

NIm-III on VP3 is 149 Å from the viral center, to some extent in the shadow of a larger protrusion of VP3 (residues 58 to 63) and the carboxy end of VP1 (282 - 286). NIm-III is in a position on VP3 corresponding to NIm-IA on VP1, that is in the loop between βB and βC. Residue K287 on VP1, associated with this site, points directly toward and is adjacent to the other residues on VP3 associated with NIm-III. The large protrusion near NIm-III, consisting of residues 282 - 286 in VP1 and 58 - 63 in VP3, has not been shown to be antigenic for HRV14 but in FMDV residues associated with the carboxy-terminal end of VP1 have shown antigenicity (Table 1). It is possible that this structural component lacks antigenicity in HRV14 because of greater rigidity (90,91), and this hypothesis will be checked when the HRV14 structure is refined and temperature factors have been determined.

ANTIGENIC SITES IN POLIO AND FMDV

Table 1 summarizes results obtained in mapping the immunogenic and antigenic surfaces of other picornaviruses. There is a dominant immunogen in poliovirus corresponding to NIm-IA of rhinovirus. Absence of a corresponding immunogen in FMDV is explained by deletion of this loop from its VP1.

The dominant antigenic site in FMDV resides in a region homologous to NIm-II in HRV14 (Table 1). The VP1 contribution to this antigen has also been shown for HRV14 (E210 of NIm-II) and for poliovirus (residues 222 - 241, Table 1). In FMDV, however, the immunogenic puff on VP2 is absent and 8 amino acids are inserted at the extreme surface of the αD-αE region of VP1. The resulting protrusion of NIm-II in FMDV would be unsupported by the puff. This protrusion may, therefore, occupy the space which in HRV14 is occupied by the puff. The inability of FMDV VP1 to elicit neutralizing antibodies (92) is consistent with the lack of structural support for this NIm-II protrusion in the absence of the neighboring VP2.

THE CANYON AS RECEPTOR BINDING SITE

The 25 Å deep canyon, circulating around each of the 12 pentamer vertices, suggests this to be the site for cell receptor binding. An antibody molecule would have difficulty in reaching the canyon floor, its entrance being blocked by the canyon rim. Thus, the residues in the deeper recesses of the canyon would not be under immune selection and could remain constant, permitting the virus to retain its ability to seek out the same cell receptors.

While retention of the canyon structure for all picornaviruses is to be expected, variation in the residues lining the canyon should be anticipated between viruses that attach themselves to different host cell receptors. That is, FMDV, polio, HRV14 and HRV2, all of which recognize different receptors, should exhibit some variation in the residues lining the canyon wall. It is thus noteworthy that not only are those parts of the carboxy-terminal ends of VP1 and VP3 which line the canyon walls some of the least conserved amino acids among picornaviruses, but also there is a deletion in HRV2 relative to HRV14 in βB at NIm-IB which is possibly associated with cell receptor carbohydrate recognition.

Since the topology of the canyon should be retained, the highly conserved sequence (MYVPPGAPNP starting at 151 of VP1 and AYTPPGARGP starting at 130 of VP3 for HRV14) in rhino, polio and FMD viruses situated in the middle of the floor of the canyon may be significant.

FMDV can be treated with trypsin causing cleavage at residues between

Table 1. Evidence for Antigenic Regions in Poliovirus and FMDV

Virus	Coat Protein	Amino Acid Number	HRV14[a] Amino Acid Equivalent	Sero-type[b]	Ref.	Method[c]	Position in HRV14 Structure
Polio	VP1	11-17	deleted	1M	50	3	Within RNA
Polio	VP1	24-40	14-30	1M	101	5	Faces RNA
Polio	VP1	70-75	60-65	1M	50	3,6	Faces RNA
Polio	VP1	61-80	51-70	1M	101	2	Faces RNA
Polio	VP1	86-103	77-97	1M	101	5	NIm-IB+NIm-IA
Polio	VP1	91-109	84-103	1M	101	2	NIm-IB+NIm-IA
Polio	VP1	100-109	93-103	1M	101	2	Part of NIm-IA
Polio	VP1	93-103	86-97	1M	50	2,3,6	NIm-IB(?)+NIm-IA[d]
Polio	VP1	93-104	86-98	1	102	4	NIm-IB(?)+NIm-IA[d]
Polio	VP1	93-100	86-93	3L	103,104	1	NIm-IB(?)+part of NIm-IA[d]
Polio	VP1	141-147	135-141	1M,S	105	3	NIm-IB
Polio	VP1	161-181	154-174	1M	101	5	Partly exposed on canyon
Polio	VP1	182-201	175-193	1M	101	2	Buried on 4th corner down
Polio	VP1	222-241	212-225	1M	101	2	NIm-II
FMDV	VP1	141-160	210-228	O_1K	106	2	NIm-II
FMDV	VP1	144-159	211-227	O_1K	87	2	NIm-II
FMDV	VP1	145-168	211-236	A12	107	8	NIm-II
FMDV	VP1	146-154	211-223	O_1K	95	7	NIm-II
FMDV	VP1	169-179	237-247	A12	96,107	8	NIm-IB central strand
Polio	VP1	270-287	254-272	1M	101	5	Partly exposed on canyon wall
FMDV	VP1	200-213	268-294	O_1K	106	2	NIm-III(?)
FMDV	VP1	200-213	268-294	O_1K	95	7	NIm-III(?)
Polio	VP2	162-173	161-170	1M	108	2,3	NIm-II
Polio	VP3	71-82	70-81	1M	108	6	NIm-III

a. Aligned by eye and by fitting to the shell domain structure.
b. 1M (type 1, Mahoney), S (Sabin), 3L (type 3, Leon), O_1K (type O_1, strain Kaufbeuern), A12 (type A, subtype 12).
c. Method key:
   1. Monoclonal antibodies raised against intact virus select for resistant mutations.
   2. Synthetic peptides induce neutralizing antibodies.
   3. Synthetic peptides prime for high titer neutralizing response.
   4. Synthetic peptide competes with monoclonal antibody to inhibit neutralization.
   5. Synthetic peptides induce antibodies which bind virus but neutralize poorly.
   6. Neutralizing antisera or monoclonal antibodies bind peptide in ELISA.
   7. Deduced from ability or inability of protein fragments to induce neutralizing antibodies.
   8. Deduced from ability or inability of neutralizing monoclonal antibodies to bind protein fragments.
d. Uncertainty due to somewhat arbitrary nature of computer alignments.

138 and 154 of VP1. This causes the virus to lose its ability to attach to cells and its ability to stimulate neutralizing antibodies (93-96). The enzymic cleavages occur in NIm-II on the αD-αE protrusion of VP1, a large loop which also forms part of the presumed host receptor binding site.

Concanavalin A is a lectin extracted from jack beans. It has been shown to compete with solubilized coxsackie B3 virus for receptors on HeLa cells (59). The lectin also interferes with infection of HRV and poliovirus (60), suggesting that there might be common structural features between concanavalin A and these viruses. Argos et al. (97) have shown that there is a remarkable similarity in the folding topology of concanavalin A and the shell domain of TBSV (and hence also to VP1 of HRV14), thus providing some rationalization of the competition experiments at the molecular level. Many animal and bacterial viruses recognize their host cells by virtue of a specific polysaccharide (98). The latter may be a part of the virus or of the host cell receptor. Concanavalin A can bind D-glucose and D-mannose pyranosides. The functional sugar binding site of concanavalin A is in a position equivalent to the amino end of βI, that is roughly the NIm-IB site. If this comparison is meaningful, it might be anticipated that the NIm-IB site, which lies on the rim of the receptor binding cavity, would be associated with polysaccharide recognition, a hypothesis that should be readily testable crystallographically.

ASSEMBLY

Assembly of picornaviruses (1-3,8-11) proceeds from 6S protomers of VP1, VP3 and VP0, via 14S pentamers of five 6S protomers, to mature virions. The final step involves inclusion of the RNA into empty capsids or partially assembled shells with simultaneous cleavage of VP0 into VP2 and VP4. Conversely, in vitro disassembly, produced by mild denaturation, proceeds via the expulsion of VP4 followed by the RNA (3).

Both the amino and carboxy ends of VP1 and $VP3_5$ are intertwined with each other. Furthermore, if VP4 and VP2 are considered as VP0, then VP0 is also intertwined with VP1 and $VP3_5$. This strongly suggests that the 6S protomer is as shown in Figure 1c. These protomers are themselves intertwined by virtue of the fivefold β-annulus formed by the amino ends of the VP3's and the proximity of the observed amino ends of VP4's to the fivefold axis. Thus, the 14S pentamers closely correlate with the observed structure, shown diagrammatically in Figure 1c. Such an assembly sequence matches that observed in plant viruses, in particular that of SBMV, where the building blocks are dimers corresponding to VP1 and $VP3_5$ and where the formation of intermediates with fivefold symmetry is considered to be a critical stage in the formation of $\underline{T} = 1$ and $\underline{T} = 3$ capsids (99). Indeed, it has now been shown (100) that the 15-protein cluster, corresponding to the one shown in Figure 1c, is conserved between $\underline{T} = 1$ and $\underline{T} = 3$ SBMV particles.

VP2, once cleaved from VP4, is globular and does not contact the other proteins extensively. There are large solvent accessible regions between VP2 and the surrounding proteins. This, as well as the extraordinarily internal heavy atom sites on VP2, is consistent with the loose binding of VP2 to the capsid. (The ability of the heavy atoms to penetrate deeply into the shell suggests cautious interpretation of data based upon the use of small chemical labels for mapping viral surfaces.) Disruption of pentamer-pentamer contacts, mediated by a slight reorientation of VP2 or its complete removal, could provide a port by which the VP4 and RNA exit. Binding of a cell receptor in the canyon adjacent to VP3 could facilitate this process, possibly accompanied by an isoelectric change.

It is remarkable how, both in plant and animal RNA viruses, there is a β-annulus type structure between the amino ends of some subunits. It is equally remarkable how the amino ends of the capsid proteins are invariably associated with the RNA. In TBSV, SBMV and STNV they are basic, whereas in HRV14 they are slightly acidic. These properties may be significant for the initial events of assembly.

ACKNOWLEDGMENTS

This paper is adapted from a manuscript submitted to Nature. We are grateful to Sharon Wilder, Kathy Shuster, Bill Boyle, Jun Tsao, Gale Rhodes, Diana Delatore and Tim Schmidt for help in data collection and presentation; to Keith Moffat, Wilfried Schildkamp, Robert Hunt, Don Bilderbeck, Aggie Sirrine and Boris Batterman at CHESS; to Hans Bartunik and Klaus Bartels at DESY; to Saul Rosen, John Steele, Tom Putnam and Paul Townsend at the Purdue University Computer Center; to Richard Colonno, Ann Palmenberg, Jeffrey Bolin, Abelardo Silva, Ignacio Fita, Celerino Abad-Zapatero, R. Usha, M. V. Hosur, Cynthia Stauffacher, Patrick Argos and J. K. Mohana Rao for helpful discussions. The work was supported by an NIH grant, NSF grants for supercomputer time and graphics, a Purdue University Showalter Foundation grant and a contribution by Merck Sharp & Dohme Co. to M.G.R. and an ACS grant to R.R.R. CHESS is supported by an NSF grant and the macromolecular diffraction facility by an NIH grant. A postdoctoral Walter Winchell-Damon Runyon fellowship supported E.A. and a predoctoral NIH training grant supported B.S.

REFERENCES

1. J. R. Putnak and B. A. Phillips, Picornaviral structure and assembly, Microbiol. Rev. 45:287-315 (1981).
2. D. V. Sangar, The replication of picornaviruses, J. Gen. Virol. 45:1-13 (1979).
3. R. R. Rueckert, On the structure and morphogenesis of picornaviruses, in: "Comprehensive Virology," H. Fraenkel-Conrat and R. R. Wagner, eds., Vol. 6, pp. 131-213, Plenum, New York (1976).
4. R. R. Rueckert, A. K. Dunker, and C. M. Stoltzfus, The structure of mouse-Elberfeld virus: a model, Proc. Natl. Acad. Sci. U.S. 62:912-919 (1969).
5. R. R. Rueckert and E. Wimmer, Systematic nomenclature of picornavirus proteins, J. Virol. 50:957-959 (1984).
6. S. McGregor, L. Hall, and R. R. Rueckert, Evidence for the existence of protomers in the assembly of encephalomyocarditis virus, J. Virol. 15:1107-1120 (1975).
7. M. F. Jacobson, J. Asso, and D. Baltimore, Further evidence on the formation of poliovirus proteins, J. Mol. Biol. 49:657-669 (1970).
8. S. McGregor and R. R. Rueckert, Picornaviral capsid assembly: similarity of rhinovirus and enterovirus precursor subunits, J. Virol. 21:548-553 (1977).
9. C. B. Fernandez-Tomas, N. Guttman, and D. Baltimore, Morphogenesis of poliovirus III. Formation of provirion in cell-free extracts, J. Virol. 12:1181-1183 (1973).
10. M. F. Jacobson and D. Baltimore, Morphogenesis of poliovirus. I. Association of the viral RNA with coat protein, J. Mol. Biol. 33:369-378 (1968).
11. C. B. Fernandez-Tomas and D. Baltimore, Morphogenesis of poliovirus. II. Demonstration of a new intermediate, the provirion, J. Virol. 12:1122-1130 (1973).
12. Y. F. Lee, A. Nomoto, B. M. Detjen, and E. Wimmer, A protein covalently linked to poliovirus genome RNA, Proc. Natl. Acad. Sci. U.S. 74:59-63 (1977).

13. D. V. Sangar, D. J. Rowlands, T. J. R. Harris, F. Brown, Protein covalently linked to foot-and-mouth disease virus RNA, Nature (London), 268:648-650 (1977).
14. V. Ambros and D. Baltimore, Protein is linked to the 5' end of poliovirus RNA by a phosphodiester linkage to tyrosine, J. Biol. Chem. 253:5263-5266 (1978).
15. F. Golini, B. L. Semler, A. J. Dorner, and E. Wimmer, Protein-linked RNA of poliovirus is competent to form an initiation complex of translation in vitro, Nature (London), 287:600-603 (1980).
16. A. B. Vartapetian, E. V. Koonin, V. I. Agol, and A. A. Bogdanov, Encephalomyocarditis virus RNA synthesis in vitro is protein-primed, EMBO J. 3, 2593-2598 (1984).
17. E. Wimmer, Genome-linked proteins of viruses, Cell, 28:199-201 (1982).
18. M. A. Pallansch, O. M. Kew, B. L. Semler, D. R. Omilianowski, C. W. Anderson, E. Wimmer, and R. R. Rueckert, Protein processing map of poliovirus, J. Virol. 49:873-880 (1984).
19. R. Hanecak, B. L. Semler, C. W. Anderson, and E. Wimmer, Proteolytic processing of poliovirus polypeptides: antibodies to polypeptide P3-7c inhibit cleavage at glutamine-glycine pairs, Proc. Natl. Acad. Sci. U.S. 79:3973-3977 (1982).
20. P. Carthew and S. J. Martin, The iodination of bovine enterovirus particles, J. Gen. Virol. 24:525-534 (1974).
21. K. Lonberg-Holm and B. E. Butterworth, Investigation of the structure of polio- and human rhinovirions through the use of selective chemical reactivity, Virology, 71:207-216 (1976).
22. T. W. Beneke, K. O. Habermehl, W. Diefenthal, and M. Buchholz, Iodination of poliovirus capsid proteins, J. Gen. Virol. 34:387-390 (1977).
23. G. A. Lund, B. R. Ziola, A. Salmi, and D. G. Scraba, Structure of the Mengo virion. V. Distribution of the capsid polypeptides with respect to the surface of the virus particle, Virology, 78:35-44 (1977).
24. K. Wetz and K. O. Habermehl, Topographical studies on poliovirus capsid proteins by chemical modification and cross-linking with bifunctional reagents, J. Gen. Virol. 44:525-534 (1979).
25. J. S. Hordern, J. D. Leonard, and D. G. Scraba, Structure of the Mengo virion. VI. Spatial relationships of the capsid polypeptides as determined by chemical cross-linking analyses, Virology, 97:131-140 (1979).
26. K. Wetz and K. O. Habermehl, Specific cross-linking of capsid proteins to virus RNA by ultraviolet irradiation of poliovirus, J. Gen. Virol. 59:397-401 (1982).
27. H. Toyoda, M. Kohara, Y. Kataoka, T. Suganuma, T. Omata, N. Imura, and A. Nomoto, Complete nucleotide sequences of all three poliovirus serotype genomes. Implication for genetic relationship, gene function and antigenic determinants, J. Mol. Biol. 174:561-585 (1984).
28. N. Kitamura, B. L. Semler, P. G. Rothberg, G. R. Larsen, C. J. Adler, A. J. Dorner, E. A. Emini, R. Hanecak, J. J. Lee, S. van der Werf, C. W. Anderson, and E. Wimmer, Primary structure, gene organization and polypeptide expression of poliovirus RNA, Nature (London), 291:547-553 (1981).
29. G. Stanway, A. J. Cann, R. Hauptmann, P. Hughes, L. D. Clarke, R. C. Mountford, P. D. Minor, G. C. Schild, and J. W. Almond, The nucleotide sequence of poliovirus type 3 leon 12 a_1b: comparison with poliovirus type 1, Nucl. Acids Res. 11:5629-5643 (1983).
30. V. R. Racaniello and D. Baltimore, Molecular cloning of poliovirus cDNA and determination of the complete nucleotide sequence of the viral genome, Proc. Natl. Acad. Sci. U.S. 78:4887-4891 (1981).
31. A. J. Makoff, C. A. Paynter, D. J. Rowlands, and J. C. Boothroyd, Comparison of the amino acid sequence of the major immunogen from three serotypes of foot and mouth disease virus, Nucl. Acids Res. 10:8285-8295 (1982).
32. A. R. Carroll, D. J. Rowlands, and B. E. Clarke, The complete nucleo-

tide sequence of the RNA coding for the primary translation product of foot and mouth disease virus, Nucl. Acids Res. 12:2461-2472 (1984).

33. S. Forss, K. Strebel, E. Beck, and H. Schaller, Nucleotide sequence and genome organization of foot-and-mouth disease virus, Nucl. Acids Res. 12:6587-6601 (1984).

34. J. C. Boothroyd, P. E. Highfield, G. A. M. Cross, D. J. Rowlands, P. A. Lowe, F. Brown, and T. J. R. Harris, Molecular cloning of foot and mouth disease virus genome and nucleotide sequences in the structural protein genes, Nature (London), 290:800-802 (1981).

35. G. Stanway, P. J. Hughes, R. C. Mountford, P. D. Minor, and J. W. Almond, The complete nucleotide sequence of a common cold virus: human rhinovirus 14, Nucl. Acids Res. 12:7859-7875 (1984).

36. P. L. Callahan, S. Mizutani, and R. J. Colonno, Molecular cloning and complete sequence determination of RNA genome of human rhinovirus type 14, Proc. Natl. Acad. Sci. U.S. 82:732-736 (1985).

37. T. Skern, W. Sommergruber, D. Blaas, P. Gruendler, F. Fraundorfer, C. Pieler, I. Fogy, and E. Kuechler, Human rhinovirus 2: complete nucleotide sequence and proteolytic processing signals in the capsid protein region, Nucl. Acids Res. 13:2111-2126 (1985).

38. A. C. Palmenberg, E. M. Kirby, M. R. Janda, N. L. Drake, G. M. Duke, K. F. Potratz, and M. S. Collett, The nucleotide and deduced amino acid sequences of the encephalomyocarditis viral polyprotein coding region, Nucl. Acids Res. 12:2969-2985 (1984).

39. R. Najarian, D. Caput, W. Gee, S. J. Potter, A. Renard, J. Merryweather, G. Van Nest, and D. Dina, Primary structure and gene organization of human hepatitis A virus, Proc. Natl. Acad. Sci. U.S. 82:2627-2631 (1985).

40. D. L. Linemeyer, J. G. Menke, A. Martin-Gallardo, J. V. Hughes, A. Young, and S. W. Mitra, Molecular cloning and partial sequencing of hepatitis A viral cDNA, J. Virol. 54:247-255 (1985).

41. P. Argos, G. Kamer, M. J. H. Nicklin, and E. Wimmer, Similarity in gene organization and homology between proteins of animal picornaviruses and a plant comovirus suggest common ancestry of these virus families, Nucl. Acids Res. 12:7251-7267 (1984).

42. H. Franssen, J. Leunissen, R. Goldbach, G. Lomonossoff, and D. Zimmern, Homologous sequences in non-structural proteins from cowpea mosaic virus and picornaviruses, EMBO J. 3:855-861 (1984).

43. G. Kamer and P. Argos, Primary structural comparison of RNA-dependent polymerases from plant, animal and bacterial viruses, Nucl. Acids Res. 12:7269-7282 (1984).

44. J. Haseloff, P. Goelet, D. Zimmern, P. Ahlquist, R. Dasgupta, and P. Kaesberg, Striking similarities in amino acid sequence among nonstructural proteins encoded by RNA viruses that have dissimilar genomic organization, Proc. Natl. Acad. Sci. U.S. 81:4358-4362 (1984).

45. P. Ahlquist, E. G. Strauss, C. M. Rice, J. H. Strauss, J. Haseloff, and D. Zimmern, Sindbis virus proteins nsP1 and nsP2 contain homology to nonstructural proteins from several RNA plant viruses, J. Virol. 53:536-542 (1985).

46. M. A. Rezaian, R. H. V. Williams, K. H. J. Gordon, A. R. Gould, and R. H. Symons, Nucleotide sequence of cucumber-mosaic-virus RNA 2 reveals a translation product significantly homologous to corresponding proteins of other viruses, Eur. J. Biochem. 143:277-284 (1984).

47. N. J. Dimmock, Mechanisms of neutralization of animal viruses, J. Gen. Virol. 65:1015-1022 (1984).

48. J. Icenogle, H. Shiwen, G. Duke, S. Gilbert, R. Rueckert, and J. Anderegg, Neutralization of poliovirus by a monoclonal antibody: kinetics and stoichiometry, Virology, 127:412-425 (1983).

49. B. Mandel, Neutralization of poliovirus: a hypothesis to explain the mechanism and the one-hit character of the neutralization reaction, Virology, 69:500-510 (1976).

50. E. A. Emini, B. A. Jameson, and E. Wimmer, Priming for an induction of

50.   anti-poliovirus neutralizing antibodies by synthetic peptides, Nature (London), 304:699-703 (1983).
51. M. Schrom, J. A. Laffin, B. Evans, J. J. McSharry, and L. A. Caliguiri, Isolation of poliovirus variants resistant to and dependent on arildone, Virology, 122:492-497 (1982).
52. E. A. Emini, S. Kao, A. J. Lewis, R. Crainic, and E. Wimmer, Functional basis of poliovirus neutralization determined with monospecific neutralizing antibodies, J. Virol. 46:466-474 (1983).
53. E. A. Emini, P. Ostapchuk, and E. Wimmer, Bivalent attachment of antibody onto poliovirus leads to conformational alteration and neutralization, J. Virol. 48:547-550 (1983).
54. B. Sherry and R. Rueckert, Evidence for at least two dominant neutralization antigens on human rhinovirus 14, J. Virol. 53:137-143 (1985).
55. B. Sherry, A. G. Mosser, R. J. Colonno, and R. R. Rueckert, Use of monoclonal antibodies to identify four neutralization immunogens on a common cold picornavirus, human rhinovirus 14, J. Virol., manuscript in preparation (1985).
56. G. Abraham and R. J. Colonno, Many rhinovirus serotypes share the same cellular receptor, J. Virol. 51:340-345 (1984).
57. R. J. Colonno, personal communication.
58. P. D. Minor, P. A. Pipkin, D. Hockley, G. C. Schild, and J. W. Almond, Monoclonal antibodies which block cellular receptors of poliovirus, Virus Res. 1:203-212 (1984).
59. D. L. Krah and R. L. Crowell, Properties of the deoxycholate-solubilized HeLa cell plasma membrane receptor for binding group B coxsackieviruses, J. Virol. 53:867-870 (1985).
60. K. Lonberg-Holm, The effects of concanavalin A on the early events of infection by rhinovirus type 2 and poliovirus type 2, J. Gen. Virol. 28:313-327 (1975).
61. C. F. T. Mattern and H. G. duBuy, Purification and crystallization of coxsackie virus, Science, 123:1037-1038 (1956).
62. F. L. Schaffer and C. E. Schwerdt, Crystallization of purified MEF-1 poliomyelitis virus particles, Proc. Natl. Acad. Sci. U.S. 41:1020-1023 (1955).
63. B. D. Korant and J. T. Stasny, Crystallization of human rhinovirus 1A, Virology, 55:410-417 (1973).
64. J. T. Finch and A. Klug, Structure of poliomyelitis virus, Nature (London), 183:1709-1714 (1959).
65. S. C. Harrison, A. J. Olson, C. E. Schutt, F. K. Winkler, and G. Bricogne, Tomato bushy stunt virus at 2.9 Å resolution, Nature (London), 276:368-373 (1978).
66. C. Abad-Zapatero, S. S. Abdel-Meguid, J. E. Johnson, A. G. W. Leslie, I. Rayment, M. G. Rossmann, D. Suck, and T. Tsukihara, Structure of southern bean mosaic virus at 2.8 Å resolution, Nature (London), 286:33-39 (1980).
67. L. Liljas, T. Unge, T. A. Jones, K. Fridborg, S. Lövgren, U. Skoglund, and B. Strandberg, Structure of satellite tobacco necrosis virus at 3.0 Å resolution, J. Mol. Biol. 159:93-108 (1982).
68. J. M. Hogle, Preliminary studies of crystals of poliovirus type I, J. Mol. Biol. 160:663-668 (1982).
69. J. W. Erickson, E. A. Frankenberger, M. G. Rossmann, G. S. Fout, K. C. Medappa, and R. R. Rueckert, Crystallization of a common cold virus, human rhinovirus 14: "isomorphism" with poliovirus crystals, Proc. Natl. Acad. Sci. U.S. 80:931-934 (1983).
70. E. Arnold, J. W. Erickson, G. S. Fout, E. A. Frankenberger, H. J. Hecht, M. Luo, M. G. Rossmann, and R. R. Rueckert, Virion orientation in cubic crystals of the human common cold virus HRV14, J. Mol. Biol. 177:417-430 (1984).
71. M. Luo, E. Arnold, J. W. Erickson, M. G. Rossmann, U. Boege, and D. G. Scraba, Picornaviruses of two different genera have similar structures, J. Mol. Biol. 180:703-714 (1984).

72. "International Tables for Crystallography," T. Hahn, ed., Vol. A, Reidel Publishing, Dordrecht (1983).
73. M. G. Rossmann and D. M. Blow, The detection of sub-units within the crystallographic asymmetric unit, Acta Crystallogr. 15:24-31 (1962).
74. P. Argos and M. G. Rossmann, A method to determine heavy-atom positions for virus structures, Acta Crystallogr. B32:2975-2979 (1976).
75. P. Argos and M. G. Rossmann, Determining heavy-atom positions using non-crystallographic symmetry, Acta Crystallogr. A30:672-677 (1974).
76. M. G. Rossmann and E. Arnold, Comparison of vector search and feedback methods for finding heavy atom sites in isomorphous derivatives, manuscript in preparation (1985).
77. P. Argos, G. C. Ford, and M. G. Rossmann, An application of the molecular replacement technique in direct space to a known protein structure, Acta Crystallogr. A31:499-506 (1975).
78. G. Bricogne, Methods and programs for the direct space exploitation of geometric redundancies, Acta Crystallogr. A32:832-847 (1976).
79. J. E. Johnson, Appendix II. Averaging of electron density maps, Acta Crystallogr. B34:576-577 (1978).
80. T. A. Jones, A graphics model building and refinement system for macromolecules, J. Appl. Crystallogr. 11:268-272 (1978).
81. M. G. Rossmann and D. M. Blow, Determination of phases by the conditions of non-crystallographic symmetry, Acta Crystallogr. 16:39-45 (1963).
82. J. E. Johnson, T. Akimoto, D. Suck, I. Rayment, and M. G. Rossmann, The structure of southern bean mosaic virus at 22.5 Å resolution, Virology, 75:394-400 (1976).
83. I. Rayment, T. S. Baker, D. L. D. Caspar, and W. T. Murakami, Polyoma virus capsid structure at 22.5 Å resolution, Nature (London), 295:110-115 (1982).
*83a. T. Unge, L. Liljas, B. Strandberg, I. Vaara, K.K. Kannan, K. Friedeorg, C.E. Nordman and P.J. Lentz, Jr., Satellite tobacco necrosis virus structure at 4.0 Å resolution, Nature, 285:373-377 (1980).
84. W. P. J. Gaykema, W. G. J. Hol, J. M. Vereijken, N. M. Soeter, H. J. Bak, and J. J. Beintema, 3.2 Å structure of the copper-containing, oxygen-carrying protein Panulirus interruptus haemocyanin, Nature (London), 309:23-29 (1984).
85. M. G. Rossmann, C. Abad-Zapatero, M. R. N. Murthy, L. Liljas, T. A. Jones, and B. Strandberg, Structural comparisons of some small spherical plant viruses, J. Mol. Biol. 165:711-736 (1983).
86. D. L. D. Caspar and A. Klug, Physical principles in the construction of regular viruses, Cold Spring Harbor Symp. Quant. Biol. 27:1-24 (1962).
87. E. Pfaff, M. Mussgay, H. O. Böhm, G. E. Schulz, and H. Schaller, Antibodies against a preselected peptide recognize and neutralize foot and mouth disease virus, EMBO J. 1:869-874 (1982).
88. H. Neurath, Evolution of proteolytic enzymes, Science, 224:350-357 (1984).
89. T. A. Steitz and R. G. Shulman, Crystallographic and NMR studies of the serine proteases, Ann. Rev. Biophys. Bioeng. 11:419-444 (1982).
90. E. Westhof, D. Altschuh, D. Moras, A. C. Bloomer, A. Mondragon, A. Klug, and M. H. V. Van Regenmortel, Correlation between segmental mobility and the location of antigenic determinants in proteins, Nature (London), 311:123-131 (1984).
91. J. A. Tainer, E. D. Getzoff, H. Alexander, R. A. Houghten, A. J. Olson, R. A. Lerner, and W. A. Hendrickson, The reactivity of anti-peptide antibodies is a function of the atomic mobility of sites in a protein, Nature (London), 312:127-134 (1984).
92. R. H. Meloen, J. Briaire, R. J. Woortmeyer, and D. Van Zaane, The main antigenic determinant detected by neutralizing monoclonal antibodies on the intact foot-and-mouth disease virus particle is absent from isolated VP1, J. Gen. Virol. 64:1193-1198 (1983).

93. D. Cavanagh, D. V. Sangar, D. J. Rowlands, and F. Brown, Immunogenic and cell attachment sites of FMDV: further evidence for their location in a single capsid polypeptide, J. Gen. Virol. 35:149-158 (1977).
94. T. F. Wild, J. N. Burroughs, and F. Brown, Surface structure of foot-and-mouth disease virus, J. Gen. Virol. 4:313-320 (1969).
95. K. Strohmaier, R. Franze, and K. H. Adam, Location and characterization of the antigenic portion of the FMDV immunizing protein, J. Gen. Virol. 59:295-306 (1982).
96. B. Baxt, D. O. Morgan, B. H. Robertson, and C. A. Timpone, Epitopes on foot-and-mouth disease virus outer capsid protein VP_1 involved in neutralization and cell attachment, J. Virol. 51:298-305 (1984).
97. P. Argos, T. Tsukihara, and M. G. Rossmann, A structural comparison of concanavalin A and tomato bushy stunt virus protein, J. Mol. Evol. 15:169-179 (1980).
98. K. Lonberg-Holm and L. Philipson, Early interaction between animal viruses and cells, in: "Monographs in Virology," J. L. Melnick, ed., Vol. 9, Karger, Basel (1974).
99. M. G. Rossmann, C. Abad-Zapatero, M. A. Hermodson, and J. W. Erickson, Subunit interactions in southern bean mosaic virus, J. Mol. Biol. 166:37-83 (1983).
100. J. W. Erickson, A. M. Silva, M. R. N. Murthy, I. Fita, and M. G. Rossmann, The structure of a $T = 1$ icosahedral empty particle from southern bean mosaic virus, Science, 229:625-629 (1985).
101. M. Chow, R. Yabrov, J. Bittle, J. Hogle, and D. Baltimore, Synthetic peptides from four separate regions of the poliovirus type 1 capsid protein VP1 induce neutralizing antibodies, Proc. Natl. Acad. Sci. U.S. 82:910-914 (1985).
102. C. Wychowski, S. van der Werf, O. Siffert, R. Crainic, P. Bruneau, and M. Girard, A poliovirus type 1 neutralization epitope is located within amino acid residues 93 to 104 of viral capsid polypeptide VP1, EMBO J. 2:2019-2024 (1983).
103. D. M. A. Evans, P. D. Minor, G. S. Schild, and J. W. Almond, Critical role of an eight-amino acid sequence of VP1 in neutralization of poliovirus type 3, Nature (London), 304:459-462 (1983).
104. P. D. Minor, G. C. Schild, J. Bootman, D. M. A. Evans, M. Ferguson, P. Reeve, M. Spitz, G. Stanway, A. J. Cann, R. Hauptmann, L. D. Clarke, R. C. Mountford, and J. W. Almond, Location and primary structure of a major antigenic site for poliovirus neutralization, Nature (London), 301:674-679 (1983).
105. S. van der Werf, C. Wychowski, P. Bruneau, B. Blondel, R. Crainic, F. Horodniceanu, and M. Girard, Localization of a poliovirus type 1 neutralization epitope in viral capsid polypeptide VP1, Proc. Natl. Acad. Sci. U.S. 80:5080-5084 (1983).
106. J. L. Bittle, R. A. Houghten, H. Alexander, T. M. Shinnick, J. G. Sutcliffe, R. A. Lerner, D. J. Rowlands, and F. Brown, Protection against foot-and-mouth disease by immunization with a chemically synthesized peptide predicted from the viral nucleotide sequence, Nature (London), 298:30-33 (1982).
107. B. H. Robertson, D. O. Morgan, and D. M. Moore, Location of neutralizing epitopes defined by monoclonal antibodies generated against the outer capsid polypeptide, VP1, of foot-and-mouth disease virus A12, Virus Res. 1:489-500 (1984).
108. E. A. Emini, B. A. Jameson, and E. Wimmer, Identification of multiple neutralization antigenic sites on poliovirus type 1 and the priming of the immune response with synthetic peptides, in: "Modern Approaches to Vaccines," R. M. Chanock and R. A. Lerner, eds., pp. 65-75, Cold Spring Harbor Laboratory, Cold Spring Harbor (1984).

THE STRUCTURE OF POLIOVIRUS AT 2.9 Å RESOLUTION:
CRYSTALLOGRAPHIC METHODS AND BIOLOGICAL IMPLICATIONS

J. M. Hogle, M. Chow, and D. J. Filman

Department of Molecular Biology
Research Institute of Scripps Clinic
10666 N. Torrey Pines Road
La Jolla, California 92037

Poliovirus is a member of the Picornavirus family, which also includes a number of biologically significant pathogens such as hepatitis A virus, foot and mouth disease virus, the coxsackie viruses, and the rhinoviruses. The poliovirus capsid is approximately 300 Å in diameter, with a mass of 8.4 million daltons. It is composed of sixty copies each of the four capsid proteins VP1, VP2, VP3, and VP4 (approximately 33, 30, 26, and 7.5 kilodaltons, respectively), arranged to form a T=1 icosahedral shell. The capsid encloses one molecule of single stranded message sense RNA of approximately 7500 nucleotides (2.5 million daltons) [1]. In nature, there are three known stable serotypes of poliovirus, and within each serotype there is a large number of strains. We have solved the structure of the Mahoney strain of type 1 poliovirus at 2.9 Å resolution [2]. Recently, Rossmann et al. [3] have reported the structure of a closely related virus, rhinovirus 14.

Most of the crystallographic methods used to solve the structure of poliovirus were based on those used earlier in determining the structures of several small spherical plant viruses [4-7]. However, several additional developments in methodology were necessary, due primarily to the increased size of the crystallographic problem. Although the dimensions of the poliovirus particle, and the unit cell constants of the crystals are similar to those of the T=3 plant viruses, less of the icosahedral particle symmetry is expressed in the crystal. Specifically, poliovirus crystals have one-half particle per asymmetric unit, compared with one-sixth in SBMV, and one-twelfth in TBSV. Since the crystallographic problem was significantly larger, procedural changes were necessary at several stages of the structure determination, including film scanning and the application of noncrystallographic symmetry constraints. In addition, phases for the data between 5 and 2.9 Å were obtained by direct phase extension, starting with symmetry-constrained 5 Å SIR phases. A similar phase extension technique also has been used in the structure determinations of hemocyanin [8] and rhinovirus [3].

The greatest significance of the poliovirus structure determination, however, is not the methodology, but rather the ability to associate structure with biological function. The picornaviruses, in general, and poliovirus, in particular, are far better characterized biologically than the plant viruses whose structures are known. The ability to grow animal viruses in cultured cells has led to the characterization of the additional insights into virus evolution, picornavirus assembly, and the recognition of viruses by neutralizing antibodies. We expect that in the future, the structure will also improve our understanding of the mechanisms of virus neutralization, of the structural properties responsible for host and tissue specificity, and of the structural factors that control neurovirulence and its attenuation.

CRYSTALLOGRAPHIC METHODS

The Mahoney strain of type 1 poliovirus was crystallized by microdialysis versus low ionic strength NaCl in 10mM PIPES buffer at neutral pH. Virus recovered from redissolved crystals was fully infectious. The crystals belong to the orthorhombic space group P2 2 2 with a=322.9, b=358.0, c=380.2 Å, and contain one-half particle per asymmetric unit. The crystals diffract to at least 2.2 Å resolution, but are sensitive to x-irradiation, so that they require cooling to obtain a useful diffraction lifetime. Consequently, high resolution data were collected at -15 degrees C, using 25% ethylene glycol in 10mM PIPES buffer at pH 7 as cryosolvent. A platinum derivative was obtained by soaking crystals in 5mM K_2PtCl_4 for 48 hours.

Data were collected by (0.5 degree) oscillation photography, using CuKα radiation from a Marconi GX-20 generator (for the native data) or a GX-18 (for the derivative data), with 100 micron focus and Franks mirror optics. Films were digitized on a 50 micron raster, and integrated with a locally developed film scanning program that uses profile fitting to resolve overlapped reflections, and two dimensional function fitting to compensate for inadequate background measurements in crowded regions of the film. Procedures for the accurate estimation of measurement errors were also devised, so that the large number of overlapped reflections could be retained in the data set. Data were reduced, and partially recorded reflections were corrected using a local implementation of the method of Winkler, Schutt, and Harrison [9]. The native data set (consisting of 3.6 million measurements of 894,000 unique reflections) yielded 737,000 usable intensities, out of 950,000 in the 2.9 Å sphere. A partial data set for the platinum derivative (consisting of 1.1 million measurements of 540,000 unique reflections) yielded 310,000 usable intensities.

The location and orientation of the poliovirus particle were initially estimated by inspection of native diffraction patterns. Packing considerations dictated that the particle center (and a particle two-fold axis) must be located along the crystallographic two-fold axis. The occurrence of body centered extinctions that extend to about 25 Å resolution indicated that the particle center could be located no further than 3 Å from Z=0.25, with particle two-fold axes rotated by no more than 3 degrees relative to the crystallographic a and b axes. The precise orientation of the particle in the unit cell (a 2 degree rotation) was determined by a one-dimensional icosahedrally locked rotation function. Inspection of native Patterson maps failed to indicate any significant deviation of the particle position from Z=0.25.

The sites of platinum substitution in the K_2PtCl_4 derivative were determined by an icosahedrally locked systematic search of the derivative difference Patterson map at 5 Å resolution. The icosahedral asymmetric unit was searched for potential sites on a 1 Å grid. Three mutually consistent sites were identified by evaluation of predicted cross-vectors: one general site, one site on the five-fold axis, and one site on the three-fold axis. To confirm these results, the general site was used to determine single isomorphous replacement (SIR) centroid phases at 8 Å resolution, and the phases were refined by iterative electron density averaging about the noncrystallographic symmetry elements. Derivative difference Fourier maps calculated with these refined phases contained significant peaks only at the input general site and at the fivefold site. Centroid phases based on these two sites were used to initialize all subsequent phase calculations between 50 and 5 Å.

Native phases were refined at 5 Å resolution by real space averaging, using the method of Bricogne[10]. During this process, the particle position and orientation, and the platinum substitution sites in the particle were repeatedly optimized, treating them as variables in a systematic search that maximized the electron density at the predicted heavy atom sites in unaveraged derivative difference Fourier maps. After the final application of SIR phases, phase refinement at 5 Å converged in eight cycles, but the progress of the refinement suggested that SIR phases would be of limited use at higher resolution. Therefore, the structure determination was extended from 5 Å to 2.9 Å resolution by the addition of 0.05 Å shells of structure factor amplitudes in every few cycles of density averaging. New terms were phased by inversion of the previous cycle's averaged electron density.

As the resolution of the averaging procedure increased, the calculation progressively became larger. If the Bricogne implementation had been used without modification, the large asymmetric unit of poliovirus, together with its high degree of noncrystallographic symmetry, would have required a calculation exceeding the available computing resources (a VAX 11/750 computer with 8 megabytes of memory, and 500 megabytes of disk storage). Repeated tape operations (that would have been required to divide the calculation into pieces of manageable size) would have been impractical, in view of the large number of cycles required by the phase extension procedure. We found, however, that substantial savings in cpu time and disk space could be realized by modifying the double sort algorithm in three ways (that are described below). With these modifications, an entire cycle of density were present) could be run without intervention.

As described by Bricogne[10], the constraints of noncrystallographic symmetry are imposed by iterative averaging of symmetry equivalent points in the electron density map. Fourier inversion of the averaged map yields constrained phases that are applied to observed amplitudes in the next iteration. The feature of the Bricogne implementation that usually makes it suitable for averaging large maps is the "double sorting" technique. Initially, "index records" containing the coordinates of corresponding points in the "source" and "destination" maps are generated, and then bin-sorted according to section number in the source map. Typically, the source map is calculated on a grid five times finer than the nominal resolution, to reduce the errors introduced by linear interpolation to acceptable levels. In the interpolation step, sections of the source map are input in sequence, and held in machine memory two at a time. As each "index record" is input, the interpolated electron density value is determined for the point specified by the "source" coordinates, whereupon an output "density record" is produced that contains the density value, along with a copy of the "destination" coordinates. After these output records have been bin-sorted according to section number in the destination map, sections of the output map are constructed in sequence. The following three modifications were implemented to increase the speed and reduce the storage requirements of this map averaging procedure.

First, Bricogne's "double interpolate" option was used. Thus, the input crystallographic asymmetric unit (calculated on a fine grid) was first averaged to produce the icosahedral asymmetric unit (also on a fine grid); the output crystallographic asymmetric unit (on a coarser grid) was then derived from it by a second interpolation. For poliovirus, with thirty-fold noncrystallographic redundancy, this reduced the size of the calculation by a factor of 2.

Second, during the interpolation steps, output density records were bin-sorted (by section in the destination map) as they were calculated. Specifically, whenever the grid size or the molecular envelope was changed, so that a new index file was generated and sorted, the number of coordinate records in each section of the destination map was determined. During interpolation, the resulting "table of contents" was used to specify the physical record number in a direct-access output file to which each bin-specific buffer of density values was written when it was filled. By eliminating the need for scratch files in the sort operation, this procedure reduced storage requirements by a factor of 3, and made it somewhat faster.

In the third, and most significant modification, 8-point linear interpolation was replaced by a routine that used previously tabulated coefficients to do 64-point nonlinear interpolation. Here, the set of coefficients looked up depends only on the fractional parts of the "source" coordinates. Interpolation is equivalent to determining by least squares the zero-order coefficient of the best-fitting function of a specified form. The function chosen for the poliovirus phase extension was the product of a second order polynomial and a gaussian. Although this function was satisfactory, the performance of other functions is currently being investigated. This modification permitted interpolation out of more coarsely sampled maps without a detectable loss in accuracy. By calculating the "source" map on a grid only three times finer than the nominal resolution, the number of points to be evaluated was reduced by a factor of 5.

The complete process of extending phases between 5 and 2.9 Å resolution required 70 cycles of density averaging, yielding a quadratic R value of 0.27 and a linear correlation coefficient of 0.85. The quality of the electron density map was such that it was possible to locate and identify almost all of the residues in all four capsid protein subunits at 3.6 Å resolution. Residue placement and side chain identifications have been confirmed by inspection of the 2.9 Å map; and a complete atomic model is now being constructed.

No electron density was present for residues 1-20 at the amino terminus of VP1, 1-7 at the amino terminus of VP2, 1-12 at the amino terminus of VP4 (all of which are in the interior of the particle) and residues 234-238 at the carboxy terminus of VP3 (on the outside surface of the virion). In addition, there was no significant unassigned density that could correspond either to the viral RNA or to VPg (a small genome-linked protein). The lack of density for the amino terminal residues and the RNA implies that the interior of the virion is spatially disordered with respect to the icosahedrally symmetric protein shell.

POLIOVIRUS STRUCTURE

Capsid Protein Subunits

The structures of the four capsid proteins are shown in Figure 1. VP1, VP2, and VP3 show an obvious structural homology. Each of the three is composed of a conserved core with variable elaborations. The cores consist of an eight-stranded antiparallel beta barrel with two

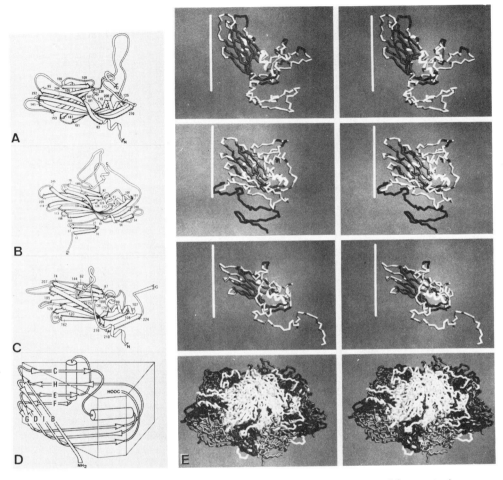

FIGURE 1. Alpha carbon models of the poliovirus capsid proteins. (A) VP1; (B) VP2 and VP4; (C) VP3; (D) diagrammatic representation of the conserved folding pattern of the major capsid proteins; (E) the organization of the major capsid proteins around the five-fold axis. The structurally conserved "cores" of the proteins each consists of a radial (back) helix, a tangential (front) helix, and an eight-stranded antiparallel beta barrel. In the leftmost panels, strands of the beta barrel are indicated by arrows, and labelled either by residue number, or by single strand designations used for the icosahedral plant viruses[29]. In A and C terminal extensions have been removed from the ribbon drawings for clarity. The center and rightmost panels are stereo pairs. In A, B and C helices and beta strands of the core are white and dark gray, respectively; variable loops and terminal extensions are medium gray, except for the sites of monoclonal release mutations, and VP4, which are black. Vertical bars extend from 110 to 160 Å along the five-fold axis (in A), or the three-fold axis (in B and C). In E, five copies each of VP1 (light gray) VP2 (medium gray) and VP3 (dark gray) are shown arranged in a pentamer.

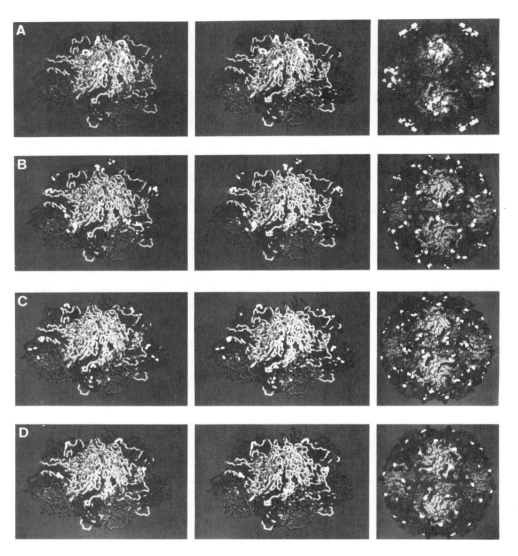

FIGURE 2. Monoclonal release mutations on the outer surface of poliovirus. VP1 is light gray; VP2 is medium gray; VP3 is dark gray. The sites of individual monoclonal release mutations are shown in white. As described in the text, these mutations are grouped by proximity into four clusters: (1) cluster 1; (B) cluster 2; (C) cluster 3; (D) cluster 4. The center and leftmost panels are stereo pairs, showing the organization of the major capsid proteins around the five-fold axis. The rightmost panels depict the outer surface of the virion. Both representations are based on alpha carbon models.

flanking helices. Four of the beta strands (B, I, D, and G in Figure 1d) make up a large twisted beta sheet that forms the front and bottom surfaces of the barrel. The remaining four strands (C, H, E, and F in Figure 1d) make up a shorter, flatter beta sheet that forms the back surface of the barrel. The strands of the front and back surfaces are connected at one end by short loops, giving the barrel the shape of a triangular wedge. Although the cores of the three large capsid proteins are very similar, the connecting loops and amino and carboxy terminal extensions are all dissimilar. In particular, VP1 has a large insertion (residues 207-237) in the loop connecting the G and H strands of the beta barrel and a relatively large loop (residues 96-104) connecting the top two strands (B and C) of the beta barrel. VP2 has a large insertion (residues 127-185) in the loop connecting the E strand of the barrel with the "back" helix. The largest insertion in VP3 (residues 53-69) occurs in the middle of the B strand of the beta barrel.

The structural description of the three large proteins in terms of cores, connecting loops, and terminal extensions is also relevant to function, since each of these units appears to play a different role in the structure of the intact virus. The cores make up the continuous shell of the virus, the connecting loops and carboxy terminal extensions make up many of the major features on the external surface of the virion, and the amino terminal extensions cover much of the internal surface of the capsid. The small protein VP4 (shown along with VP2 in Figure 1b) is similar in many respects to the amino terminal extensions of VP1 and VP3. Like them, it is rather extended in structure, (except for a short two-stranded antiparallel beta sheet near its amino terminus), and it occupies a totally internal position, forming a significant portion of the internal surface of the capsid. Since VP4 is covalently linked to VP2 until very late in the assembly of the virions, it may be considered to be the detached amino terminal extension of VP2.

Virion Structure

The arrangement of the major capsid proteins in the virion is depicted in Figures 1e and 2. Five copies of VP1 are clustered around the particle five-fold axis, with the narrow ends of the beta barrels closest to the axis. A pronounced tilt of the beta barrels outward along the five-fold axis results in the exposure of the top three connecting loops at the narrow end of the barrel, and the formation a prominent surface protrusion extending to 160 Å radius. VP2 and VP3 alternate around the particle three-fold axis, also with the narrow ends of the barrels closest to the axis. The tilts of the barrels of VP2 and VP3 are less pronounced. As a result, only the top two loops at the narrow end of the barrels of VP2 and VP3 are exposed, and the surface protrusion at the three-fold extends to a lesser radius (150 Å). This three-fold protrusion is ringed by two sets of outward projections (or promontories). The larger projection is dominated by the large insertion in VP2 (residues 127-185) with contributions from the carboxy terminus of VP2, the large insertion in VP1 (residues 207-237) and residues 271-295 near the carboxy terminus of VP1. This large projection extends to a radius of 165 Å. The smaller projection is formed by the insertion in VP3 (residues 53-70) and extends to 155 Å radius. The peaks at the five-fold axes are surrounded by broad deep valleys, while the three-fold plateaus are separated by saddle surfaces at the two-fold axes.

SIGNIFICANCE

Virus Evolution

As it had been predicted from earlier sequence comparison, the structures of the Mahoney strain of type 1 poliovirus and of rhinovirus 14 are strikingly similar. In fact, the "cores" of the capsid proteins are nearly identical, and the major structural differences occur in loops that are exposed on the surface of the virions (especially the top loop [residues 96-104] in VP1, and the large insertions in VP1 and VP2). The degree of sequence and structural similarity clearly indicates that the two viruses are very closely related, and points to the danger of classifying viruses within families on the basis of pathology, or physical characteristics such as acid stability or permeability to ions.

A much more surprising observation is that poliovirus (and hence the rhinovirus) capsid proteins are similar in structure to the capsid proteins of the icosahedral plant viruses whose structures are known. The capsid proteins of the plant viruses have an identical folding pattern (illustrated in Figure 1d), and a similar organization into cores (that pack together to form the capsid shell), connecting loops (that decorate the outer surface), and amino terminal strands (that are located on the inner surface of the capsid, and play a role in directing assembly). Overall, the packing of capsid proteins in poliovirus is similar to the packing of proteins in the "S-domains" of the T=3 plant viruses (although it differs significantly in detail, since the constraints imposed by the quasi-equivalent packing of chemically identical subunits have been relaxed). Although these structural similarities strongly indicate an evolutionary relationship, the exact nature of the relationship is unclear. Despite the structural similarity, plant viruses and picornaviruses differ significantly in their gene order, their mechanisms for the control of gene expression, and in details of the assembly process. Therefore, alternatives to direct descent from a common ancestor must be considered. One alternative possibility is that these viruses have "borrowed" a common protein and its message from their respective hosts.

Assembly

The poliovirus structure has yielded important insights into the process of picornavirus assembly. The amino terminal extensions of VP1, VP2 (including VP4), and VP3 form an intricate network on the inner surface of the capsid. Within pentamers (Figure 1e), these interactions are particularly extensive, but between adjacent pentamers, they are less extensive. This confirms the importance of pentameric intermediates in capsid assembly, as originally proposed from the physical characterization of assembly intermediates isolated from infected cells. It also suggests that the amino terminal arms may play a role in controlling the assembly process, as they do in the T=3 plant viruses. One especially interesting portion of the network is a structure formed at the five-fold axis. Five copies of the amino terminus of VP3 wrap around the axis, forming a highly twisted tube stabilized by parallel beta interactions. This structure can form only upon pentamer formation, and it may serve to direct and/or stabilize the formation of the pentameric intermediate. An analogous structure (called a "beta annulus") has been observed in the T=3 plant viruses. Like the twisted tube in poliovirus, the beta annulus appears to play a critical role in directing the assembly of T=3 capsids. In the T=3 plant viruses, however, the beta annuli occur at three-fold axes, rather than at five-fold axes, underscoring the differences in the assembly of T=3 as opposed to T=1 virions.

The locations of the amino and carboxy termini of the capsid proteins provides information about the process of virion assembly. All virally encoded proteins are translated together as a 220 kilodalton polyprotein that is subsequently processed by viral proteases. One product of an early cleavage is the precursor P1-1a containing the capsid proteins in the order VP4-VP2-VP3-VP1. This precursor is cleaved to yield VP0, VP3, and VP1 in an event that appears to be linked to the early stages of assembly. The final cleavage of VP0 into VP4 and VP2 occurs late in assembly, after the encapsidation of RNA[11]. In the mature virion, the carboxy terminus of VP4 is located very close to the the first residue that is ordered at the amino terminus of VP2 (residue 8). The seven missing residues could easily bridge the gap. It is significant, however, that both termini are located in the interior of the virion, where they are inaccessible to exogenous proteases. Unless a substantial conformational rearrangement occurs subsequent to the final cleavage, this cleavage must be (in some sense) "autocatalytic".

In contrast to the proximity of the termini generated by the maturation cleavage, the amino and carboxy termini generated by the earlier cleavages of VP2 from VP3, and VP3 from VP1 are quite distant from one another. Indeed, the amino termini are on the inner surface of the capsid, while the carboxy termini are on the outside. The termini must, therefore, undergo substantial rearrangement subsequent to the cleavages. The apparent role of the amino terminal extensions in directing and stabilizing the formation of pentamers thus suggests that the proteolytic processing and subsequent rearrangements play an important role in the control of virion assembly.

Antigenic Sites

The antigenic sites of poliovirus have been identified by two techniques: the mapping of mutants resistant to neutralizing monoclonal antibodies[12-18], and the identification of synthetic peptides capable of eliciting or priming for neutralizing antibodies.[16,19-23] The antigenic sites identified by monoclonal release mutations are shown in Figure 2. Residues that contribute to these sites are indicated in Figure 1. (The distribution of these sites is similar to that observed in rhinovirus 14).[3] As illustrated in Figure 2, the sites occur on the exterior of the capsid, in four clusters defined by proximity. Cluster 4 is sufficiently close to clusters 2 and 3 that it actually might be part of either; and recent data indicates that residues from clusters 3 and 4 can form parts of the same neutralizing site [24]. Each of the clusters, except for cluster 4, includes amino acids from several discrete parts of a polypeptide, and two of the clusters are composed of amino acids from more than one polypeptide chain. In several instances, it has been confirmed experimentally that these linearly separated residues can form parts of the same conformational determinant (since mutations in different sites can result in escape from neutralization by the same monoclonal antibody). An intriguing, but as yet unexplained finding is that whereas most of the escape mutations from type 1 poliovirus map into clusters 2 and 3, nearly all of the escape mutations in type 3 poliovirus map into cluster 1 (specifically, to residues 89-104 in the top loop of VP1).

All sites of monoclonal release mutation identified thus far are located on the surface of the virus, in highly exposed loops that are readily accessible to antibody binding. These sites are expected to accommodate point mutations locally, without disrupting structure elsewhere in the virion. Thus, in poliovirus, there is no need to propose that the mutations act by disrupting spatially distant antibody binding sites. The location of antibody binding sites on exposed loops at the surface of an antigen has also been observed in the haemagglutinin and neuraminidase of influenza virus [25-26]. The decoration of a virus surface with exposed (and mutable) loops seems likely to be a common mechanism by which animal viruses escape immune surveillance.

Synthetic peptides capable of eliciting or priming for a neutralizing response have been identified for portions of all four clusters. In several cases, two peptides that are distant from one another in the sequence, but map into the same cluster, were both capable of eliciting a neutralizing response. This observation suggests that the distinction between conformational (discontinuous) and sequential (continuous) antigenic sites may be somewhat artificial since an immunogenic peptide (normally thought to indicate a sequential determinant) may form a portion of a conformational site.

Several of the peptides that are capable of eliciting a neutralizing response do not map in or near the clusters of monoclonal release mutations. Surprisingly, these peptides (three from the amino terminal extension of VP1 and one from the bottom loop of the closed end of the barrel of VP1) are on the interior of the mature virion. The ability of these buried peptides to induce a neutralizing response may prove to be important in understanding the conformational changes that may be involved in neutralization.

FUTURE IMPLICATIONS

Knowlege of the structure of poliovirus has begun to provide insights into the evolution and assembly of the virus, and into the relationship between virus structure and immune recognition. We believe that the potential of the structure is just beginning to be realized. In particular, the structure will be useful for the design and interpretation of experiments exploring several new areas, including the mechanisms of neutralization, and the structural factors responsible for host and tissue tropism and the attenuation of neurovirulence. Finally, we anticipate that the structure will be crucial for the design and interpretation of site directed mutagenesis experiments, using infectious DNA copies of the genome to construct mutants in the capsid region.

ACKNOWLEDGEMENTS

This work was supported by NIH Grant AI-20566 to J. M. Hogle, and in part by a grant from the Rockefeller Foundation and by NIH Grant AI-22346 to D. Baltimore (Whitehead Institute for Biomedical Research). D.J. Filman was supported in part by Training Grants NS-07078 and NS-12428. Preliminary crystallographic studies which led to this work were supported by NIH Grant CA-13202 to S. C. Harrison. We would like to thank Richard Lerner, David Baltimore, and Stephen Harrison for their interest and support throughout various phases of this work; and Joseph Icenogle, Carl Fricks, Michael Oldstone and Ian Wilson for helpful comments and discussions. Ribbon drawings in Figure 1 were produced with the assistance of E. Getzoff. Stereo pairs in Figures 1 and 2 were produced with the assistance of D. L. Bloch, using the program RAMS2[27]. Molecular surface representations in Figure 2 were calculated by A. J. Olson, using the programs AMS and RAMS[27-28].

This is contribution no. 4217-MB from the Department of Molecular Biology, Research Institute of Scripps Clinic, La Jolla, California.

REFERENCES

1. R.R. Rueckert, On the structure and morphogenesis of picornaviruses, in: "Comprehensive Virology", vol. 6, H. Fraenkel-Conrat and R. R. Wagner, eds., Plenum (1976).
2. J.M. Hogle, M. Chow, and D.J. Filman, Three-dimensional structure of poliovirus at 2.9 Å resolution. Science, 229:1358 (1985).
3. M.G. Rossmann, E. Arnold, J.W. Erickson, E.A. Fankenberger, J.P. Griffith, H.-J. Hecht, J.E. Johnson, G. Kamer, M. Luo, A.G. Mosser, R.R. Rueckert, B. Sherry, and G. Vriend, Structure of a human common cold virus and functional relationship to other picornaviruses. Nature, 317:145 (1985).
4. S.C. Harrison, A.J. Olson, C.E. Schutt, F.K. Winkler, and G. Bricogne, Tomato bushy stunt virus at 2.9 A resolution. Nature, 276:368 (1978).
5. C. Abad-Zapatero, S.S. Abdel-Mequid, J.E. Johnson, A.G.W. Leslie, I. Rayment, M.G. Rossmann, D. Suck, and T. Tsukihara, Structure of southern bean mosaic virus at 2.8 Å resolution. Nature, 286:33 (1980).
6. L. Liljas, T. Unge, T.A. Jones, K. Fridborg, S. Lovgren, U. Skoglund, and B. Strandberg, Structure of satellite tobacco necrosis virus at 3.0 Å resolution. J. Mol. Biol., 159:93 (1982).
7. J.M. Hogle, A. Maeda, and S.C. Harrison, Structure and assembly of turnip crinkle virus, I: x-ray crystallographic structure analysis at 3.2 Å resolution. J. Mol. Biol. 191:625-638 (1986).
8. W.P.J. Gaykema, W.G.J. Hol, J.M. Vereijken, N.M. Soeter, H.J. Bak, and J.J. Beintema, 3.2 Å structure of the copper-containing, oxygen-carrying protein Panulirus interruptus haemocyanin. Nature, 309:23 (1984).
9. F.K. Winkler, C.E. Schutt, and S.C. Harrison, The oscillation method for crystals with very large unit cells, Acta Cryst., A35:901 (1979).
10. G. Bricogne, Methods and programs for direct-space exploitation of geometric redundancies. Acta Cryst., A32:832 (1976).
11. M.A. Pallansch, O.M. Kew, B.L. Semler, D.R. Omilianowski, C.W. Anderson, E. Wimmer, and R. R. Rueckert, Protein processing map of poliovirus. J. Virol., 49:873 (1984).
12. E.A. Emini, B.A. Jameson, A.J. Lewis, G.R. Larsen, and E. Wimmer, Poliovirus neutralization epitopes: analysis and localization with neutralizing monoclonal antibodies. J. Virol., 43:997 (1982).
13. R. Crainic, P. Couillin, B. Blondel, N. Cabau, A. Boue, and F. Horodniceanu, Natural variation of poliovirus neutralization epitopes. Infect. Immun., 41:1217 (1983).
14. P.D. Minor, G.C. Schild, J. Bootman, D.M.A. Evans, M. Ferguson, P. Reeve, M. Spitz, G. Stanway, A.J. Cann, R. Hauptmann, L.D. Clarke, R.C. Mountford, and J.W. Almond, Location and primary structure of a major antigenic site for poliovirus neutralization. Nature, 301:674 (1983).
15. P.D. Minor, D.M.A. Evans, M. Ferguson, G.C. Schild, G. Westorp, and J.W. Almond, Principal and subsidiary antigenic site of VP1 involved in the neutralization of poliovirus type 3. J. Gen. Virol., 65:1159 (1985).
16. M. Ferguson, D.M.A. Evans, D.I. Magrath, P.D. Minor, J.W. Almond, and G.C. Schild, Induction by synthetic peptides of broadly reactive, type-specific neutralizing antibody to poliovirus type 3. Virology, 143:505 (1985).
17. D.C. Diamond, B.A. Jameson, J. Bonin, M. Kohara, S. Abe, H. Itoh, T. Komatsu, M. Arita, S. Kuge, A. Nomoto, A. D. M. E. Osterhaus, R. Crainic, and E. Wimmer, Antigenic variation and resistance to neutralization in poliovirus type 1. Science, 229:1090 (1985).
18. P. D. Minor, personal communication.

19. M. Chow, R. Yabrov, J. Bittle, J. Hogle, and D. Baltimore, Synthetic peptides from four separate regions of the poliovirus type 1 capsid protein VP1 induce neutralizing antibodies. Proc. Natl. Acad. Sci. USA, 82:910 (1985).
20. E. A. Emini, B.A. Jameson, and E. Wimmer, Priming for and induction of anti-poliovirus neutralizing antibodies by synthetic peptides. Nature, 304:699 (1983).
21. E. A. Emini, B.A. Jameson, and E. Wimmer, Identification of a new neutralization antigenic site on poliovirus coat protein VP2. J. Virol., 52:719 (1984a).
22. E.A. Emini, B.A. Jameson, and E. Wimmer, Identification of multiple neutralization antigenic sites on poliovirus type 1 and the priming of the immune response with synthetic peptides, in: "Modern Approaches to Vaccines", R. M. Chanock, and R. A. Lerner, eds., Cold Spring Harbor Laboratory (1984b).
23. B.A. Jameson, J. Bonin, M.G. Murray, E. Wimmer, and O. Kew, Peptide-induced neutralizing antibodies to poliovirus, in: "Vaccines 85", R.A. Lerner, R.M. Chanock, and F. Brown, eds., Cold Spring Harbor Laboratory (1985).
24. P.D. Minor, M. Ferguson, D.M.A. Evans, and J.P. Icenogle, Antigenic structure of polioviruses of serotypes 1, 2, and 3. J. Gen. Virol. 67:1283-1291 (1986).
25. D.C. Wiley, I.A. Wilson, and J.J. Skehel, Structural identification of the antibody-binding sites of Hong Kong influenza haemagglutinin and their involvement in antigenic variation, Nature, 289:373 (1981).
26. P.M. Colman, J.N. Varghese, and W.G. Laver, Structure of the catalytic and antigenic sites in influenza virus neuraminidase, Nature, 305:41 (1983).
27. M.L. Connolly, Depth-buffer algorithms for molecular modelling. J. Mol. Graphics, 3:19 (1985).
28. M.L. Connolly, Analytical molecular surface calculation. J. Appl. Crystallogr., 16:548 (1983).
29. M.G. Rossmann, C. Abad-Zapatero, M.R.N. Murthy, L. Liljas, T.A. Jones, and B. Strandberg, Structural comparisons of some small spherical plant viruses. J. Mol. Biol., 165:711 (1983).

THE STRUCTURE OF COWPEA MOSAIC VIRUS AT 3.5 Å RESOLUTION

C.V. Stauffacher, R. Usha[+], M. Harrington, T. Schmidt,
M.V. Hosur and J.E. Johnson

Department of Biological Sciences, Purdue University
West Lafayette, IN 47907

[+]Present address: Department of Organic Chemistry, Indian
Institute of Science, Bangalore, India

INTRODUCTION

There are nine groups of icosahedral, positive stranded RNA plant viruses (Francki et al., 1985) established on the basis of immunological relationships, genomic size, the number of RNA molecules composing the genome and the stability of the capsid. Eight of these groups are remarkably similar in both structural and functional properties while one group, the comoviruses, is quite different from the others (Goldbach and vanKammen, 1985). Cowpea mosaic virus (CPMV), the type member of the comovirus group, has only two properties in common with any of the other plant virus groups. The genome of CPMV is bipartite and a small protein (VpG) is linked to the 5' terminus of each RNA molecule (Bruening, 1977; Daubert et al., 1979). In all its other properties CPMV is unique among the plant viruses. The RNA molecules of CPMV are polyadenylated at their 3' termini while the genomes of the other virus groups are not. CPMV proteins are generated by the processing of two polyproteins, produced by translation of single open reading frames on each RNA molecule. The proteases that cleave the polyproteins are virally encoded and are initially part of the translation product of the larger of the two RNA molecules. The other plant virus groups utilize subgenomic RNA molecules as the messengers for protein synthesis with no posttranslational processing. The CPMV genome codes for a total of eight proteins, three (including the capsid proteins) derived from the small RNA molecule, and five from the large RNA. The other plant virus groups normally contain the genes for only four proteins. The T=1 icosahedral capsid of CPMV is composed of two proteins, one large (42 Kd) and one small (24 Kd), which separately encapsulate each of the genomic RNAs. The RNA molecules of other spherical plant viruses are encapsulated in either T=1 or T=3 icosahedral shells (Caspar and Klug, 1962) but these shells contain only a single type of capsid protein.

While CPMV differs from the other plant viruses in the properties discussed above, all of these properties, with the exception of the bipartite genome, are common to the animal picornaviruses (Rueckert, 1985). In addition to these qualitative similarities, recent sequence comparisons have revealed homology between the large RNA of CPMV and the 3' proximal region of the poliovirus RNA (Franssen et al., 1984). These comparisons (Figure 1) suggest that CPMV and the picornaviruses may have the same gene order if the small RNA of CPMV is considered the 5' end of the total CPMV genome. The physical properties of CPMV particles, protein and RNA molecules are listed in Table I.

We undertook the study of CPMV using X-ray crystallography to compare its structure with typical plant viruses (e.g. southern bean mosaic virus (SBMV), Abad Zapatero et al., 1980) and picornavirus structures (e.g. human rhinovirus 14 and poliovirus, this volume). We present here a 3.5 Å electron density map of CPMV from which we have traced nearly the entire polypeptide chain of each protein subunit. Our results show that CPMV is like the picornaviruses in the quaternary structure of the capsid, but is similar to plant virus subunits in its tertiary structure.

METHODS

Virus Production

The Bil mutant of the yellow strain of CPMV was propagated in blackeyed peas. The initial innoculum was kindly supplied by Professor George Bruening, University of California, Davis. The virus was prepared using the method described by Siler et al. (1976) and separated into the four components listed in Table I.

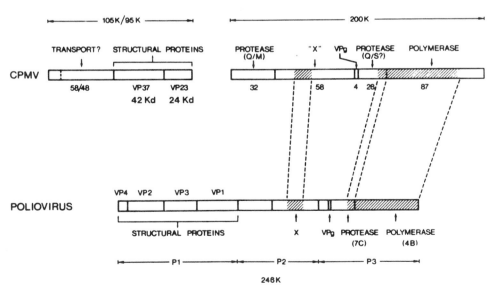

Figure 1: Comparison of the genome organization of CPMV and poliovirus. M-RNA (left) and B-RNA (right) of CPMV are shown aligned with the RNA of poliovirus. The molecular weight and function are marked for each gene product. Homologous regions in the two genomes are shaded. (From Franssen et al., 1984)

Table I

Cowpea Mosaic Virus
Icosahedral Particles

Spherically Averaged Diameter	284 Å
5-5 Dimension	308 Å
3-3 Dimension	272 Å
2-2 Dimension	254 Å
Average Thickness of Protein Shell	39 Å

Capsid Proteins

	# of Amino Acids	M.W.
Large Subunit (VP37)	374	41,720
Small Subunit (VP23)	213	23,930

(60 copies of each subunit in the capsid)

Virus Components

Particle	%RNA	M.W.	S	Buoyant Density
Top	0	3.94×10^6	58	1.297
Middle	24%	5.16×10^6	98	1.402
Bottom (upper)	34%	5.98×10^6	118	1.422
Bottom (lower)	34%	5.98×10^6	?	1.470

RNA

Component	M.W.	# of Ribonucleotides	5'	3'	# of Proteins
M-RNA	1.22×10^6	3482	VpG(4Kd)	polyA(100)	3
B-RNA	2.02×10^6	5889	VpG(4Kd)	polyA(87)	5

Crystallization

All three nucleoprotein components, as well as the empty capsid, form isomorphous crystals and can even cocrystallize suggesting that no differences exist in their capsid structures. The results in this paper were obtained from a study of cubic crystals of CPMV. These crystals differ from the hexagonal form previously reported (White and Johnson, 1980; Johnson and Hollingshead, 1981). Cubic crystals displaying rhombic dodecahedral morphology (Figure 2) were prepared by vapor diffusion (McPherson, 1982). The reservoir solution was 0.4M ammonium sulfate, 2% polyethylene glycol (PEG) 8000 (w/v) and 0.05M potassium phosphate adjusted to pH 7.0. The virus solution was prepared at 35 mg/ml in 0.05M potassium phosphate buffer (pH 7.0) using one of the isolated virus components. Purified middle component was consistently used in these crystallizations. The solution in the crystallization well was prepared by mixing 25 μl of virus solution with 25 μl of reservoir solution. Crystals normally appeared in one week and often grew to a size in excess of 1mm. The number of crystals and their rate of growth was greatly increased by seeding. This was performed by crushing a single crystal in 5 μl of 0.05M phosphate buffer. The slurry of microcrystals was injected in the first well of virus solution and agitated. Then 5 μl from this well was transferred to the next well and the process repeated for all nine wells in the plate. The number of crystals in each well was usually proportional to the dilution. Crystals grew in one to two days when the wells were seeded.

Figure 2: I23 cubic crystals of CPMV. The long axis of the rhombic dodecahedron for the largest crystal shown measures approximately 1mm.

Heavy atom derivative preparation

An ethylmercuryphosphate (EMP) derivative of CPMV was prepared by cocrystallization. A 1 mM solution of EMP (M.W. 362.5) was prepared in 0.4M ammonium sulfate, 0.05M potassium phosphate solution (pH 7.0). A second solution of 0.4M ammonium sulfate, 5% PEG 8000 (w/v) and 0.05M potassium phosphate (pH 7.0) was prepared for mixing with the virus solution. The solution in the well consisted of 25 μl of virus solution (40 mg/ml) in 0.05M phosphate buffer (pH 7.0) mixed with 25 μl of the buffer containing 5% PEG and 28 μl of the EMP solution. The reservoir solution was the same as that used for the native crystals and the wells were serially seeded as previously described. Rhombic dodecahedral crystals with a pale yellow color grew in approximately one week.

Crystal data

Crystals obtained by these methods are either P23 or I23 cubic crystals. Morphologically these two forms are indistinguishable. The primitive crystals diffract to a maximum resolution between 5.0 Å and 3.5 Å and display significant diffuse scattering. The body centered crystals diffract to beyond 2.8 Å resolution and show little diffuse scattering. The lattice constants of the two forms are identical within experimental error (a = 317.3(1) Å). Figure 3a shows a 4.5° precession photograph of the primitive crystals and Figure 3b shows a 0.5° oscillation photograph of an I23 crystal. The I23 crystal can be converted to the P23 form on extended exposure in the X-ray beam. Approximately 90% of the crystals are initially in the primitive form, as indicated by the diffraction patterns we have collected, necessitating more extensive data collection to complete a data set consisting of only body centered patterns. The I23 cubic crystals contain two virus particles per unit cell at the tetrahedral special positions 0 0 0 and 1/2 1/2 1/2. Since the virus particles have icosahedral symmetry there is five-fold noncrystallographic symmetry in the crystals.

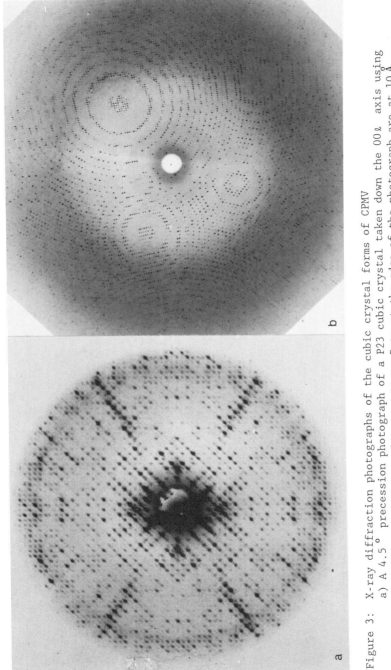

Figure 3: X-ray diffraction photographs of the cubic crystal forms of CPMV
a) A 4.5° precession photograph of a P23 cubic crystal taken down the 00ℓ axis using an Elliott rotating anode generator. Data at the edge of the photograph are at 10 Å resolution.
b) A 0.5° oscillation photograph of an I23 cubic crystal taken in a four minute exposure at the Cornell Synchrotron (CHESS). The diffraction pattern in this film extends to 2.8 Å.

Table II

Data Collection and Processing
Cowpea Mosaic Virus I23 Cubic Crystals

I. Native (3.5Å)
   (Data Collected on Rotating Anode)

Number of Photographs	25
Number of Measurements >2 Sigma	53,743
Number of Unique Measurements >2 Sigma	38,654
Scaling R Factor	10.5%

II. Derivative (3.5Å) Ethyl Mercury Phosphate
    (Data Collected at LURE, Orsay, France)

Number of Photographs	25
Number of Measurements >2 Sigma	115,929
Number of Unique Measurements >2 Sigma	49,784
Number of Scaling R Factor	13.7%

III. Fitting Native and Derivative Data

Resolution Ranges (Å)	15.0	7.5	5.0	3.75	3.0
Abs. Difference	162.4	75.7	68.6	78.9	80.7
Number	359	4091	9346	14713	2661
Average F	549.9	539.1	462.1	495.8	427.8
Mean Error	24.4	29.3	38.4	46.0	52.4
R Factor	18.7%	13.8%	14.8%	15.9%	19.1%

Data collection

Native data were collected using a rotating anode generator employing the oscillation method (Arndt and Wonnacott, 1977). Crystals were mounted with the 10$\bar{1}$ axis coincident with the spindle. In this orientation an angular range of 23° is required for a complete data set. For the 3.5 Å resolution data collection, 1.0° oscillation photographs were taken with an overlap of .1° and exposure times of twenty hours. Five native photographs were also recorded at the LURE synchrotron, Orsay France. These photographs were taken in a similar manner but required only a thirty minute exposure.

The data from the EMP derivatized virus were all collected at LURE. Thirty minute, 1.0° exposures were also used for these films. Native and derivative films were processed using the program developed by Rossmann (1979). Each data set was scaled and post refined (Rossmann et al., 1979) with results shown in Table II. The native and derivative data sets were locally scaled in thirteen resolution ranges to eliminate systematic differences in the data sets. Significant changes in intensity between native and derivative are seen to 3.5 Å.

Heavy atom position determination, refinement, and single isomorphous replacement phasing.

A difference Patterson map for the EMP derivative was computed to 5.5 Å resolution at a grid spacing of 1.5 Å. The map was systematically searched within one icosahedral asymmetric unit for heavy atom sites with interatomic vectors consistent with icosahedral symmetry (Argos and Rossmann, 1976). The search grid was sampled at 2 Å intervals within radial limits of 90 Å and 150 Å. The search revealed a single site where the vector density was two times higher than any other position in the map. The section containing the largest peak is shown in Figure 4. The heavy atom position was refined and single isomorphous replacement (SIR) phases were computed. Difference electron density maps using these phases showed no additional EMP sites. Table II shows that the derivative data set collected at LURE was much more complete than the native rotating anode data set. After calculating SIR phases based on the native and derivative data we used the derivative data as if it were native for calculating the maps since phase refinement and extension by real space averaging is more powerful when a complete or nearly complete data set is used (Rayment, 1983). As a result our electron density map contains the heavy atom as part of the density, but this has not caused any problems during the structure solution. Figure 5a shows the appearance of the SIR phased 5.5 Å map. This electron density was then averaged using the five-fold noncrystallographic symmetry (Bricogne, 1976; Johnson, 1978). The resultant map in Figure 5b clearly revealed an envelope of the protein shell indicating that the SIR phases were of good quality. This envelope agrees well with the projected shape of the CPMV particle obtained from electron microscopy studies of two dimensional crystals (Figure 6 and Hollingshead and Johnson, 1980).

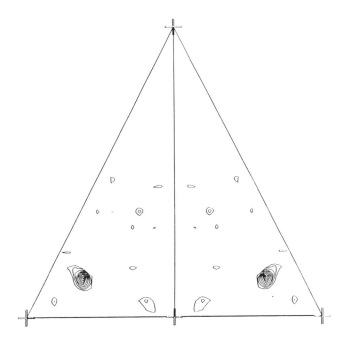

Figure 4: Patterson vector search for the EMP derivative.
The results of the Patterson search are represented as a vector density map of the icosahedral asymmetric unit. The triangular section shown here is defined by the icosahedral fivefold at the top, the twofold at the bottom center and the two adjacent threefold axes at the corners. Patterson symmetry requires that the vector peaks be mirrored across the line connecting the fivefold and twofold axes. The highest peak in the map is the peak of nine contours shown in this section. The next highest peak in the map is half this size.

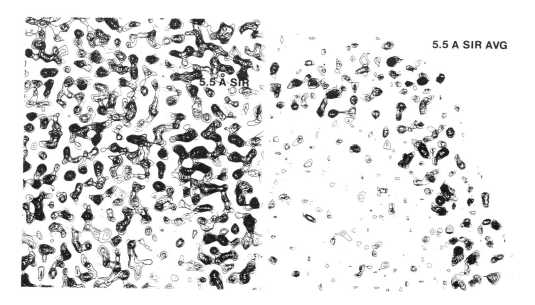

Figure 5: Effect of averaging about the noncrystallographic symmetry axis on the CPMV electron density map. A stack of six sections (11 Å total thickness) of the 5.5 Å electron density map for CPMV is shown. The map is sectioned along the z coordinate of the I23 cell. The origin of the map is located at the lower left hand corner of each map, which is the center of a CPMV particle. Twofold axes run along each edge and the noncrystallographic fivefold at an angle of 32° from the vertical axis.
a) 5.5 Å SIR phased electron density map of CPMV.
b) The same map averaged once over the noncrystallographic fivefold symmetry.

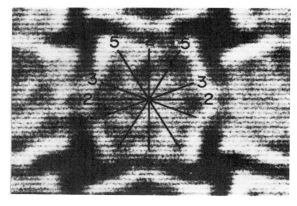

Figure 6: A single CPMV particle selected from a computer filtered image of a two dimensional array of the virus. The micrograph shows the characteristic shape of CPMV with a bulge at the fivefold (Crowther et al., 1974). Icosahedral axes are superimposed on the particle image. The upper right hand quarter of the particle is comparable to the electron density shown in Figure 5b.

Phase Refinement and Extension

A process of cycling the electron density maps with constraints of noncrystallographic symmetry and a molecular envelope was used for both phase refinement and extension. First a map was calculated with phases and weights from previous cycles using $\omega F_{obs} e^{i\alpha}calc$ as coefficients where

$$\omega = \exp\left(-\frac{(|F_{obs} - F_{calc}|)}{F_{obs}}\right).$$

All data with $F/\sigma(F) > 1.0$ were included in the calculation of the map. This map was masked with a simple geometrical envelope representing the protein capsid shell. "Solvent" outside the mask was flattened, and the map averaged over the non-crystallographic fivefold symmetry. This averaged electron density was Fourier transformed to produce calculated structure factors. New weights and phases were extracted from this data and the cycle repeated until convergence.

The first cycles of this process were used to improve the SIR phases for CPMV at 5.5 Å resolution. Eight cycles of phase refinement were performed giving a final phase change of less than 5° and an R factor of 26.7%. Phase extension to 3.5 Å was then done in seventeen increments following the procedure used successfully for rhino 14 virus (this volume). In the first cycle of each increment of phase extension the electron density map was Fourier transformed to a resolution radially extended by two lattice points in reciprocal space beyond the converged resolution. For example, in the extension from 5.5 Å, data to 5.32 Å were calculated and thereafter included in the cycling. The reflections in this additional shell of data were at first very weak, but present due both to the envelope and the averaging process. Four to five cycles of real space averaging were then performed until phases, R factor, and correlation coefficient converged.

One additional change was made in the phase extension cycles from 4.0 Å to 3.5 Å. The observed data set is approximately 80% complete using the criteria that $F/\sigma(F) > 1.0$. Rayment (1983) has suggested that including an estimate for the missing data by using F_{calc}s increases the power of the averaging method. During the phase extension from 4.0 Å to 3.5 Å resolution the scaled F_{calc} values were used in the calculation of the electron density map to be averaged. Although this made little visible difference in the map itself, the addition of calculated data significantly improved the overall R factor and correlation coefficient through all resolution ranges. A summary of the phase extension results is shown in Figure 7. The final R factor at 3.5 Å resolution was 21.1% and the correlation coefficient was .80 for 51,839 reflections.

RESULTS

The 3.5 Å electron density map of CPMV shows a clear contrast between protein and solvent. The maximum particle dimension is 308 Å near the fivefold icosahedral axes. The particle is 267 Å near the threefold axes and 260 Å at the twofold axes. These dimensions agree well with the particle size and shape derived from the filtered electron micrographs of two dimensional CPMV particle arrays (Figure 6 and Table I). The thickness of the protein shell varies between 25 Å and 40 Å which agrees well with the results obtained from solution X-ray scattering studies of CPMV empty capsids (Schmidt et al., 1983).

The electron density in the icosahedral asymmetric unit of the CPMV map (the volume defined by a fivefold axis and two adjacent threefold axes)

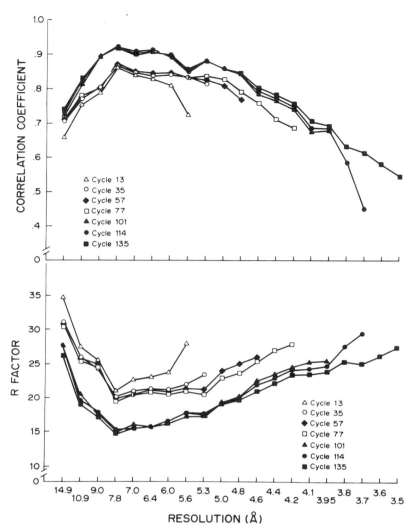

Figure 7: Statistics for the phase extension from 5.5 Å to 3.5 Å. Statistics are given in shells of resolution for the final cycle in representative steps of the phase extension.

a) Correlation coefficients (C)
$$C = \frac{\sum_h (<F_o> - |F_o(h)|)(<F_c> - |F_c(h)|)}{[\sum_h (<F_o> - |F_o(h)|)^2 (<F_c> - |F_c(h)|)^2]^{1/2}}$$

b) R factor (R)
$$R = \frac{\sum_h |(|F_o(h)| - |F_c(h)|)|}{\sum_h |F_o(h)|}$$

where $|F_o(h)|$ and $|F_c(h)|$ are the observed and calculated structure factor amplitudes for reflection h, $<F_o>$ and $<F_c>$ are the mean observed and calculated amplitudes for the shell of data, and the sum is taken over all reflections h in the shell.

displays a striking degree of quasi-threefold symmetry (Figure 8). In this figure the sections of the electron density map are cut perpendicular to the icosahedral twofold axis positioned midway between the icosahedral threefold axes. The quasi-threefold axis is at the centroid of the triangle. This volume element contains the density for one large and one small protein subunit, a total molecular weight of 66 Kd.

It was immediately apparent in the 3.5 Å resolution map that the electron density distribution in CPMV closely resembled that of the T=3 icosahedral plant viruses (e.g. southern bean mosaic virus (SBMV), Abad Zapatero et al., 1980). Indeed, at all levels of our structural analysis there are straight forward relationships between the two viruses. This

Figure 8: 3.5 Å resolution electron density map of CPMV. Ten sections of the electron density map are shown, representing an 11 Å thick slice of the CPMV structure. The large triangle defines the icosahedral asymmetric unit as described in Figure 4. Two of the β strands and their connecting turns are marked for each of the three domains, as well as the top of the loop containing αB. The symbols represent comparable strands in SBMV (see Figure 9) as follows:

βB-βC
βD-βE
βF-βG
βH-βI
αB loop

similarity compels us to present our results with respect to the SBMV structure, since a nomenclature for this and the related virus structures of tomato bushy stunt virus (TBSV) (Harrison et al., 1978) and satellite tobacco necrosis virus (STNV) (Liljas et al., 1982) is well established. Figure 9c is a diagrammatic representation of the subunit positions in the SBMV icosahedral asymmetric unit. The A subunits are clustered at the fivefold axes while the B and C subunits are clustered at the icosahedral threefold axes. The three domains of CPMV are located roughly at the positions of the A, B and C subunits in SBMV, and we refer to the domains by these letters. In SBMV each domain is an individual subunit and each subunit has the same amino acid sequence. In CPMV each domain has a different amino acid sequence and two of the domains are covalently linked. The electron density of each domain is roughly wedge shaped with the narrow end near the icosahedral axes and the wide end near the centroid of the triangle. The wedge of density for the B and C domains is roughly tangential to the virus sphere while the wedge for the A domain is canted upwards at the fivefold axis. This leads to the large protrusions at the fivefold axes seen in electron micrographs of the virus (Figure 6). The B and C domains of CPMV can be approximately superimposed on each other by a rotation of $112°$ about the quasi-threefold axis when this axis is assumed parallel to the adjacent icosahedral twofold axis. However, due to the difference in orientation of the A domain it is impossible to superimpose it onto the B or C domain by a simple rotation about the quasi-threefold axis.

The electron density in each of the three domains was immediately recognizable as the β barrel structure observed in other virus capsid proteins. Figure 9b shows the fold of SBMV with eight antiparallel strands (βB-βI) folded into a barrel connected by loops containing helices αA-αE. Two of the helices observed in SBMV (αA and αB) were observed in the other plant virus structures while αC, αD and αE were not. The CPMV map clearly showed the eight β strands and the four short turns connecting them near the icosahedral threefold axes for the B and C domains and near the fivefold axes for the A domain. A molecular model has been built for the B and C domains using an Evans and Sutherland PS330 and the Frodo model building program (Jones, 1982). A polyalanine backbone was used initially to fit the main chain density and the side chains identified in this fit were tentatively correlated with the known amino acid sequence (Van Wezenbeek et al., 1984).

The C domain exhibits all the common features of the virus barrel structure. The eight stands of the barrel (beginning with βB) occur in the same order of primary sequence as seen in other plant and animal viruses. Only 130 residues were required to construct this domain, compared to the SBMV shell domain which required 200 residues. The TBSV shell domain and the STNV structure required 165 and 172 residues respectively. This domain is clearly the "essential barrel" with no excursions from the secondary structural elements described above (Figure 9a). In addition to helices αA and αB which exist in all plant viruses, there is a short (1 1/2 turn) helix connecting βG and βH which replaces the long connecting loop containing αD and αE in SBMV. The carboxy terminus of this β barrel runs to the interior of the shell and loops under the barrel. The B domain is also clearly organized as a β barrel. The heavy atom, bound in this domain, lies between βF and βG in the interior of the virus producing a greater separation between these strands than is observed in the C domain. Both αA and αB are also present, but poorly defined, in this domain. The major difference between the B and C domain is an insertion of 21 residues between βG and βH very similar to one of the helical insertions observed in SBMV (αD in figure 9b). The strands in the A domain can be assigned in the same order of primary sequence as in the B and C domain. Because the long barrel axis of the A subunit points more radially from the particle than the axes of the B and C barrels, the βB-βC turn and βH-βI turn in this domain point out into the solvent. The first turn forming contacts with neighboring A subunits at

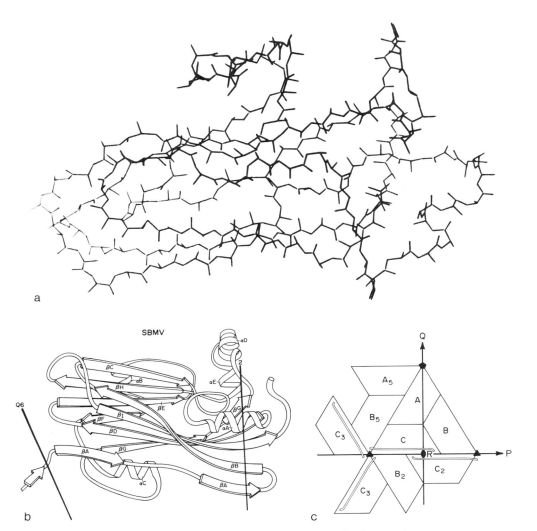

Figure 9: Comparison of CPMV domain C to a typical plant virus.
a) Polyalanine backbone fit to the main chain of the C domain of CPMV. This view clearly shows the strands and turns of the "essential barrel" structure of this domain. The carboxy terminus can be seen extending under the barrel at the right.
b) The fold of the polypeptide chain of one subunit of the SBMV capsid protein. This view corresponds to a 90° rotation about the large axis of the barrel in 9a.
c) Positions of the subunits of SBMV in the icosahedral asymmetric unit and neighboring subunits in the capsid.
The descriptions in the text of the structure of the CPMV domains are referred to the SBMV diagrams shown in b) and c).

the pentamer is the βD-βE turn, in contrast to the B-C contacts at the threefold axes which are formed by βH-βI. There are other features that differ when the A domain is compared to B and C but their nature is still being examined.

DISCUSSION

The structural features of CPMV that we see in the 3.5 Å resolution map are a combination of those of the known plant viruses and the animal picornaviruses. The T=3 plant viruses, SBMV and TBSV, use one type of protein subunit arranged with quasi-threefold symmetry to form the icosahedral asymmetric unit of the capsid. In the picornaviruses the three major proteins of the T=1 shell each fold into β barrels which occupy positions similar to those of the quasi-equivalent subunits in the T=3 viruses. The structure of CPMV clearly shows that the two capsid proteins (VP23 and VP37) also produce three distinct β barrel domains in the icosahedral asymmetric unit. The covalent linkage of two of these barrels does not disrupt this general capsid forming structure. In the CPMV structure these domains are arranged like the picornaviruses with the major axes of two of the barrels tangential to the sphere and one canted upwards at the fivefold axis. On the other hand, the folding in each of the individual domains of CPMV is clearly comparable to the plant viruses and does not include the extensive excursions found in the picornavirus capsid proteins.

The assignment of these three domains to the two capsid proteins in CPMV can now be cautiously made. It is most likely that two of the β barrel domains are formed from the large protein (VP37) and the third domain from the small protein (VP23). The amino terminus of the β barrel in the C domain (which corresponds to βB in SBMV) appears to start at residue 100 of the VP37 sequence. Twelve residues (110 to 121 in the sequence) fit the well defined side chains in the turn and strands of βB and βC. Following the sequence of VP37 from this point shows that the turn forming residues and β strands seem to fall correctly, although shifts in the fit will be necessary. The B domain contains a definitive marker for sequence fitting in the mercury heavy atom, which will most likely bind to a cysteine. Using a characteristic pattern of bulky side chains near that site, the heavy atom position can be tentatively assigned to cysteine 323 in VP37. This would then assign the B domain to the carboxy terminal half of VP37, which is consistent with the sequence alignment for domain C to the amino terminal half of VP37. A potential connection between these two β barrel domains exists at the interior of the shell where the carboxy terminus of the C domain loops under the barrel and connects to density leading to the B domain. This connection would link domain C in one icosahedral asymmetric unit with the B domain in the fivefold related position located to the left in Figure 9c. More detailed interpretation of the structure waits on the completion of the side chain fitting and the extension of the map to higher resolution.

Comparing the gene order of CPMV and poliovirus in Figure 1, and the arrangement of their capsid proteins in the structure, can now provide insight into the relationship between CPMV and the picornaviruses. The capsid protein of the picornaviruses is synthesized as a polyprotein precursor containing VP4, VP2, VP3 and VP1. CPMV capsid protein is also originally produced with VP37 and VP23 covalently linked. The polyproteins of both viruses are cleaved before assembly, but the cleavage in CPMV produces only two proteins rather than the four found in the picornaviruses. In the assembled picornavirus VP2 occupies the C position, VP3 the B position and VP1 the A position. VP4 is on the internal portion of the capsid and extends from near the amino terminus of the VP2 molecule toward the fivefold axis. Our analysis of CPMV suggests that VP37 takes the same position in the capsid structure as does VP2 and VP3, and that VP23 corresponds in position to VP1. Although we have not as yet seen a structure similar to that of VP4 in the CPMV electron density map, there is a significant amount of density in the interior of the capsid, and the 100 residues in VP37 at the amino terminus that precede the β barrel in our fit may form a structure that can be related to that picornavirus protein. This correspondence in the structural positions of the capsid proteins as well as their order in the RNA genome suggests that the VP37 represents the uncleaved form of three picornavirus capsid proteins (VP4, VP2, VP3). The details of the relationship between the covalently linked domains in CPMV may reveal important information on the role and mechanism of cleavage in the picornaviruses. In this context it is also interesting that the VP37 protein subunit formed by the connection of domains B and C we have suggested produces, with the VP23 subunit, a structural unit which corresponds to the group of proteins in rhino virus that Rossmann et al. (this volume) have suggested is the protomer for capsid assembly in the picornaviruses.

The 3.5 Å resolution structure of CPMV has shown us that this virus is a mixture of the structures of the T=3 plant viruses and the animal picornaviruses. Clearly a plant virus in the folding of the β barrel domains, CPMV still mimics in replication strategy and now, structural organization, the picornaviruses. The differences between these structures may then emphasize the changes that suit these viruses to their hosts in the plant and animal kingdoms, as well as provide some interesting clues to virus evolution.

ACKNOWLEDGEMENTS

We would like to thank Eddie Arnold, Gert Vriend, Greg Kamer and Michael Rossmann for many stimulating and helpful discussions. We thank Lucy Winchester for her help in preparation of the manuscript and the Purdue University Computing Center for their help and support. Most of the data used for this study were collected at LURE, Orsay, France. We express deep appreciation for the use of this facility and the enthusiastic cooperation we received from the staff at LURE, especially Roger Fourme and Richard Kahn. This work was supported by project Grants R01 AI18764 from NIH, DMB-8511084 from NSF and grants for computing DMB-8416890 and the Evans and Sutherland graphics system DMB-8320527 from NSF.

REFERENCES

Abad Zapatero, C., Abdel-Meguid, S.S., Johnson, J.E., Leslie, A.G.W., Rayment, I., Rossmann, M.G., Suck, D. and Tsukihara, T., 1980, Nature, 286:33-39.
Argos, P. and Rossmann, M.G., 1976, Acta Cryst., B32:2975-2979.
Arndt, U. and Wonnacott, A. (eds.), 1977, The Rotation Method in Crystallography. North Holland Publishing, New York, U.S.A.
Bricogne, G., 1976, Acta Cryst. A32:832-847.
Bruening, G., 1977, Plant covirus systems: two component systems 55-141. In Comprehensive Virology, Vol 11. H. Fraenkel-Conrat and R.R. Wagner (ed.), Plenum Press, New York U.S.A.
Caspar, D.L.D. and Klug, A., 1962, Cold Spring Harbor Symposium on Quantitative Biology, 27:1-24.
Crowther, R.A., Geelen, J.L.M.C., and Mellema, J.E., 1974, Virology, 57:20-27.
Daubert, S.D. and Bruening, G., 1979, Virology Journal, 98:246.
Francki, R.I.B., Milne, R.G. and Hatta, T., 1985, Atlas of Plant Viruses. CRC Press, Boca Ratan, Florida U.S.A.

Franssen, H., Leunissen, J., Goldback, R., Lomanossoff, G. and Zimmern, D. 1984, EMBO Journal, 3:855-861.
Goldback, R. and vanKammen, A., 1985, Structure replication and expression of the bipartite genome of cowpea mosaic virus. In Molecular Plant Virology, J.W. Davis (ed.) CRC Press, Boca Raton, Florida U.S.A. (in press).
Harrison, S.C., Olson, A.J., Schutt, C.E., Winkler, F.K. and Bricogne, G. 1978, Nature, 276:368-373.
Hollingshead, C. and Johnson, J.E., 1980, unpublished results.
Johnson, J.E., 1978, Acta Cryst., B34:575-577.
Johnson, J.E. and Hollingshead, C., 1981, J. Ultrastructure Res., 74:223-231.
Jones, T.A., 1982, FRODO: A graphics fitting program for macromolecules, 303-317. In Computational Crystallography, D. Sayre (ed.) Oxford University Press, New York.
Liljas, L., Unge, T., Jones, T.A., Fridberg, K., Lovgren, S. Skogland, U. and Strandberg, B., 1982, J. Mol. Biol., 159:93-108.
McPherson, A., 1982, Preparation and Analysis of Protein Crystals. Wiley, New York U.S.A.
Rayment, I., 1983, Acta Cryst., A39:102-116.
Rossmann, M.G., 1979, J. Appl. Cryst., 12:223-238.
Rossmann, M.G., Lesli, A.G.W., Abdel-Meguid, S.S., and Tsukihara, T., 1979, J. Appl. Cryst., 12:570-581.
Rueckert, R.R., 1985, Picornaviruses and Their Replication, 705-738. In Virology, B.N. Fields (ed.) Raven Press, New York.
Schmidt, J., Johnson, J.E. and Phillips, W., 1983, Virology, 127:65-73.
Siler, D.J., Babcock, J. and Bruening, G., 1976, Virology, 71:560.
VanWezenbeck, P., Verver, J., Harmsen, J., Vos, P. and vanKammen, A., 1983, EMBO J. 2:941-946.

ADENOVIRUS ARCHITECTURE

Roger M. Burnett, Michael M. Roberts and Jan Van Oostrum

Department of Biochemistry and Molecular Biophysics
College of Physicians and Surgeons - Columbia University
630 West 168th Street, New York - New York 10032

The adenovirus family consists of related viruses infectious to several different mammalian and avian hosts (Ginsberg, 1984). The simple icosahedral shape of the virion (Fig. 1), with the characteristic fibres projecting from the capsid vertices, conceals a complex organisation. The particle contains approximately 2700 copies of at least 10 different polypeptides (van Oostrum and Burnett, 1985) and a single linear copy of double stranded DNA. Electron micrographs show 252 morphological units, which correspond to the major coat proteins penton and hexon. Penton, a complex between penton base AND fibre, lies at each of the 12 vertices where it is surrounded by 5 peripentonal hexons. The 20 facets and 30 edges of the icosahedral capsid are formed from 240 trimeric hexons, which contain the majority of the immunological determinants distinguishing the different adenoviral species. Dissociation of the virus releases first the 12 penton and their 60 surrounding peripentonal hexons to free 20 planar group-of-nine hexons (Fig. 1). The size of the virion, which is twentyfold larger than the RNA viruses whose structure are emerging currently (as reported elsewhere in these proceedings), has so far prohibited a direct X-ray crystallographic structure analysis of the complete particle. Our approach to understanding adenovirus has been through conbining structure determination for hexon with complementary studies of the virion and group-of-nine hexons using electron microscopy and biochemistry. Structural studies on hexon have been reviewed recently (Burnett, 1984).

STRUCTURE OF HEXON, THE MAJOR COAT PROTEIN

Hexon is the major coat protein of adenovirus, accounting for 60 % of the virion protein (van Oostrum and Burnett, 1985). Free adenovirus type 2 (ad2) hexon trimers (Grutter and Franklin, 1974), purified from infected KB cells, form cubic crystals in space group $P2_13$ with a cell edge of 150.5 ± 0.1 Å. The threefold molecular and crystal axes coincident so that one ad2 hexon polypeptide of 967 residues lies in the asymmetric unit. Averaging methods of phase determination are therefore inaplicable for hexon. And earlier tracing in an electron density map at 2.9 Å resolution (Burnett et al., 1985), calculated using multiple isomorphous replacement phases from five derivatives, provided ten separate segments of unknown sequence. We have now established the connectivity of these segments and identified the residues within them.

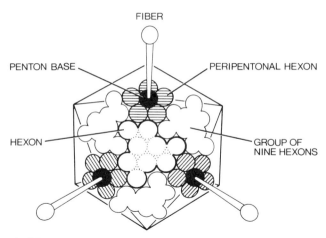

Figure 1. A symbolic representation of the adenovirus capsid showing the major external components. The penton is a complex of the fibre and penton base. Dissociation of the virion releases the pentons, peripentonal hexons, and then the planar groups-of-nine hexons from each facet. Illustration by John Mack from Burnett (1984)

This, together with a modest rearragement of the original tracing, has revealed the topology of the polypeptide (Roberts et al., 1985). Of the total of 967 amino acids, the 859 that have been located account for all prominent electron density in the map. The backbone has been fitted to the mini-map and its coordinates measured. A schematic view of one subunit is shown in Figure 2. The ad2 hexon monomer (Mr 109,077) is currently the longest polypeptide for which a three-dimensional crystal structure has been reported.

Hexon has a bulky pseudo-hexagonal base, which allows the molecules to close-pack in the adenovirus capsid, and a spiky triangular top. The base contains a central cavity and forms two different vertical external contact faces. These alternate around the molecule to facilitate hexon-hexon interactions in the capsid. Three towers, forming the triangular top, are above the basal interfaces between neighboring subunits.

Secondary structure is limited in hexon, which the tracing reveals to contain only 8 % α-helix and 22 % β-sheet, located mainly in the base (Fig. 2). Two eight-stranded "Greek key" β-barrels form domains P1 and P2, which define the basal corners. The β-barrels are identical in their topoloty to those found in the RNA plant viruses (Olson et al., 1983 ; Rossmann et al., 1983), rhinovirus and poliovirus (these proceedings). In the RNA viruses, there is only one β-barrel per polypeptide. The topological similarity of the two β-barrels within the hexon polypeptide is emphasized by the occurence of other structural features. One example is that each domain contains an α-helix, followed by a loop to the tower, situated between the fifth and sixth strands. The tower domain, T, is composed of three upper loops, γ_1 to γ_3, each arising from a different subunit (Fig. 2). The extensive and interwoven subunit interface explains hexon's extreme resistance to denaturation. Regions of sequence heterology between ad2 and ad5 hexon lie predominantly on the surface at the top of the molecule. We are currently refining the 2.9 Å model of ad2 hexon, and are collecting 1.9 Å resolution synchrotron data for both ad2 and ad5 hexon. Since the latter crystallizes isomorphously, we will be able to visualize the difference Fourier maps based on the ad2 hexon phases.

ORGANIZATION OF THE OUTER ADENOVIRUS CAPSID

The crystallographic results indicate that hexons are designed to form close-packed arrays, but do not reveal how the coat protein is organized in the adenovirus capsid. At the start of this investigation groups-of-nine hexons were known to lie on a p3 net (Crowther and Franklin, 1972), but their orientation with respect to each other was unknown. Moreover, there was no information concerning the spatial relationship of the peripentonal hexons to the group-of-nine hexons, and it was even postulated that there could be different chemical classes of hexons within the virion (Shortridge and Biddle, 1970). In order to extend the detailed model of hexon to higher levels of organization, we have used the accurate molecular model to interpret electron micrographs of various capsid fragments.

Our first results came from correspondance analysis of digitized electron micrographs of negatively strained left- and right-handed groups-of-nine hexons. Sets of related images were averaged, and the resulting pictures showed a correlation with the height to which negative stain rose around the hexons. The exact orientation of the hexons was then deduced by comparison with the hexon model to establish the relationship of the hexons within the p3 net of the group-of-nine. The orientation of the remaining 60 peripentonal hexons, with respect to those within the group-of-nine, then was obtained from a novel capsid fragment (van Oostrum and Burnett, 1984). The "quarter capsid" contains a central penton surrounded by five capsid facets. Its image was rationally fivefold averaged about the vertex symmetry axis, which lay normal to the electron microscope grid. This work showed that all 12 facet hexons are symmetrically distributed on the p3 net (Fig. 3) and confirmed a model for the organization of hexons within the capsid (Burnett, 1984 ; 1985).

The release of groups-of-nine, rather than complete facets, after disruption of the virion is not consistent with the symmetrical distribution of hexons within the facet. A minor capsid protein, polypeptide IX, is known to stabilize the capsid and a model has been proposed for its distribution (Fig. 3), which explains the dissociation of the virion as well as that of the group-of-nine (Burnett, 1984). We have obtained supporting evidence for this model from the molecular

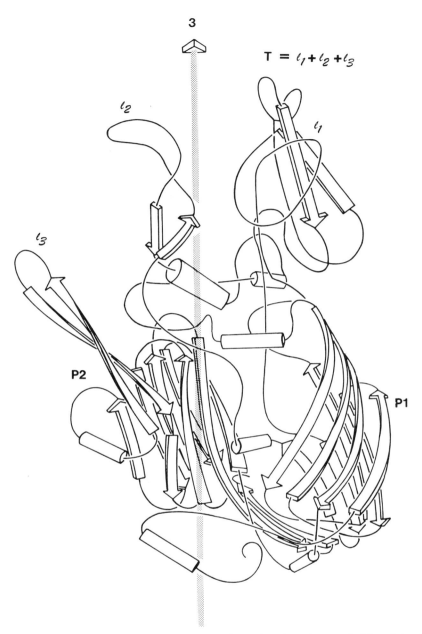

Figure 2. A sketch of one hexon subunit, showing its relationship to the threefold molecular axis. The molecule has been simplified by restricting the length of connections between the secondary structural features, which are therefore over-emphasized. Two small strands from loop ℓ_2, with their symmetry related conterparts, from a β-sheet bonded formation about the threefold axis. This causes a tight constriction in the internal molecular cavity at this point due to the presence of β-bulgess in alterning strands. Pedestal domains P1 and P2 form corners of the pseudo-hexagonal molecular base, and each of the three upper loops (ℓ_1, ℓ_2 and ℓ_3) contributes to one of hexon's three towers.

composition of the virion. Stoichiometric ratios for the structural polypeptides were obtained by radiolabeling the viral proteins, using the methionine content given by the known DNA sequence (van Oostrum and Burnett, 1985). Twelve copies of polypeptide IX were found per facet, with their locations restricted to within the group-of-nine. This study also resolved the controversy concerning the subunit composition of the penton complex by revealing pentameric penton bases and trimeric fibres. The biochemical analysis also revealed microherogeneity in the penton base polypeptides, which are present in approximately two copies of the full-length polypeptide and three copies of a slightly shorter polypeptide. A chemical mechanism for overcoming the symmetry mismatch within the complex would be provided of binding of fibre to the three shorter polypeptides of the base. The pentameric nature of the base was in accord with the fivefold symmetry revealed by rational frequency analysis of the vertex region in image of the quarter-capsid.

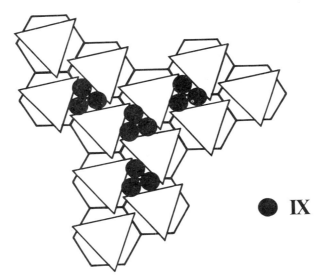

Figure 3. Distribution of protein IX in a capsid facet formed from 12 hexon trimers arranged on a p3 net. The protein is buried in the large, but not the small, cavities created between the hexon molecules. This confers stability upon the hexons within the facet, but does not stabilize the peripentonal hexons or the capsid edges. The distribution of protein IX therefore explains the preferential loss of peripentonal hexons upon disruption of the captis, and the non-random dissociation of the group-of-nine itself.

ARCHITECTURAL PRINCIPLES

The morphology of many mammalian viruses was originally explained by Caspar et Klug (1962). The morphological features were thought to arise from clustering of identical polypeptides about the local symmetry elements of hexagonal lattices. For adenovirus, the triangulation number (T) of 25 required the association of 1500 polypeptides to give 240 hexameric and 12 pentameric groups. However, since later work revealed that adenovirus contains 240 trimeric hexons and 12 pentons the theory cannot apply to adenovirus. Another recently discovered exception is polyoma, in which pentamers and found at local sixfolds of the supposedly T=7d shell (Rayment et al., 1982).

Our model for the overall arrangement of hexons in adenovirus (Burnett, 1984 ; 1985) is illustrated Figure 4. A quartet of hexons repeats with exact 532 symmetry to form the icosahedral shell, leaving holes at the vertices into which pentons may be inserted. The packing of the pseudohexagonal hexon in four topographically distinct locations conserves no less than 2/3 of its inter-hexon chemical bonds. Since each facet of adenovirus is a small crystalline array, only one additional topographically related location is necessary to achieve a capsid of infinite size. By extension, a similar simplification may be achieved for polyma by treating the pentamers as structure units and dividing them into two classes : 60 arranged with exact icosahedral symmetry, and 12 at the vertices.

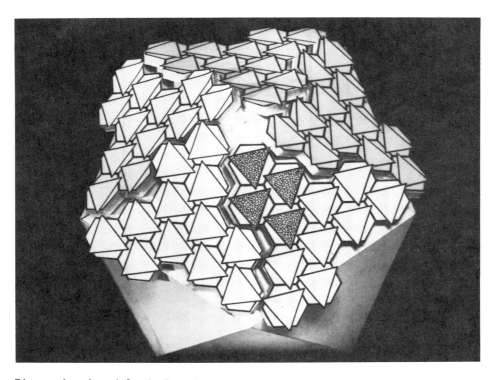

Figure 4. A model of the adenovirus capsid, showing the arrangement of hexons. The shaded asymmetric unit contains hexons in each of the four topographically distinct locations. Exact icosahedral symmetry relates the 60 assymetric units within the icosahedral hexon shell. The bonds made by hexon in each of its distinct locations are very similar.

The new approach, in which hexons are treated as the capsid building blocks, brings further simplifications. Accurate assembly of large capsids is explained by the small number of differently bonded multimeric structure units needed when each can close-pack. The earlier approach, with each polypeptide contact assumed "quasi-equivalent", failed to explain sharp capsid facets and led to a myriad of different bonding states for structure units. In adenovirus, we now see how the "crystal-like" facets are formed with the consequence that hexon need participate in only 4 rather than 25 different bonding states.

Hexon is an example of a multimeric, or "complex", structure unit (Burnett, 1985). Accurate assembly using these units, which must be highly stable to act as building blocks, requires relatively weak interactions. The final capsid strength is achieved by the addition of further components, such as polypeptide IX in adenovirus, or calcium ions in the plant viruses.

ACKNOWLEDGMENTS

We are grateful to our collaborators Drs. Minijeh Mohraz and P.R. Smith of the Department of Cell Biology, New York University Medical School, for their invaluable contributions to the electron microscopic studies. The investigation has been supported by Public Health Service grant AI-17270 from the Institute of Allergy and Infectious Diseases, grand PCM-84-18111 from the National Science Foundation, and by an Irma T. Hirschl Career Scientist Award to R.M.B.

REFERENCES

1. R.M. Burnett, Structural Investigations of Hexon, the Major Coat Protein of Adenovirus, In "Biological Macromolecules and Assemblies. A. McPherson and F.A. Jurnak, eds. Vol. 1, Virus Structure. John Wiley and Sons, New York. pp. 337-385 (1984).
2. R.M. Burnett, The Structure of the Adenovirus Capsid. II. The Packing Symmetry of Hexon and its Implications for Viral Architecture, J. Mol. Biol. 185:125-14" (1985).
3. R.M. Burnett, M.G. Grutter and J.L. White, The Structure of the Adenovirus Capsid. I. An Envelope Model of Hexon at 6 Å Resolution, J. Mol. Biol. 185:105-123 (1985).
4. D.L.D. Caspar and A. Klug, Physical Principles in the Construction of Regular Viruses, Cold Spring Harbor Symp. Quant. Biol. 27:1-24 (1962).
5. R.A. Crowther and R.M. Franklin, The Structure of the Groups of Nine Hexons from Adenovirus, J. Mol. Biol. 68:181-184 (1972).
6. H.S. Ginsberg, editor, The Adenovirus, Plenum Press, New York (1984).
7. M.G. Grutter and R.M. Franklin, Studies on the Molecular Weight of the Adenovirus Type 2 Hexon and its Subunit, J. Mol. Biol. 89:163-178 (1974).
8. A.J. Olson, G. Bricogne and S.C. Harrison, Structure of Tomato Bushy Stunt Virus IV. The Virus Particle at 2.9 Å Resolution, J. Mol. Biol. 171:61-93 (1983).
9. I.K. Rayment, T.S. Baker, D.L.D. Caspar and W.T. Murukami, Polyoma Virus Capsid Structure at 22.5 Ångstrom Resolution. Nature (London) 295:110-115.
10. M.M. Roberts, J.L. White, M.G. Grutter and R.M. Burnett, The Three-Dimensional Structure of Adenovirus Type 2 hexon at 2.9 Å Resolution, Submitted for Publication (1985).
11. M.G. Rossmann, C. Abad-Zapatero, M.R.N. Murthy, L. Liljas, T.A. Jones and B. Strandberg, Structural Comparison of some Small Spherical Plant Viruses, J.Mol. Biol. 165:711-736.
12. K.F. Shortridge and F. Biddle, Arch. Ges. Virusforsch. 29:1-24 (1984).
13. J. van Oostrum and R.M. Burnett, Architecture of the Adenovirus Capsid, Annal. N.Y. Acad. Sci. 435:578-581 (1984).

14. J. van Oostrum and R.M. Burnett, Molecular Composition of the Adenovirus Type 2 Virion, J. Virol 56:439-448 (1985).

7 PROTEIN - NUCLEIC ACIDS

A BACTERIOPHAGE REPRESSOR/OPERATOR COMPLEX AT 7 Å RESOLUTION

S. C. Harrison and J. E. Anderson
Department of Biochemistry and Molecular Biology
Harvard University
7 Divinity Avenue, Cambridge, MA 02138

Possible mechanisms for sequence-specific affinity of proteins for DNA include: (1) 'direct readout' of base sequence by interactions of residues in the protein with functional groups on the DNA bases; (2) 'indirect readout' by interactions with backbone conformations sensitive to base sequence; (3) dependence of binding on long-range structural characteristics of DNA - e.g. twist, bend, or superhelicity; (4) induction of cooperative transitions in DNA, such as B -> A or B -> Z. The crystal structure of a bacteriophage repressor/operator complex, determined at 7 Å resolution, permits us to examine these mechanisms in the context of a particular example (Anderson et al., 1985). The protein in this complex is the amino-terminal domain (residues 1-69 and denoted R1-69) of the repressor encoded by coliphage 434, a close relative of bacteriophage λ. The DNA is a 14 base-pair synthetic oligonucleotide, with a self-complementary sequence (Fig. 1) that contains a consensus of the six repressor binding sites (operators) in the phage.

Structure of the Complex at 7 Å

Analysis of packing in the crystals of the R1-69/DNA complex and qualitative evaluation of helix-like features of the diffraction pattern showed that the DNA segments pack end-to-end, parallel to the body diagonal of the I422 unit cell (a = b = 166.4 Å, c = 139.4 Å: Anderson et al., 1984). There is a non-crystallographic threefold screw axis relating stacked repressor/DNA complexes, and these stacked rods extend from cell to cell through the crystal. The pitch of the threefold screw is 137 Å; this repeat contains 42 base pairs (three 14mers) and four turns of the pseudocontinuous DNA helix. Accurate data, sufficient to measure differences due to substitution of -Br for -CH_3, were collected with the aid of a Xentronics detector (Durbin et al., 1985). SIR phase determination, followed by phase refinement using non-crystallographic symmetry, gave a clearly interpretable 7 Å map (Fig. 2).

The map has strong ribbons of density, representing B-DNA backbone, and rod-like features showing α-helical segments in the protein. Two of these rods precisely fit the helix-turn-helix structure found in three other sequence-specific DNA binding proteins, including the repressor of phage λ (Anderson et al., 1981; McKay and Steitz, 1981; Pabo and Lewis, 1982). Moreover, when helices 2 and 3 of λ repressor are superimposed on these rods of density, two additional rods lie very close to the positions of helices 1 and 4 of λ repressor N-terminal domain. Thus, the N-terminal domains of 434 and λ repressors have similar folded structures. The map suggests, however, that the C-terminal segment (residues 60-69) of

```
O_R1      A C A A G A A A G T T T G T
          T G T T C T T T C A A A C A

O_R2      A C A A G A T A C A T T G T
          T G T T C T A T G T A A C A

O_R3      A C A A G A A A A A C T G T
          T G T T C T T T T T G A C A

O_L1      A C A A G G A A G A T T G T
          T G T T C C T T C T A A C A

O_L2      A C A A T A A A T A T T G T
          T G T T A T T T A T A A C A

O_L3      A C A A A C T C C A T T G T
          T G T T T G A G G T A A C A

14mer     A C A A T A T A T A T T G T
          T G T T A T A T A T A A C A
```

Fig. 1. The sequence of the 14-base-pair synthetic operator ('14-mer') used for crystallization, compared to the six subsites on O_R and O_L of the phage 434.

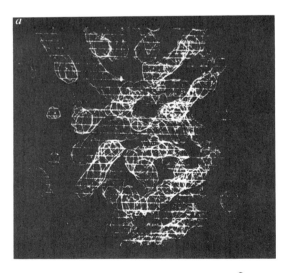

Fig. 2. A part of the electron density map at 7 Å resolution, viewed normal both to the non-crystallographic 3_1 and to the operator dyad. The positions of Br atoms are shown as stippled spheres. Model B-DNA is superimposed on DNA electron density.

R1-69 does not follow the same direction as helix 5 of λ repressor N-terminal domain.

The helix-turn-helix motif in the complex lies against DNA (Fig. 3) in an orientation similar to the one suggested for λ cro by Ohlendorf et al. (1982) and for λ repressor by Pabo and Lewis (1982). Strong contacts with DNA backbone appear to be formed by the N-termini of helices 2, 3, and 4, by the loop between helices 3 and 4, and perhaps by the very C terminus of the domain. These backbone contacts position the domain on the DNA helix so that helix 3 lies in the major groove, in contact with base pairs 1-4 of the operators. The loop between helices 3 and 4 might approach base pair 5. Fig. 1 shows that base pairs 1-4 correspond to a consensus sequence found in all 434 operators.

The backbone of DNA in the complex can be fit quite well by a standard B model with 10.5 base pairs/turn, except for a small deviation in the 3' terminal 3 to 4 nucleotides. This deviation correlates with a protein contact described below. The individual 14mers (operators) are aligned with their axes at about $10°$, and an approximate model suggests that the 5' adenines may stack directly on each other, leaving the 3' termini exposed. As shown in Fig. 4, the N terminus of helix 2 in R1-69 contacts the backbone of an adjacent 14mer, at the first phosphate from the 3' end. We believe that this contact may also be made when an operator is embedded in continuous DNA, since ethylation of the phosphate at position '-1' (Fig. 4) strongly interferes with binding of R1-69 (Bushman et al., 1985). The relative tilt of the stacked 14mers in the crystal is required for close approach of this phosphate and the N terminus of helix 2, and continuous DNA may bend in a corresponding way when repressor binds. The bend would be gentle (radius of curvature approximately equal to 100 Å).

We have built a detailed model for R1-69, based on homology with λ repressor, in order to examine in a preliminary way possible contacts of amino-acid side chains with groups in DNA (Fig. 5). Of the residues in the part of the domain that lies in the major groove, gln 28, gln 29 and gln 33 appear respectively to approach base pairs 1, 2/3 and 4 - precisely the part of the operator sequence conserved in all sites. The disposition of gln 29 in this model is interesting, since the Cβ contacts the methyl group of T12 (paired to A3), while its amide group could H-bond to functions on base pair 2 (CG). Changes in the base sequence at one position would probably influence the most favorable configuration for this gln and hence its interaction with the bases at the other position. This proposed interaction illustrates how the effect of a sequence change at one base pair may depend on the identity of an adjacent base pair and that the correlation can be mediated by features of the protein structure rather than by DNA conformation.

Determinants of Specificity

The way in which side chains of α3 appear to contact conserved base pairs is consistent with the notion that these interactions are critical determinants of binding specificity. Models invoking the importance of the homologous α helix in λ cro and in CAP first directed attention to this possibility (Anderson et al., 1981; McKay and Steitz, 1981), and subsequent studies of λ repressor and of 434 repressor have provided further evidence

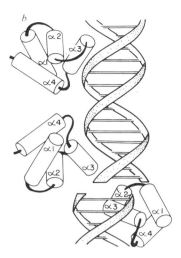

Fig. 3. Schematic diagram showing most of the protein. The map indicates that the polypeptide chain ceases after α4 to follow the fold of the λ repressor N-terminal domains, but it cannot of course be traced at 7 Å. Thus only about 75% of the protein is shown in this diagram.

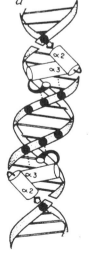

Fig. 4. Protein-DNA backbone contacts. Filled symbols in B indicate phosphates that cannot be ethylated without blocking binding (Bushman et al., 1985). Open symbols indicate contacts suggested by the map but not detected in the ethylation interference experiment.

322

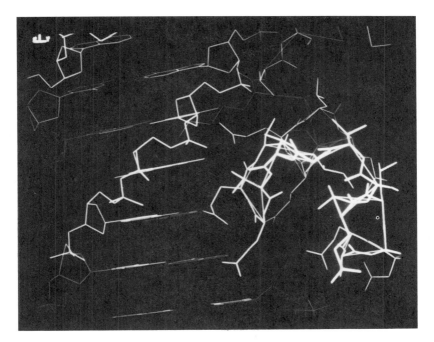

Fig. 5. Model of α3 with side chains, showing the basis for proposed protein-DNA contacts (see text).

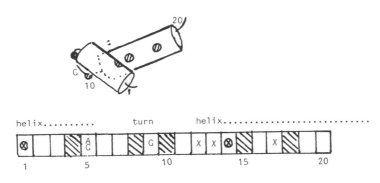

Fig. 6. The helix-turn-helix motif: conserved characteristics of residues at indicated positionns. Hatched boxes are hydrophobic residues (sometimes thr). Contacts among these hydrophobic side chains depend on their relative sizes. They form a hydrophobic 'core' in the pocket between α2 and α3. The ala or gly at position 5 appears to be a nearly invariant feature, because longer side chains would collide with backbone between residues 10 and 11. The gly at position 9 forms the tight turn between α2 and α3. An active mutant of λ repressor has been isolated with glu at position 9 (A. Hochschild, personal communication), showing that gly is not absolutely essential. Crosses show side chains that contact backbone (1,14) or bases (12, 13, 1) in the model described here.

in this direction (Pabo and Lewis, 1982; Nelson et al., 1983; Wharton and Ptashne, 1985). A more precise model must await the higher resolution analysis of the 434 repressor/operator complex. The evidence, summarized above, for extensive backbone contacts, suggests that tight binding will occur at those sequences for which the 'fit' of α3 in the major groove is consistent with optimal backbone-binding geometry. A good fit in the major groove will depend on complementary van der Waals surfaces and full realization of hydrogen bonds, in the same way that stable protein/protein interfaces depend on such interactions. Water molecules may also participate. Note that in this view, pairwise matching between particular residues and particular base pairs is less a central feature than overall match of the two structures. The binding energy may be provided largely by backbone contacts, counteracted at all but the optimal sequences by unfavorable major-groove interactions (poor van der Waals contacts, unsatisfied H-bonds, 'holes' left by expelled waters).

Does backbone conformation play a role in sequence specificity? Again, a clear answer must await higher resolution. A robust effect of base sequence on conformation at a phosphate could influence hydrogen bonding to that phosphate and alter affinity, perhaps by 1-1.5 kcal (a 10-fold effect). This level is significant on the scale of differences between O_R1 and O_R3, but not on the scale of operator vs. non-specific binding. Indirect readout of sequence through its effect on local backbone conformation might therefore modulate affinity, but it appears unlikely to be a principal determinant of specificity.

Other kinds of backbone effects may actually play a more significant role. Two sorts are suggested by the present model. One involves bending of the DNA helix. If 434 repressor bends DNA as suggested above, then some sequences flanking the operator may be more readily bent in the correct way than others. A second backbone effect involves overall twist, as it determines the relative azimuth of half sites to either side of the operator dyad. There are clearly interactions between the two R1-69 domains bound to a single 14mer. The twist of the DNA helix imparted by the central base pairs may then determine how consistent are the protein/protein interactions with the required orientation of the two domains about the DNA axis.

<u>Acknowledgment</u>

This work was supported by NIH Grant GM-29109 (to SCH and M. Ptashne). We thank M. Ptashne and members of his laboratory for ongoing collaboration.

REFERENCES

Anderson, W.F., Ohlendorf, D.H., Takeda, Y., and Matthews, B.W. (1981). Nature 290, 754-758.

Anderson, J.E., Ptashne, M., and Harrison, S.C. (1984). PNAS 81, 1307-1311.

Anderson, J.E., Ptashne, M., and Harrison, S.C. (1985). Nature 316, 596-601.

Bushman, F.D., Anderson, J.E., Harrison, S.C., and Ptashne, M. (1985). Nature 316, 651-653.

Durbin, R., Burn, R., Moulai, J., Metcalf, P., Freymann, D., Anderson, J.E., Harrison, S.C., and Wiley, D.C. (1985). Science, in press.

Nelson, H.C.M., Hecht, M.H., and Sauer, R.T. (1983). Cold Spring Harbor Symp. Quant. Biol. 47, 441-449.

Ohlendorf, D.H., Anderson, W.F., Fisher, R.G., Takeda, Y., and Matthews, B.W. (1982). Nature 298, 718-723.

Pabo, C.O. and Lewis, M. (1982). Nature 298, 443-447.

Wharton, R. and Ptashne, M. (1985). Nature 316, 601-605.

REFINED STRUCTURE OF DNase I AT 2 Å RESOLUTION

AND MODEL FOR THE INTERACTION WITH DNA

Christian Oefner and Dietrich Suck

European Molecular Biology Laboratory
Postfach 102209, D-6900 HEIDELBERG, FRG

ABSTRACT

The structure of bovine pancreatic deoxyribonuclease I (DNase I) has been refined at 2 Å resolution using the restrained parameter LS refinement method of Hendrickson and Konnert. The conventional R-factor for the 16,104 reflections with $I \geqslant 3(I)$ from 6.0 to 2.0 Å resolution is 0.157. Bond lengths and angles of the refined structure are close to ideal values, the r.m.s. deviations being 0.023 Å and 1.4°, respectively. The r.m.s. deviation of short non bonded contacts from the sum of the van der Waals radii is 0.18 Å. The average atomic thermal vibration parameter, B, for the 2,031 atoms of the enzyme is 11.8 $Å^2$. This low B-value is a consequence of the large hydrophobic core of DNase I formed by the two central, tightly packed ß-pleated sheets, which are surrounded by 8 helices and extended loop-regions. 377 water molecules have been located and refined, many of them are forming water-clusters. The protein sequence as determined chemically by the groups of Moore and Stein has to be corrected at 5 positions, the major correction being an insertion of 3 amino acid residues at R27. Two N-acetylglucosamine- and 5 mannose-residues have been located in the carbohydrate side-chain, which is branched at the mannose in position 3.

Based on the 3D-structure and the active site geometry determined from the binding of Ca^{2+}-thymidine-3', 5'-diphosphate in the crystal, we have derived a model for interactions of DNase I with ds-DNA. According to this model, the DNA binds at the slightly concave surface between the two ß-sheets with the exposed loop R70 - N71 - S72 - Y73 - K74 interacting in the minor groove of B-DNA. The hydroxyl group of Y73 is hydrogen-bonded to either O2 of a pyrimidine or N3 of a purine base. The positively charged residues R9, R70, R108, R38 and K74 are interacting with phosphates opposing each other in the minor groove. The cleavage patterns observed in oligonucleotides can be nicely explained in terms of this model.

INTRODUCTION

Deoxyribonuclease I (DNase I) from bovine pancreas is an endonuclease of molecular weight 30,400 degrading double-stranded DNA through hydrolysis of the P-O3'-bond (for a review, see Laskowski, 1971; Moore, 1981). The enzyme attacks DNA in a double-hit mechanism, i.e.

only one strand is cut at a time. Accordingly, only single-strand nicks and no reduction in molecular weight are observed in the early stages of the reaction. Divalent cations are an absolute requirement for enzymatic activity, which shows a maximum in the presence of Ca^{2+} and Mg^{2+} and Mn^{2+} (Price, 1975). DNase I is a glycoprotein with a carbohydrate unit of 8 - 10 sugar residues attached to Asn 18 (Price et al., 1960a), but as in the case of RNase B, the carbohydrate seems not to be essential for activity.

DNase I shows little base- or sequence-specificity, the cleavage patter and the frequency of cutting is, however, clearly sequence-dependent showing that DNase I recognizes variations in the DNA backbone (Bernardi et al., 1973 ; Lomonossoff et al., 1981 ; Drew and Travers, 1984). The enzyme bas been used as a biochemical tool in a variety of reactions and has proved particularly useful as a probe for the helical periodicity of DNA on the nucleosome and in solution (Prunell et al., 1979 ; Rhodes and Klug, 1980 ; Klug and Lutter, 1981). Limited digestion by DNase I - 'DNase I footprinting' - is routinely use to detect protected regions in DNA-protein complexes (Galas and Schmitz, 1978). For a correct interpretation of these experiments, knowledge of the 3D-structure of DNase I and its mode of interaction with DNA is essential. We have recently solved the structure of DNase I at 2.5 Å resolution by X-ray structure analysis (Suck et al., 1984). Here we report on the rafinement at 2Å resolution and the resulting DNase I model, which shows excellent sterochemistry. We also describe a model for the interaction with double-stranded DNA, which we have derived on the basis of the refined structure of DNase I.

MODEL BUILDING AND REFINEMENT

The starting point for the model building of DNase I was a set of approximte C_α co-ordinates read directly from a 2.5 Å-MIR-electron density map (Suck et al., 1984). These positions were used as guide-points for the fitting procedure with an Evans and Sutherland PS2 colour graphics display system. A modified version of the interactive computer graphics program FRODO (Jones, 1978) was used. The initial model was built by using the known sequence of DNase I (Liao et al., 1973) and assuming ideal geometry for the peptide bonds and the amino acid side chains. The starting R-value ($R = \Sigma\ ||F_o| - |F_c|| / \Sigma\ |F_o|$) for this model comprising 2073 non-hydrogen atoms was 45.5 % in the resolution range 6.0 to 2.5 Å.

The refinement of DNase I was carried out using the Hendrickson-Konnert restrained, reciprocal least-squares procedure (Hendrickson and Konnert, 1980). The computing time (CPU-time) on a VAX 11/780 was about 5 hours for one cycle of refinement including ≃ 16.000 reflections from 6.0 - 2.0 Å and ≃ 2.400 atoms using individual temperature factors. The progress of refinement, which was carried out in 10 stages, is illustrated in the table. At the end of each stage, $2|F_o| - |F_c|$, α_c and partial difference Fourier maps were calculated, to check for errors in the model and, if necessary, correct them manually. The weighting factors used for the stereochemical restraints were essentially those recommended by Hendrickson and Konnert (1980). The structure factor weight was set to $1/2\ ||F_o| - |F_c||$ in the early stages and was later made a function of resolution. Throughout the whole refinement, the weights were balanced in such a way to keep a good stereochemistry.

The refined 2.5 Å model was used to extend the resolution to 2 Å. During this extension it became obvious, that the chemically determined sequence (Liao et al., 1973)) had to be corrected at several places, the major correction being an insertion of the tripetide Ile - Val - Arg at position 27. Other changes include replacement of Thr 14 by a Ser and G224 - P225 by P22 - G225. With the corrected sequence and after successive inclusion of solvent molecules the R-factor dropped much faster than in the early stages and is presently at 15.7 % for the 16,100 reflections with I $\geq 3\sigma_o$(I) in the resolution range from 6.0 to 2.0 Å. Details of the refinement procedure will be described elsewhere (Oefner and Suck, manuscript in preparation). The mean values of bond lengths and angles of the final model are very close to ideal values, the r.m.s. deviations being 0.023 Å and 1.4°, respectively. The same is true for non-bonded interactions with an r.m.s. deviation of 0.18 Å from the sum of the van der Walls radii. As an additional indication of the quality of the structure and the improvement of the model during refinement we show the Ramachandran plot of main-chain conformational angles at the biginning and the end of refinement (figure 1).

STRUCTURE OF DNase I

A schematic view of the 3D-structure of DNase I along with the C_α-backbone is given in figure 2. DNase I is a compact structure of approximate dimensions 45 x 40 x 35 Å with the carbohydrate moiety protruding by about 15 Å. The central core of the molecule consists of two 6-stranded ß-pleated sheets which are tightly packed and are surrounded by three longer and five shorter helices and several loop regions. This 4-layer or sandwich-type structure of DNase I gives rise to the formation of three hydrophobic cores, one in between and two on either side of the sheets. The extended hydrophobic area is a very rigid part of the molecule, as is evident from the low isotropic temperature factors (see figure 3). On one side of the molecule, a shallow groove is formed at the surface between the two sheets. As will be discussed later, this is the location of the active site and this is where according to our model one strand of the DNA double-helix is bound.

As pointed out in our 2.5 Å paper (Suck et al., 1984), the topology of the two ß-sheet is very similar. The two central strands run parallel, all others are antiparallel. If we cut the molecule after residue 120, the resulting halves can be superimposed quite nicely by an approximate 2-fold rotation. The r.m.s. deviation for 54 corresponding C_α-positions is 1.2 Å, suggesting that DNase I might have undergone gene-duplication. It is interesting to note, that this internal symmetry is not detectable in the sequence.

As is usually the case in α,β-proteins, all α-helices of DNase I are exposed to the solvent and show corresponding amphiphatic character. Their length varies from one to five turns, the longest one consisting of 19 residues. As can be seen from figure 2, this helix has a kink and actually consists of two α-helical segments separated by one turn of 3_{10}-helix in the middle. Formation of a salt-bridge and tight packing against the hydrophobic surface formed by one of the ß-sheet seem to be reason for the 22°-kink of this helix.

Several factors are contributing to the several stability and rigidity of DNase I, as reflected in the low average temperature factor of 11.8 Å^2. The molecule has a relatively high percentage of secondary structure, 60 % of all residues are involved in either α-helices or ß-sheets. In addition, there are several salt-bridges and two S-S-bridges, one of which (C170 - C206) is essential for activity. Adding to the structural rigidity are also two Ca^{2+}-binding sites, which stabilize loop-regions in the molecule (figure 2). In the absence of Ca^{2+}, DNase I becomes extremely sensitive against the action of proteases (Price et al., 1969b). These "structural" Ca^{2+}-sites are distinct from the Ca^{2+} binding-site at the active center discussed below.

TABLE 1: Progress of refinement

Refinement Survey

Stage no.	1	2	3	4 A	4 B	4 C	5 A	5 B	6 A	6 B	7	8	9	10
No. of cycles	10	22	30	11	6	19	8	3	11	4	10	11	17	11
No. of parameters	6221	6221	6104	6065	6065	6065	7484	2495	9629	9629	9465	9953	10086	9974
No. of atoms	2073	2073	2034	2021	2021	2021	2494	2494	2407	2407	2366	2488	2521	2493
No. of solvent atoms	–	–	–	–	–	–	376	376	311	311	268	383	405	377
$<B> (\text{Å}^{-2})$	10.0	7.0	5.8	10.0	10.0	10.0	11.0	*	*	*	*	*	*	*
r.m.s shift in coordinates in a given stage (Å)	0.45	0.41	0.42			0.83	0.38			0.35	0.18	0.26	0.26	0.20
No. of F-data	7016	7016	7016	8451	12346	16104	16661	16661	16661	16517	16517	16517	16104	16104
$F's >$	10σ	10σ	10σ	3σ	3σ	3σ	3σ	3σ	3σ	3σ	3σ	3σ	3σ	3σ
Resol. range (Å)	6 – 2.5	6 – 2.5	6 – 2.5	6 – 2.5	6 – 2.2	6 – 2	10 – 2	10 – 2	10 – 2	8 – 2	8 – 2	8 – 2	6 – 2	6 – 2
$<|F_o| - |F_c|>$	195.7	172.4	160.2	133.2	130.5	112.6	115.4	108.2	95.4	89.7	82.8	69.8	60.6	58.9
R(initial)	0.445	0.373	0.349	0.361	0.324	0.332	0.388	0.304	0.324	0.251	0.237	0.236	0.193	0.160
R(final)	0.368	0.324	0.301	0.284	0.317	0.300	0.304	0.285	0.251	0.238	0.212	0.183	0.161	0.157

* Individual isotropic temperature factors

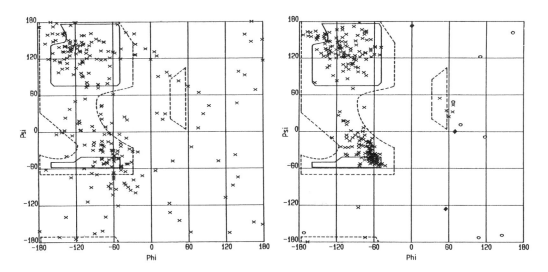

FIGURE 1: Ramachandran-plot of the starting (left) and the final model

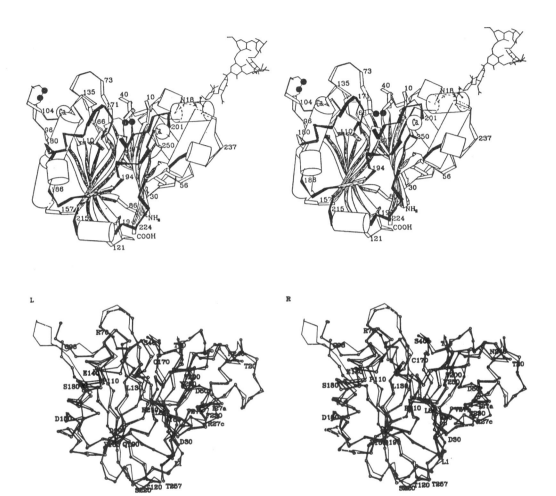

FIGURE 2: (top) Schematic stereo representation of the 3D-structure of DNase I. α-Helices are represented by cylinders, β-strands by arrows. The longest helix has a 22° kink and is therefore drawn as two separate cylinders. Protruding to the upper right is the carbohydrate unit attached to N18 consisting of two N-acetylglucosamine and five mannose residues. The S-S-bridges connecting C170/C206 and C98/C101 are indicated by black dots. The two structurally important Ca^{2+}-binding sites are marked. (bottom) Stereo view of the C_α-backbone of DNase I (same view as above). Superimposed in light lines is the starting model before refinement. Residues I27a, V27b and R27c were missing in the chemically determined sequence.

There are two regions in the enzyme molecule showing elevated mobility (figure 3) : one is of the region of the short S-S-bridge (C98 - C101) and the two residues E99 and S100 in this flexible loop are actually not visible in the electron density map. It should be mentioned here that this disulfide-bridge is not essential for activity. The other region with higher B-values is the exposed loop R70 - N71 - S72 - Y73 - K74, which according to our binding-model interacts with the minor groove of ds-DNA.

Not surprisingly, the carbohydrate site-chain shows increased temperature factors as well, since it is protruding from the enzyme surface into a large solvent channel extending throughout the whole crystal in the c^*-direction. The average B-values are ranging from 34 Å^2 for the N-acetylglucosamine residues. Two N-acetylglucosamine and five mannose residues have been located, showing a branching point at the mannose in position 3 (see figure 2.)

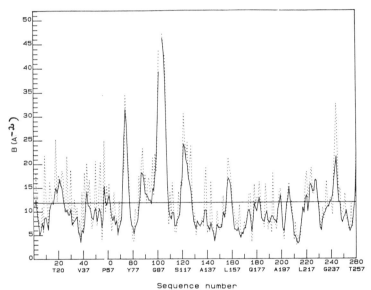

FIGURE 3: Variation of the isotropic temperature factor B with the sequence. The numbering scheme in the bottom line corresponds to the original, chemically determined sequence.

ACTIVE SITE AND MODEL FOR THE BINDING OF DOUBLE-STRANDED DNA

The di-p-nitrophenylester of thymidine-3', 5'-diphosphate is a substrate for DNase I (Liao, 1975). Like the degradation of DNA, the hydrolysis of this small substrate requires divalent cations and is inhibited by carboxymethylation of His 131. Ca^{2+}-thymidine-3', 5'-diphosphate (Ca-pTp) was diffused into DNase I crystals and its binding site determined from a 4 Å difference Fourier as described in our 2.5 Å-paper (Suck et al., 1984). Ca-pTp binds near the essential His 131 at the surface of the protein in a shallow groove between the two β-pleated sheets (figure 4). Based on the geometry of the active site and using the position of the Ca^{2+}-ion bound at the active site in the presence of the nucleotide, we have derived a model for the interaction of DNase I with double-stranded DNA in its B-form (figure 5). Given the constraints imposed by the Ca^{2+}-ion, the disposition of the charged residues and the charged residues and the van der Waals surface of the protein around the active site, the docking was rather straightforward and only the orientations of some exposed side-chains of DNase I were altered in order to optimize the interactions with DNA. The basic features of our model are the rather tight fit of the exposed loop R70 - N71 - S72 - Y73

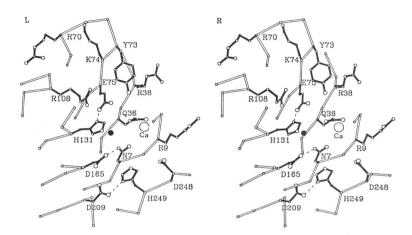

FIGURE 4: Active site of DNase I. The position of the Ca^{2+}-ion found in the Ca-pTp complex is indicated. Also indicated is a water molecule which is in hydrogen bonding contact with N3 of the essential His 131 and which might act as the attacking nucleophile in the hydrolysis reaction of DNase I.

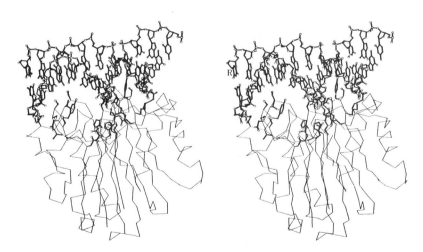

FIGURE 5: Stereoscopic view of DNase I/DNA interaction. DNase I is represented by its C_α-positions and in addition the side-chains, which are in the contact region with DNA, are included. The catalytic site Ca^{2+}-ion is indicated by an open circle. The exposed loop R70 to K74, in particular Y73, is interacting in the minor groove of B-DNA. Phosphates from both DNA-strands opposing each other in the minor groove make electrostatic interactions with positively charged side chains (R9, R109, R70, R38, K74) on either side of the exposed loop. Approximately 5 base-pairs away from the cutting site, additional mainly van der Waals type interactions occur in the region of S171 and T174.

- K74 in the minor groove of B-DNA and electrostatic interactions of phosphates from both DNA-strands with positively charged side-chain on either side of this loop. The DNA strand being cut is bound in the shallow groove between the two sheets, following the direction of the sheets. A somewhat similar similar situation has been found in the EcoRI-DNA complex, possibly indicating a general correspondance of ß-sheet structure displaying a left-handed twist with a right-handed DNA double-helix (J. Rosenberg, personal communication). The scissile phosphodiester bond is anchored to the enzyme through ghe Ca^{2+}-ion, two neighbouring phosphates in the 5'-direction make ionic interactions with R108 and R70, the phosphate following in the 3'-direction interacts with R9. Two phosphates of the other DNA-strand, directly across the minor groove, are boune to K74 and R38. Further, mainly van der Waals type interactions, involving S171 and T174 occur 5 to 6 base pairs away from the cutting site. It is the combination of these interactions - tight fit in minor groove, ionic interactions with phosphates opposing each other in the minor groove and van der Waals contacts a few base-pairs away from the cutting-site - which provides the precise alignment of the DNA-substrate on the surface of the enzyme.

Y73, which is located in the exposed loop mentioned above, penetrates the minor groove such that it is in hydrogen-bonding distance from either O2 of pyrimidines or N3 of purines three base-pairs away from the cleavage site. These two atoms are in equivalent positions relative to the helix axis and therefore, in a first approximation, this interaction will not lead to purine/pyrimidine discrimination. However, the NH_2-groups in position 2 of a guanosine will have an unfavourable steric effect and thereby forcing the tyrosine into a different position. Local variations in the cutting frequencies of DNase I could at least in part be explained by this interaction in the minor groove, since it will also be influenced by changes in the local twist angle, the propellor twist, the roll and tilt angle.

As mentioned above, the fit of the exposed loop R70 to K74 in the minor groove is rather tight and therefore the width of the helix groove will be a critical factor for the binding. X-ray studies on short double-stranded DNA pieces indicate, that the groove width can indeed vary quite substantially with the sequence (Fratini et al., 1982 ; McCall et al., 1985). The sequence-dependent variation of the DNase I cleavage pattern and cutting frequencies can therefore nicely be explained in terms of our model by a variation of the helix groove width, which directly affects the binding. Drew and Travers (1984) studied the DNase I cleavage pattern of a 160 base-pair DNA fragment and found that the locally averaged probabilities of cleavage were usually higher in regions with random base-sequence than in those with runs of A/T or G/C. In agreement with these digestion data our model predicts optimal binding for a minor groove width around 12 - 13 Å, a value occurring in normal, i.e. random-sequence, B-DNA.

DNase I digestion studies of self-complementary oligonucleotides show that the enzyme cleaves near the 3'-end or the centre of a DNA-strand but not, or at very low rates near the 5'-end (Lomonossoff et al., 1981 ; Drew, 1984). The reason for this is obvious from our model : Full binding and accurate alignment of the DNA is only possible with a complete minor groove allowing for interactions with phosphates from both strands. In case of cutting near the 3'-end this is the case and the protein binds across the minor groove with both strands. If, however, a phosphodiester-bond 4 to 5 bonds inward from the 5'-end is cut, contacts only to one of the strands are possible. One can use this argument, to exclude binding across the major groove : Exactly the opposite behaviour - namely cutting all the way to the 5'-end and no cutting near the 3'-end - would be expected from that kind of model.

Among other biochemical results in support of our model we would like to mention the oligopeptide antibiotic netropsin, which is known to bind in the minor groove at A/T rich sequences, preventing cleavage by DNase I at these sites (Kopka et al., 1985). According to our model, netropsin-binding in the minor groove directly interferes with the binding of the exposed loop R70 to K74 of DNase I. Indirect evidence for binding across the minor groove stems form a comparison of DNase I cleavage patterns of a hair-pin loop structure with a double-stranded piece of DNA of identical sequence (Drew, 1984).

Very recently we have obtained single crystals of complexes between various self-complementary octanucleotides and DNase I diffracting to 2 Å resolution (Frost and Suck, unpublished results). When checked by HPLC analysis, some of these crystals showed degraded oligonucleotides besides the parent oligonucleotides suggesting that the contain a nicked DNA double-strand bound to the enzyme. The X-ray structure analysis of these complex crystals will provide further insight into the mechanism of cleavage by DNase I and how it is influenced by the DNA sequence.

REFERENCES

1. G. Bernardi, S.D. Ehrlich and J.P. Thiery, Nature New Biol. 246:36-40 (1973).
2. H.R. Drew, J. Mol. Biol. 176:535-557 (1984).
3. H.R. Drew and A.A. Travers, Cell 37:491-502 (1984)
4. A.V. Fratini, M.L. Kopka, H.R. Drew and R.E. Dickerson, J. Biol. Chem. 257:14686-14707.
5. J.D. Galas and A. Schmitz, Nucleic Acids Res. 5:3157-3170 (1978).
6. W.A. Hendrickson and J.H. Konnert, In "Computing in Crystallography (Diamond, R., Ramaseshan, S. and Venkatesan, K., eds.) pp. 13.01-13-32, Indian Academy of Sciences, Int. Union of Crystallography, Bangalore (1980).
7. T.A. Jones, J. Appl. Cryst. 11:268-272 (1978).
8. A. Klug and L.C. Lutter, Nucleic Acids Res. 9:4267-4283 (1981).
9. M.L. Kopka, C. Yoon., D. Goodsell, P. Pjura and R.E. Dickerson, J. Mol. Biol. 183:553-563 (1985).
10. M.S. Laskowski, in The Enzymes 3rd edition (Boyer, P.D., Ed.) Academic Press NY, Vol. IV, 289-311. (1971).
11. T.H. Liao, J. Biol. Chem. 250:3721-3724 (1975).
12. T.H. Liao, J. Salnikow, S. Moore and W.H. Stein, J. Biol. Chem. 248:1489-1495 (1973).
13. G.P. Lomonossoff, P.J.G. Butler and A. Klug, J. Mol. Biol. 149:745-760 (1981)
14. M. McCall, T. Brown and O. Kennard, J. Mol. Biol. 183:385-396 (1985).
15. S. Moore, in The Enzymes 3rd edition (Boyer, P.D., Ed.) Academic Press NY, Vol. XIV, 281-296. (1981).
16. P.A. Price, J. Biol. Chem. 250:1981-1986 (1975).
17. P.A. Price, S. Moore and W.H. Stein, J. Biol. Chem. 244:924-928 (1969a).
18. P.A. Price, T.Y. Liu, W.H. Stein and S. Moore, J. Biol. Chem. 244:917-923 (1969b).
19. A. Prunell, R.D. Kornberg, L.C. Lutter, A. Klug, M. Levitt and F.H.C. Crick, Science 204:855-858 (1979).
20. D. Rhodes and A. Klug, Nature 286:573-578 (1980).
21. D. Suck, C. Oefner, and W. Kabsch, EMBO J. 3:2423-2430 (1984).

RIBONUCLEASES A AND T_1 COMPARABLE MECHANISMS OF RNA CLEAVAGE WITH DIFFERENT ACTIVE SITE GEOMETRIES

W. Saenger, R. Arni, M. Maslowska, A. Pähler and
U. Heinemann

Institut für Kristallographie, Freie Universität Berlin
Takustr 6, D-1000 Berlin 33, FRG

There are a variety of enzymes which cleave the phosphodiester link in ribo- and deoxyribonucleic acids. They exhibit different catalytic activities, different mechanisms of cleavage, and different three-dimensional structures. The best known examples are DNase I which acts upon single and double stranded DNA[1], staphylococcal nuclease which cleaves P-O bonds in RNA and DNA single strands[2], and the two RNases A and T_1 which cut at the 3'-end of pyrimidine and guanosine nucleotides respectively[3-5]. Although the two RNases have different molecular topology, the mechanism of hydrolysis is similar and suggestive of a comparable active site geometry. Since high resolution crystal structures are available (1.5Å for RNase A and 2.0Å for RNase T_1)[3-5], a study of the arrangement of the functional amino acids in the active sites of the two enzymes is of interest.

<u>Overall topologies of RNases A and T_1 are different</u> Both enzymes are small with 124 and 104 amino acids, respectively, in a single polypeptide chain. In RNase A, the main topological feature is a three-stranded, antiparallel ß-pleated sheet folded in the shape of a ∨, with the active site residues located at the inner tip of the ∨. Three short α-helices are at the N-terminus and between ß-strands, as illustrated schematically in Figure 1.

The main structural features of RNase T_1 are a long α-helix at the N-terminus and a fourfold antiparallel pleated sheet located above the helix, as shown in Figure 2. The active site residues are anchored in this sheet.

<u>Base recognition sites in RNases A and T_1</u> are different and involve interactions of bases with protein main-chain and side-chain atoms. RNase A is specific for pyrimidine bases C and U. The recognition site is a shallow pocket with amino acid Thr45 at the bottom. It forms hydrogen bonds with its γ-hydroxyl group (as acceptor) to N_3H of U and (as donor) to N_3 of C. A second hydrogen bond involving the peptide NH of Thr45 recognizes the O_2 common to U and C, see Figure 3. Discrimination of purine bases occurs on geometrical grounds because the polypeptide chain around Thr45 would interfere sterically with bases larger than pyrimidines.

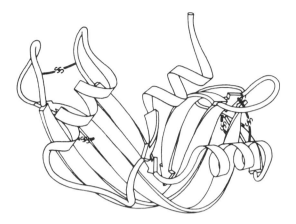

Fig. 1. Schematic diagram of the polypeptide folding in RNase A. Disulfide bridges are indicated by ss. Taken from ref. 3.

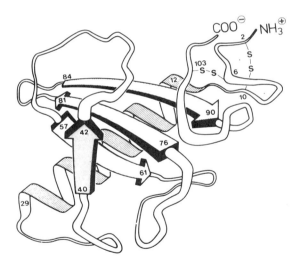

Fig. 2. Schematic diagram of RNase T_1 in the complex with the inhibitor guanosine-2'-phosphate (omitted from this picture).
Taken from ref. 4.

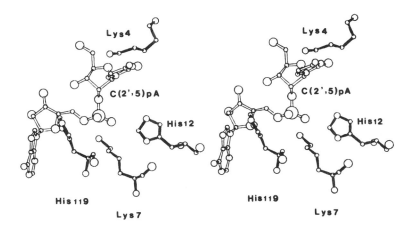

Fig. 3. The active site geometry of RNase S (RNase A with one peptide bond cleaved) in complex with C(2',5)pA. Amino acid residues His12, His119, Lys7 and/or Lys41 participate in phosphodiester cleavage. Stereo view drawn after coordinates in ref. 6 and in the Brookhaven Data File.

In RNase T_1, guanine is recognized by hydrogen bonds to main-chain functional groups: $(Asn43)NH$---O_6(guanine) and N_1H(guanine)---$O=C(Asn44)$. This interaction is augmented by stacking forces, with guanine sandwiched between side-chains of Tyr42 and Tyr45 (Figure 4).

<u>Mechanisms of RNA hydrolysis are comparable</u>   In both RNases A and T_1, the cleavage of the phosphodiester bond occurs at the 3'-side of the recognized nucleotide in a two-step process, which is illustrated for RNase T_1 in Figure 5. In the first step, the proton is removed from O_2', and transferred to His12 in RNase A or to Glu58 in RNase T_1. The O_2' then attacks the phosphate group; O_5' is released in an S_N2-type mechanism and protonated by the imidazole of His92 (RNase T_1) or His119 (RNase A). A pentacovalent phosphorus atom occurs in the transition state, and a cyclic 2',3'-phosphate is formed as product of the transesterification. This reaction step is reversible since no hydrolysis has taken place. In the second step, a water molecule is activated by the previously deprotonated His119 in RNase A or His92 in RNase T_1 so that HO^- attacks the cyclic phosphate and O_2' is released in a reversal of the S_N2 process and subsequently protonated by His12 in RNase A and by Glu58 in RNase T_1. Evidence for this sequence of events was obtained by isolation of the cyclic 2',3'-phosphate intermediate, and by the use of substrates with thiophosphate groups which allowed to follow stereochemically the course of the reaction[8].

<u>The active site residues of RNases A and T_1 are comparable</u>   Based on chemical[9] and spectroscopic[10] data and on X-ray structure analyses, it is clear that the residues directly involved in phosphodiester hydrolysis are:

RNase A: His12, His119, Lys41 and/or Lys7
RNase T_1:   Glu58, His92, Arg77.

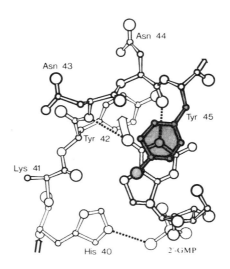

Fig. 4. Guanine recognition by RNase T_1. There are two hydrogen bonds to main-chain functional groups, (Asn43)NH---O_6(guanine) and (guanine)N_1H---O=C(Asn44), and guanine is sandwiched between side chains of Tyr42 and Tyr45. Taken from ref. 4.

In both enzymes, the positive charges of Lys or Arg appear to be essential to neutralize the negative charge of the substrate and possibly to lower the energy of the charged transition state complex. The histidines and the glutamic acid act as proton donors and acceptors in a general acid/base catalysis.

Since the pH optimum for the RNase T_1 reaction is close to 7.5 (ref.11) we should expect the amino acid side chains involved in proton transfer processes to have pK values in the same range. While this is not unusual for the imidazole groups of histidine residues in a protein, it is clearly outside the range of pK values normally observed for the carboxylate groups of glutamic acid residues. To enable Glu58 to participate in acid-base catalysis, this residue must have an unusually high pK value in the enzyme-substrate complex.

This view is supported by potentiometric titration experiments which showed that the pK of one carboxylate group (probably Glu58) shifts from 4.9 to 7.8 on binding of the inhibitor guanosine-2'-phosphate[12]. The shift appears to be caused by the close proximity of two negatively charged groups in the active site of the complex, see Figure 6. Unfortunately, nothing is known about the shift of pK of Glu58 if the substrate RNA or guanosine-3'-phosphate are bound to the active site of RNase T_1. Even if the shift were not as dramatic as observed with binding of guanosine-2'-phosphate, it would tend to make Glu58 less acidic compared with glutamic acid in aqueous solution.

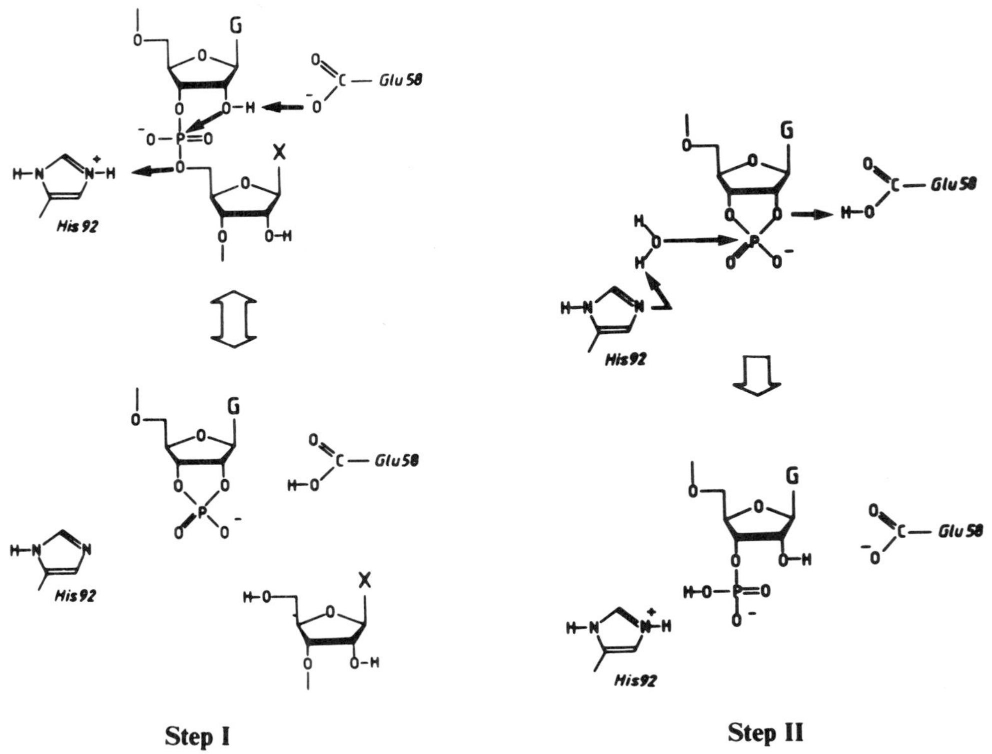

Fig. 5. Schematic illustration of the reaction mechanism of RNase T_1, illustrating the two-step hydrolysis of RNA. For RNase A, a similar mechanism holds, with guanine G replaced by cytosine C or uracil U, and Glu58, His92 replaced by His12, His119. Taken from ref. 7.

The architectures of the active sites in RNases A and T_1 are not comparable If the distances between the Cα atoms of the active site residues are compared, it strikes that there are considerable differences, as illustrated in Table 1 . The CαHis12---CαHis119 distance in RNase A, 9.2Å, contrasts the CαGlu58---CαHis92 distance of 13.3Å in RNase T_1. Because the tertiary structures of RNases A and T_1 will not change significantly upon binding of substrates as could be shown by X-ray analyses of free and inhibitor-complexed enzymes[3,13] it is improbable that the Cα---Cα distances of the active site residues His, Glu in the two RNases can be made equivalent. The same holds for the positively charged Lys and Arg residues. In RNase A, the distances between Cα atoms of Lys41 to His12 (13.0Å) and to His119 (15.4Å) are much greater than the comparable distances to Arg77 in RNase T_1 (6.7Å to Glu58 and 8.8Å to His92), so that a superposition of the Cα atoms is impossible.

Similar arguments hold if the distances between functional groups of the amino acid residues located in the active sites of RNases A and T_1 are compared, Table 1. The distance between His92-Cγ and Glu58-Cγ in RNase T_1, 8.7Å is comparable to the distance between His12-Cγ and His119-Cγ in RNase A, 8.4Å, because these pairs of side chains have to occupy similar positions relative to the common substrate, a phosphodiester bond connecting two ribose moieties, in order to catalyze the

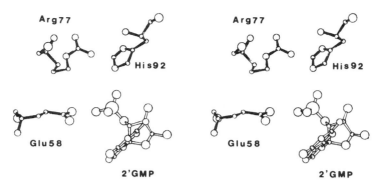

Fig. 6. Stereo view of the active site geometry of RNase T_1 in complex with guanosine-2'-phosphate. If the "substrate" guanosine-3'-phosphate is modeled into the active site, Glu58 is in hydrogen bonding contact with O_2,H, His92 hydrogen bonds to the phosphate and Arg77 has no obvious direct contact to the substrate[7].

Table 1. Distances (Å) between selected atoms of active site residues in RNase A and T_1. Distances between α-carbons above the diagonals, distances between side chain atoms below.

RNase T_1

	Glu58 Cα/Cδ	His92 Cα/Cζ	Arg77 Cα/Cγ
Glu58-Cα/Cδ	–	13.3	6.7
His92-Cα/Cγ	8.7	–	8.8
Arg77-Cα/Cζ	6.9	5.4	–

RNase A

	His12 Cα/Cγ	His119 Cα/Cγ	Lys41 Cα/Nζ
His12-Cα/Cγ	–	9.2	13.0
His119-Cα/Cγ	8.4	–	15.4
Lys41-Cα/Nζ	9.9	10.0	–

same reaction. In contrast, however, all the other distances involving the positively charged groups of Arg and Lys side chains differ so much (by several Å units) that they cannot be comparable, even if conformational adjustments are allowed for.

The obvious conclusion therefore is that the active site geometries of RNase T_1 and RNase A are different although the mechanisms of RNA hydrolysis are comparable.

This finding is in contrast to the serine proteases where structurally unrelated enzymes belonging to the mammalian trypsin and the bacterial and fungal subtilisin families have Ser, His, Asp in their active sites. The rms distances between the C_α atoms of these active site residues is only 0.9 Å between the families (trypsin and proteinase K) compared to 0.6 Å within the subtilisin family (subtilisin novo and proteinase K[14]). Based on these numbers, it appears that the serine proteases are indeed related by convergent evolution where two different enzymes were optimized to arrive at the same active site geometry. In contrast, the RNases of the mammalian RNase A family and of the bacterial and fungal RNase T_1 family are evolutionary unrelated because their three-dimensional structure and their active site geometry are both different. They are independent solutions to the problem of catalyzing an essentially identical reaction.

ACKNOWLEDGEMENTS

The authors gratefully acknowledge financial support by Deutsche Forschungsgemeinschaft (Sonderforschungsbereich 9) with a fellowship to M.M., and by Bundesministerium für Forschung und Technologie with a fellowship to R.A.

REFERENCES

1. D. Suck, C. Oefner, and W. Kabsch, Three-dimensional structure of bovine pancreatic DNase I at 2.5 Å resolution, EMBO J. 3:2423-2430 (1984).
2. F.A. Cotton, E.A. Hazen, and M.J. Legg, Staphylococcal nuclease: Proposed mechanism of action based on structure of enzyme-thymidine 3',5'-bisphosphate-calcium ion complex at 1.5 Å-resolution, Proc. Natl. Acad. Sci. (USA) 76:2551-2555 (1979).
3. A. Wlodawer, Structure of bovine pancreatic ribonuclease by X-ray and neutron diffraction, pp. 393-439, in "Biological Macromolecules and Assemblies, Vol. II. Nucleic Acids and Interactive Proteins", F.A. Jurnak and A. McPherson, eds., John Wiley & Sons, New York (1985).
4. U. Heinemann and W. Saenger, Specific protein-nucleic acid recognition in ribonuclease T_1-2'-guanylic acid complex: an X-ray study, Nature 299:27-31 (1982).
5. R. Arni, Ph.D. Thesis Freie Universität Berlin, to be submitted (1986).
6. S.Y. Wodak, M.Y. Liu, and H.W. Wyckoff, The structure of cytidylyl(2'-5')adenosine when bound to pancreatic ribonuclease S J. Mol. Biol. 116:855-875 (1977).
7. U. Heinemann and W. Saenger, Mechanism of guanosine recognition and RNA hydrolysis by ribonuclease T_1, Pure & Appl. Chem. 57:417-422 (1985).
8. F. Eckstein, Nucleoside phosphothioates, Ann. Rev. Biochem. 54:367-402 (1985).
9. K. Takahashi, The structure and function of ribonuclease T_1. XXI. Modification of histidine residues in ribonuclease T_1 with iodoacetamide, J. Biochem. (Tokyo) 80:1267-1275 (1976).
10. R. Fülling and H. Rüterjans, Proton magnetic resonance studies of ribonuclease T_1. Assignment of histidine-27 C2-H and C5-H proton resonances by a photooxidation reaction, FEBS Lett. 88:279-282 (1978).

11. F. Egami, T. Oshima, and T. Uchida, Specific interactions of base-specific nucleases with nucleosides and nucleotides, Mol. Biol. Biochem. Biophys. 32:250-277 (1980).
12. S. Iida and T. Ooi, Titration of ribonuclease T_1, Biochemistry 8:3897-3901 (1969).
13. C. Hill, G. Dodson, U. Heinemann, W. Saenger, Y. Mitsui, K. Nakamura, S. Borisov, G. Tischenko, K. Polyakov and S. Pavlovsky, The structural and sequence homology of a family of microbial ribonucleases, Trends Biochem. Sci. 8:364-369 (1983).
14. A. Pähler, A. Banerjee, J.K. Dattagupta, T. Fujiwara, K. Lindner, G.P. Pal, D. Suck, G. Weber and W. Saenger, Three-dimensional structure of fungal proteinase K reveals similarity to bacterial subtilisin, EMBO J. 3:1311-1314 (1984).

STRUCTURAL STUDIES OF ECO RV ENDONUCLEASE AND OF ITS COMPLEXES WITH

SHORT DNA FRAGMENTS

Fritz K. Winkler, Raymond S. Brown, Kevin Leonard and
John Berriman

European Molecular Biology Laboratory
Meyerhofstr. 1, Postfach 10.2209, 6900 Heidelberg
Federal Republic of Germany

INTRODUCTION

Many bacteria restrict the expression of foreign DNA introduced through phage infection, conjugation or transformation. At the molecular level such host-controlled restriction is the result of an endonuclease activity which cuts foreign DNA and a modification activity which protects the host's DNA against this cleavage.
Of the three types of restriction-modification systems usually distinguished the type II systems are the simplest and consist of two separate enzymes, an endonuclease and a methylase which recognize the same sequence of typically 4 to 6 base pairs. The endonucleases cleave double stranded DNA at these recognition sites in the presence of Mg^{2+} ions. Their remarkable specificity and its modulation by various factors make them attractive systems for structurally oriented studies of protein nucleic acid interactions (for a recent review see Modrich and Roberts[1]).

THE SPECIFICITY OF TYPE II RESTRICTION ENDONUCLEASES

Of the approximately 400 known type II endonucleases only the Eco RI enzyme has been studied in great detail. Quantitative data on its accuracy under normal buffer conditions have recently been measured by Pingoud and Alves[2]. They find that the canonical site (-GAATTC-) is attacked 3000 and 9000 times faster than the two most labile non-canonical sites which are present on the DNA fragments used in their studies. In the presence of spermidine cleavage at non-canonical sites becomes even slower[3]. The opposite trend, i.e. enhanced cleavage at non-canonical sites, is observed under non-optimal buffer conditions[4,5]. These so-called * conditions are high pH, low ionic strength, the presence of polar organic solvents or the replacement of Mg^{2+} by Mn^{2+}. Together this means that the accuracy of Eco RI endonuclease, expressed as the relative cleavage rate at the canonical site versus that at the most labile non-canonical site(s), ranges from 10 to 100 under * conditions to 10^6 or more under optimal conditions. Qualitatively similar observations have been made with most type II endonucleases so far examined[3]. Understanding the structural basis of this remarkable specificity requires the determination of the three-dimensional structures of enzyme DNA complexes. Such complexes with DNA carrying a canonical site (cognate DNA) can only be studied in the absence of Mg^{2+}. It is therefore important that binding studies[6,7] have shown a strong preference for canonical sites in the absence of Mg^{2+} meaning that the formation of specific interactions is not dependent on the presence of Mg^{2+}. First results of the crystal structure analysis of a complex between Eco RI endonuclease and a tridecameric DNA fragment have been reported by Frederick et al.[8]. Although the protein DNA interface has not yet been analysed in detail major deviations of the DNA fragment from a B-DNA like conformation are observed. This raises interesting questions such as whether bound non-cognate DNA fragments are similarly or less distorted or wether the enzyme undergoes large conformational changes too. In addition to the structure of a complex with cognate DNA one therefore also needs to know the structure of the free enzyme and that of a complex with non-cognate DNA. In the following account of our work with Eco RV endonuclease and small DNA fragments we show that this system is well suited to provide such information in the near future.

STRUCTURAL STUDIES OF ECO RV ENDONUCLEASE AND OF ITS COMPLEXES WITH DNA FRAGMENTS

Eco RV endonuclease, a dimeric enzyme of 2 x 29000 dalton, cleaves double stranded DNA at the sequence 5'-GAT/ATC and produces blunt ended fragments[9,10]. Its DNA derived amino acid sequence (245 residues) shows no significant homology to that of Eco RI endonuclease[11] and the same holds true for the two corresponding methylases. Large amounts of pure RV endonuclease have become available through the construction of an inducible overproducing strain[12] and have made it possible to grow crystals of the enzyme and of its complexes with various short DNA fragments.

Studies by X-ray Crystallography

Four crystal forms of Eco RV endonuclease have been obtained by precipitation with polyethylene glycol 4000[13] two of which diffract to about 2 Å resolution. They both have space group $P2_12_12_1$ with one dimer in the asymmetric unit and show similar cell dimensions of 58.2 (59.9), 71.7 (74.5) and 130.6 (121.8) Å for a, b and c of crystal form A and (B) respectively. Complexes of the enzyme with short self-complementary oligonucleotides were crystallized under rather similar conditions. Those with cognate DNA had to be crystallized in the absence of Mg^{2+} to prevent cleavage. Careful characterization of washed redissolved cocrystals by UV absorption and HPLC showed the presence of 1:1 complexes (1 RV dimer per DNA double strand) and no cleavage of the DNA fragments. Crystallographic parameters of these cocrystals are given in Table 1 (full details of their growth and characterization will be given elsewhere[14]). As can be seen structure determinations appear feasible for both types of complexes.

Thus far our efforts have concentrated on determining the structure of the free enzyme partly since we hope to be able to use molecular replacement methods to solve the structures of its complexes with DNA fragments. Data to 2.5 Å resolution were collected for native and Pb-derivative form A crystals by the oscillation method using synchrotron radiation at the EMBL outstation at DESY in Hamburg. The films were processed using A. Wonacott's film processing system MOSCO and good merging R-factors of 0.055 and 0.075 for native and derivative intensities respectively were obtained.

Table 1

Crystallographic parameters of complexes of Eco RV endonuclease with self complementary deoxy-oligonucleotides
(series I-n for cognate DNA fragment and series II-n for non-cognate DNA fragments)

Form	Oligonucleotide Sequence (5'-3')	Space-group	Unit cell parameters a	b	c (Å)	Diffraction limit (Å)	Crystal morphology
I-1	GGATATCC	n.d.				3.2	small prisms
I-2	GGGATATCCC and GCGATATCGC	$C222_1$	59.7	77.1	367.0	2.7	thick plates
I-3[a]	CCGATATGGC	n.d.	54	64	~160	3.0	very thin platelets
II-1	CGAATTCG	I222	84.0	146.1	74.0	3.2-4.0	plates
II-2	CGAATTCG	$C222_1$	84.0	127.3	74.0	3.2-4.0	plates
II-3	CGAATTCG	$P2_12_12_1$	90.3	61.7	251.3	4.5	prisms
II-4	CGAGCTCG	$P2_1$	65.5	78.4 $\beta = 106°$	71.3	3.0	thick plates

a) These crystals were only investigated by electron microscopy and the c-periodicity has been estimated by tilting experiments.

Two major Pb sites, already identified in a 6 Å resolution difference Patterson synthesis calculated with diffractometer data, were refined using centric terms only. A low resolution diffractometer data set was also collected for a $Au(CN)_2^-$-derivative which could not be used for phasing beyond 6 Å resolution but proved very helpful to determine the position and orientation of the noncrystallographic twofold and the correct enantiomorph. Molecular replacement techniques as described by Bricogne[15] were used to produce electron density maps at 6 and 2.5 Å resolution. Initial best phases were derived from the isomorphous differences of the Au- (to 6 Å only) and the isomorphous and anomalous differences of the Pb-derivative. The molecular envelope traced in the first averaged 6 Å map was updated after a few cycles at low resolution and used without further change at higher resolution. Figure 1 shows a few sections of the final unaveraged 2.5 Å electron density map obtained after 10 cycles of molecular replacement. Two α-helices of 4 to 5 turns related by the molecular twofold are clearly recognizable but show poorly defined side chain density. Altogether the map has proved difficult to interpret and no chain tracing has yet been achieved. Different ways to improve the map are currently underway. Apart from continuing the search for more derivatives we are trying to extend the molecular replacement to crystal form B for which a 2.5 Å data set has been collected and processed. As the initial heavy atom phases might be rather poor beyond 3 Å resolution the procedure is also being repeated at somewhat lower resolution. Position and orientation of the molecular twofold were derived from the refined coordinates of the two sets of major sites occupied in the two derivatives. Although they appear well determined, - the two heavy atom-heavy atom vectors are almost at right angles and are about 22 Å apart along the twofold -, we cannot exclude that they are somewhat in error.

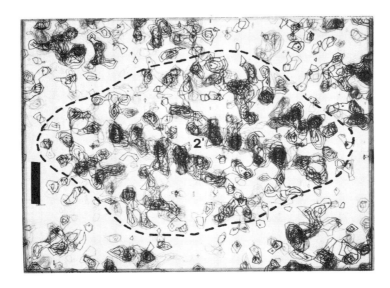

Fig. 1 2.5 Å electron density map
Eight sections spaced at 1 Å of the final unaveraged density map are shown. The view down the noncrystallographic twofold axis shows 2 symmetry related α-helices lying approximately in the plane and forming part of the dimer contact near the twofold in the center. The molecular boundary is indicated (- - -) and the bar represents 10 Å.

The presumed twofold symmetry of RV endonuclease DNA complexes leaves only few possibilities to dock a DNA fragment onto the protein. A balsa wood model of the dimer derived from the 6 Å electron density map and a simple wire model of the backbone of double stranded B-form DNA were used for this purpose. Only the arrangement illustrated in Fig. 2 satisfied a number of properties that such a complex is expected to have. The RV dimer as shown in this side view is about 60 by 60 Å across and 30 to 35 Å deep (see Fig. 5b for a view down the twofold). The bottom part of the molecule forms a tight dimer contact but its pointed end does not show any complementarity to DNA. At the top the dimer splits into two lobes leaving a depression around the twofold axis. The crude DNA model fits quite well into this depression when the major groove faces the protein. At the same time the two scissile phosphodiester bonds are in reasonable proximity to protein density. Contacts with the sugar phosphate backbone would be almost exclusively on the 3'-side of the scissile bond and could extend over 4 to 5 backbone moieties on each side (see Fig. 5b). Although we believe that this overall arrangement is correct, conformational changes of the DNA and of the protein could yield a much tighter complex in which the two lobes might have moved more into the major groove. These questions can only be answered by solving the structures of the complexes with DNA fragments. For the cocrystals with the decamer GGGATATCCC a 3 Å data set has been collected and processed and we hope to solve its structure soon after that of the native enzyme.

Fig. 2   Balsa wood model derived from 6 Å map
A side view of an Eco RV dimer is shown. A possible binding mode of DNA is illustrated by a wire model of a B-form double helix about 14 base pairs long. The scissile phosphodiester bonds (marked P) are near density protrusions at the top of the molecule. The bar represents 20 Å.

Studies by Electron Microscopy

The first cocrystallization attempts carried out with the cognate DNA decamer CCGATATCGG yielded crystalline platelets too thin to be studied by X-rays but very suitable for imaging in the electron microscope. Electron diffraction patterns of frozen hydrated or glucose dried cocrystal specimens extended to 3 Å resolution (Fig. 3). The hk0 projection shown has almost perfect p2gg plane group symmetry but many specimens show violations of the absencies along b*.

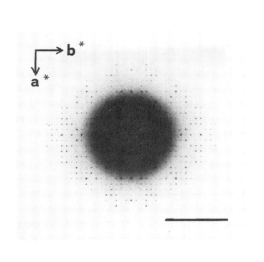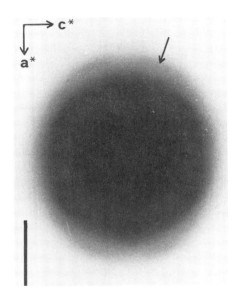

Fig. 3 (left) Electron diffraction of Eco RV-CCGATATCGG cocrystals
Thin cocrystals dried in glucose yield diffraction patterns extending to about 3 Å resolution. The bar indicates 0.2 Å^{-1}.

Fig. 4 (right) Electron diffraction of Eco RV-CGAATTCG cocrystals
Thin cocrystals with this non-cognate DNA octamer dried in glucose yield only poor diffraction patterns (hidden in the central black area due to overexposure) but show faint arcs at about 3.4 Å resolution centered at about +/- 20° from a*. The bar indicates 0.2 Å^{-1}.

We believe that a unit cell translation along c, estimated to be 140 to 160 Å from tilting experiments, comprises two molecular layers which are related by a twofold screw axis along b. Thin crystals with an odd number of layers would then not show strict absencies along b*. A two-dimensional image reconstruction from negatively stained specimens showed rather featureless, ellipsoidal density peaks. The double layer structure causes overlap in this projection and we have not yet unambiguously resolved the molecular packing. These crystals appear very suitable to develop methods of three-dimensional image reconstruction from unstained specimens of multilayered crystals.

Crystalline platelets of the complex with the non-cognate DNA octamer GGAATTCC were examined by X-ray diffraction and electron microscopy. They proved much more fragile and diffraction patterns from glucose dried specimens extended at best to about 6 Å resolution. Negatively stained specimens of multilayered crystals show c2mm plane group symmetry but single layers with p2gg symmetry could be generated by diluting the crystal mother liquor prior to staining. Fig. 5a shows the result of a two-dimensional image reconstruction from such single layers. The molecules are well resolved and the projected density suggests that we are looking down the molecular twofold axis. This was confirmed by a subsequent three-dimensional reconstruction which yielded molecular dimensions very comparable to those obtained in the X-ray analysis. Due to the low resolution of about 30 Å of this reconstruction no detailed features could be compared and no density was seen where DNA is expected to be bound. This could be due to positive staining of the nucleic acid but also due to loss of the DNA on dilution of the mother liquor needed to generate single layers.

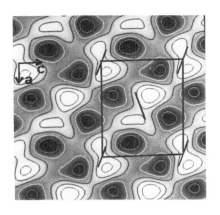 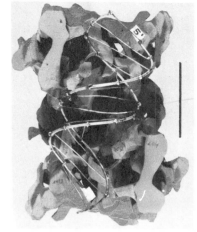

a b

Fig. 5 Orientation of DNA helix axis in RV endonuclease DNA complexes.
a) Two-dimensional image reconstruction from single layers of negatively stained RV-CGAATTCG cocrystals. Two RV dimers viewed down the twofold are clearly recognizable in one unit cell (no p2gg symmetry has been imposed). For one choice of the orientation of the DNA helix axis (marked by lines for some of the dimers) good agreement is observed to the orientation postulated from model building.
b) View down the molecular twofold of the balsa wood model of Fig. 2. The long dimension of the projected dimer is approximately aligned with the corresponding dimension in the image reconstruction. The bar represents 20 Å.

Since the DNA can hardly be involved in intra-layer contacts this appears easily possible. Some information on the orientation of the DNA in these cocrystals could however be derived from the diffraction of multilayered crystals dried in glucose. As shown in Fig. 4 faint arcs at about 3.4 Å resolution can be seen and they are centered at about 20° to the a axis of the crystal (the lattice is not visible due to overexposure necessary to bring out the arcs). Assuming that these are due to the DNA base stacking the helix axis can be chosen at + or -20° from a with respect to a particular dimer. One of these choices gives quite good agreement with the orientation derived from model building. In both cases the angle between the helix axis and the line connecting the centers of the projected monomer densities is around 45°. Analysis of the molecular packing in these cocrystals has told us that the space group of crystal form II-2 is I222 (and not $I2_12_12_1$), that form II-3 represents simply a different stacking of the layers along the b-axis and will also help the X-ray crystallographic analysis.

351

Acknowledgement

We thank M. Gait (MRC, Cambridge), J. van Boom (University of Leiden) and H. Blöcker and R. Frank (GBF, Braunschweig) for their generous gifts of some of the oligonucleotides used in the cocrystallization experiments.

References

(1) P. Modrich and R.J. Roberts, in: "Nucleases", M. Linn and R.J. Roberts, eds., Cold Spring Harbor Laboratory, New York (1982).
(2) A. Pingoud and J. Alves, personnel communication.
(3) A. Pingoud, Eur. J. Biochem. 147 : 105 (1985).
(4) B. Polisky, P. Greene, D. Garfin, B. McCarthy, H. Goodman and H. Boyer, Proc. Natl. Acad. Sci. USA 72 : 3310 (1975).
(5) R.C. Gardner, A.J. Howarth, J. Messing and R.J. Sheperd, DNA 1 : 109 (1982).
(6) B.J. Terry, W.E. Jack, R.A. Rubin and P. Modrich, J. Biol. Chem., 258 : 9820 (1983).
(7) A.D. Frankel, G.K. Ackers and H.O. Smith, Biochemistry 24 : 3049 (1985).
(8) C.A. Frederick, J. Grable, M. Melia, J. Samudzi, L. Jen-Jacobsen, B.-C. Wang, P. Greene, H.W. Boyer and J.M. Rosenberg, Nature 309 : 327 (1984).
(9) G.V. Kholmina, B.A. Rebentish, Y.S. Skoblov, A.A. Mironov, Y. Yankovskii, Y.I. Kozlov, L.I. Glatman, A.F. Moroz and V.G. Debabov, Dokl. Akad. Nauk. SSSR 253 : 495 (1980).
(10) I. Schildkraut, C.D.B. Banner, C.S. Rhodes and S. Parekh, Gene 27 : 327 (1984).
(11) L. Bougueleret, M. Schwarzstein, A. Tsugita and M. Zabeau, Nucleic Acids Res. 12 : 3659 (1984).
(12) L. Bougueleret, M.L. Tenchini, J. Botterman and M. Zabeau, Nucleic Acids Res. 13 : 3823 (1985).
(13) A. D'Arcy, R.S. Brown, M. Zabeau, R. Wijnaendts van Resandt and F.K. Winkler, J. Biol. Chem. 260 : 1987 (1985).
(14) F.K. Winkler, J. van Boom, H. Blöcker and R. Frank, in preparation for J. Mol. Biol.
(15) G. Bricogne, Acta Cryst. A32 : 832 (1976).

ERRORS IN DNA HELICES: G.T MISMATCHES IN THE A, B AND Z FORMS OF DNA

Dov Rabinovich

Department of Structural Chemistry

The Weizmann Institute of Science, Rehovot 76100, Israel

DNA replication is perhaps the most specific of all enzymic processes. Genetic experiments in DNA replication of E. Coli indicate an error range of one mistake in 10 million to 100 billion nucleotide polymerized. This remarkable fidelity is essential in order to preserve the genetic identtity of organisms. Yet, a few errors must be permitted for species evolution and adaptation to changing environments. Mutagenesis appears also to be important in determining the life cycle of individual organisms, as it is central to many pathological processes.

This fidelity is believed to result from three consecutive discrimination steps[1]. In the first, which occurs at the growing point of the DNA chain the four dNTP's (deoxynucleotide triphosphate) compete for pairing with the template base. This insertion step is thermo-dynamically controlled and depends on the difference in free energy between the correct and incorrect base pairing. The difference may depend solely on the nature of the base pairs[2] or may be amplified by DNA polymerases and other proteins. Typical frequencies of misinsertions depend on the type of mismatch, e.g., 10^{-4} and 10^{-7} for G.T and C.T, respectively[3]. The final fidelity is achieved by two additional steps of proofreading and post-replicative mismatch repair mechanism. The rare mismatches that pass through this screening may lead to spontaneous substitution mutations.

Two hypotheses have been proposed to describe the nature of the mismatches and to account for misinsertion frequencies. The first, guided by the concept of a rigid and monotonous DNA duplex invoked the imino and enol tautomers to explain the pairing of non-complementary bases[4,5]. This hypothesis is stereochemically attractive as it conserves the Watson-Crick geometry and seems to account for misinsertion frequencies. The second hypothesis assumes non Watson-Crick wobble pairing of the bases in their major tautomeric forms[6].

G.U wobble pairs have been observed by single-crystal X-ray analyses in tRNA[7,8,9]. NMR studies sprovide evidence for wobble pairing of G.T, G.A, A.C and C.T in major tautomer forms[10]. Theoretical calculations also indicate that wobble pairs can be accomodated in DNA duplexes[11,12]. However, until very recently there exists no X-ray evidence for mismatches in DNA.

We have undetaken a systematic comparative X-ray study at near-atomic resolution of the structures of mispaired synthetic DNA fragments and their parent duplexes in order to answer some of the following questions: What are the combinations of mispaired bases that can form stable duplexes in the solid state? Do they form wobble pairs in their major tautomer forms or mimic Watson-Crick pairing by utilizing rare tautomers? What effect have these errors on the geometry of neighbouring nucleotides and on the global helical parameters? Are the stabilities of the affected duplexes form composition and sequence dependent? To what extent can the differences in free energy between the correct and incorrect pairing account for the observed misinsertion frequencies?

The apparently most stable G.T mismatch was chosen as the first candidate for the study. The polymorphism of DNA structures allows one to investigate this mismatch in the A, B and Z forms of DNA. Of the number of 'self-complementary' fragments with G.T mispairs which were synthesized by the solid-phase phosphotriester method[13] and crystallized, four were subjected to X-ray structure analysis. Starting coordinates for the analyses were obtained utilizing ULTIMA[14] and refined by CORELS[15]. The relevant X-ray data are listed in the Table.

The analyses of the four structures provide partial answers to our questions:

(1) The G.T mismatch can be accomodated in crystallizable DNA duplexes in the A, B and Z forms. The B-DNA mismatch is perhaps the most interesting as this form is the most biologically relevant.

(2) All G.T mismatches adopt the wobble pair geometry with both the guanine and thymine in their major forms. The two hydrogen bonds from the N1 and O6 of the G to respectively, the O2 and N3 of the Tm are in almost perfect geometry.

(3) The dispositions of the G.T with respect to the G.C pair of the parent duplexes are form and sequence dependent and occur in such a manner as to cause minimal loss of stacking interaction.

(4) The deformations caused by the errors are localized and affect mainly the sugar and phosphate moieties of the mismatches and the flanking nucleotides. They are also form and sequence dependent. It appears that the smallest perturbations occur in the A form, whereas in the Z one, where the mismatches are at the boundaries, appreciable distortions are observed in the deoxyribose moiety. The effects on the global helical parameters are quite negligible.

(5) Since the refinement of the three structures has not yet been completed and solvent molecules and counter-ions have been only partially located, it is too early to estimate the change in free energy between the parent and affected duplexes and therefore to address ourselves to the question of relative stability versus misinsertion frequencies.

This work has been carried out in collaboration with Olga Kennard, Geoff Kneale and Tom Brown at the University Chemical Laboratory, Cambridge, U.K.

TABLE 1

Duplex	Form	Cell dimensions			S.G.	Z	Res.	Ref.	R
GGGGCTCC	A	45.20	45.20	42.97	$P6_1$	12	2.25	1954	14^{16}
GGGGTCCC	A	44.59	44.59	42.37	$P6_1$	12	2.25	2460	14^{17}
CGCGAATTTGCG	B	24.87	40.39	66.20	$P2_12_12_1$	8	3.00	1395	18
TGCGCG	Z	17.97	30.73	45.11	$P2_12_12_1$	8	1.50	2925	15

REFERENCES

1. Loeb, A.L. and Kinkel, T.A., Ann. Rev. Biochem., 51:429 (1982).
2. Goodman, M.F., Hopkins, R.L., Watanabe, S.M., Clayton, L.K. and Guidotti, S., in: 'Mechanistic Studies of DNA Recombination', Eds. Alberts, B. and Fox, F.F., Academic Press, New York, p.685 (1980).
3. Fersht, A.R., Knill-Jones, J.W. and Tsui, W.-C., J.Mol.Biol. 156:37 (1982).
4. Watson, J.D. and Crick, F.H.C., Nature, 171:737 (1953).
5. Topal, M.D. and Fresco, J.R., Nature, 263:285 (1976).
6. Crick, F.H.C., J. Mol. Biol., 19:548 (1976).
7. Ladner, J.E., Jack, A., Robertus, J.D., Brown, R.S., Rhodes, D., Clark, B.F.C. and Klug, A., Nucl. Acids Res., 2:1629 (1975).
8. Quigley, G.J., Seeman, N.C., Wang, A.H.-J., Suddath, F.L. and Rich, A., Nucl. Acids Res., 2:2329 (1975).
9. Sussman, J.L. and Kim, S.-H., Biochem. Biophys. Res. Commun., 68: (1976).
10. Patel, D.J., Koslowski, S.A., Ikuta, S., and Itakura, K., Fed. Proc., 43:2663 (1984).
11. Chuprina, V.P. and Poltev, V.I., Nucl. Acids Res., 11:5205 (1983).
12. Rein, R., Shibita, M., Garduno-Juarez, R. and Kieber-Emmons, T., in: 'Structure and Dynamics: Nucleic Acids and Proteins', Eds. Clementi, E. and Sarma, R.H., Academic Press, New York (1983).
13. Gaits, M.J., Matthes, H.W.D., Singh, M., Sproat, B.S. and Titmus, R.C., Nucl. Acids Res., 10:6243 (1982).
14. Rabinovich, D. and Shakked, Z., Acta Cryst., A40:195 (1984).
15. Sussman, J.L., Holbrook, S.R., Church, G.M. and Kim, S.-H., Acta Cryst., A33:800 (1977).
16. Brown, T., Kennard, O., Kneale, G. and Rabinovich, D., Nature, 315:605 (1985).
17. Kneale, G., Brown, T., Kennard, O., and Rabinovich, D., Sbumitted to J. Mol. Biol. (1985).

8 PROTEINS

RELATION BETWEEN FUNCTIONAL LOOP REGIONS AND INTRON POSITIONS IN
α/β DOMAINS

Carl-Ivar Brändén

Department of Molecular Biology
Swedish University of Agricultural Sciences
Uppsala Biomedical Center
Box 590, S-751 24 Uppsala, Sweden

INTRODUCTION

It is generally recognised that protein molecules are arrangend into small folding units domains. We now have a sufficiently large sample of three dimensional structures to realise that there is only a limited number of different folding patterns for such domains (Richardson, 1981). The most common type of these patterns are the α/β domains which comprise around 25% of all known domain structures. The central fold of these domains is very simple consisting of a number of hydrogen bonded parallel strands joined by helices. The strands all have their NH_2 ends at the same edge of the resulting ß-sheet. The sheet thus has a polarity with an NH_2-edge and a COOH edge. Loop regions, which are usually relatively short, join the strands to the helices at both ends.

There are two main motifs, A and B, these α/ß structures. Type A occurs when eight consecutive strands are arranged adjacent to each other. These strands are closed into a barrel structure with eight helices on the outside as illustrated by the TIM-barrel in Fig. 1. Type B structures have a change in strand order along the sheet. The resulting structure is an open twisted sheet of paralled strands with helices on both sides as illustrated by the nucleotide binding fold in Fig. 2. There are a number of variations of this general motif with different number of strands and different changes in strand order (Brändén, 1980).

There are two essential enigmas associated with these structures. The first is the fact that these similar folds occur for a large number of completely unrelated amino acid sequences. In spite of many different approaches we still cannot identify the sequence pattern that gives rise to this fold. The second is he observation that in all these structures, with no exception, the active site is at the carboxyl edge of the sheet. Active site residues are found in the loop regions that connect the strands with the helices at this edge of te sheet. In this paper I will present evidence that intron positions for the corresponding genes are frequently found in the same loop regions and rarely at other places. This provides an evolutionary mechanism for functional variations at this edge of the sheet within the same structural motif and thus an explanation why the active sites are at the same edge in all enzymes. I will also show that this is consistent with the hypothe-

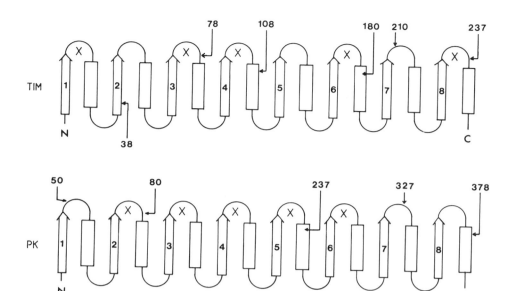

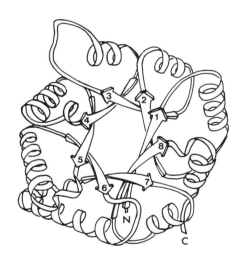

Fig. 1. Schematic diagramms of the α/β barrel structure. The strands of the parallel β-sheet is labelled 1-8. Helices are denoted by rectangles in the top part of the diagram. Residue numbers at intron positions within the barrel are given for TIM (triosophosphate isomerase) and PK (pyruvate kinase) and their positions in relation to the fold are indicated by arrows. Loop regions which provide active site residues are marked by a cross. These are at the carboxyl edge of the sheet. The bottom diagram is drawn after J. Richardson (1981).

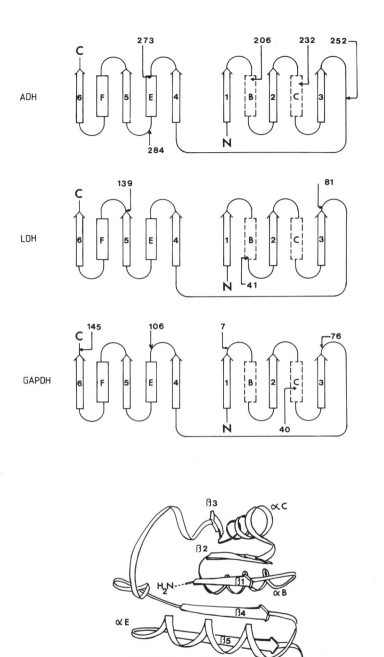

Fig. 2. Schematic diagrams of the nucleotide binding fold. The strands of the parallel β-sheet are labelled 1-6 and the helices B,C,E and F. Residue numbers at observed intron positions are given for ADH (alcohol dehydrogenase), LDH (lactate dehydrogenase) and GAPDH (glyceraldehyde phosphate dehydrogenase). Their positions in relation to the fold are indicated by arrows. The bottom diagram is drawn after B. Furugren.

sis that the genes for these proteins have been built up at an early evolutionary stage from gene pieces coding for a small structural unit comprising an α-helix followed by a ß-strand, an αß-module. Implications for synthesis at the gene level of enzymes "<u>de novo</u>" with this fold will also be discussed.

FUNCTIONAL RESIDUES AND INTRON POSITIONS IN TYPE A BARREL

The type A barrel structure has been described in triosephosphate isomerase (Banner et al., 1975), phosphogluconate aldolase (Lebioda et al., 1982), xylose isomerase (Carrel et al., 1984), α-amylase (Matsuura et al., 1980) in one domain of pyruvate kinase (Stuart et al., 1979) and glycolate oxidase (Lindqvist μ Branden, 1985). It has also been reported to be present in glucose isomerase (G. Petsko private communication), muconolactone isomerase (H.W. Wyckoff private communication), triethanol aminedehydrogenase (S. Matthews private communication) and ribulose-1,5-bisphosphate carboxylase (Schneider et al., 1986). It is thus apparent that this stable eight-stranded barrel structure is of frequent occurence and is used for a variety of different functions. The positions of active site residues have been located for triose phosphate isomerase, phosphogluconate aldolase, pyruvate kinase and glycolate oxidase. All these residues in all four strutures are within the loop regions at the carboxyl edge of the sheet.

Intron positions have been located for triosephosphate isomerase and pyruvate kinase by Gilbert and coworkers (Gilbert, 1985, Lonberg μ Gilbert, 1985). These positions are related to the schematic fold of the barrel structure in Fig. 1. Ten out of the eleven intron positions are either in the loop regions at the carboxyl edge of the sheet or close to loops at the beginning of the following helix. All active site residues and the vast majority of the intron positions are thus either in these loop regions or close the them. When examined in detail it is found that active site residues and intron positions in general do not coincide but are separated by a few residues

FUNCTIONAL RESIDUES AND INTRON POSITIONS IN TYPE B STRUCTURES

Functional residues are found in the loop regions at the carboxyl edge of the sheet in all type B structures, even though these structures are found in such functionally diverse proteins as carboxypeptidase, subtilisin, arabinose binding proteins, kinases and the coenzyme binding domain of dehydrogenase (Bränden, 1980). Here we will only compare the last class of domains, because there is no variation in the folds and they bind NAD in the same way using the same loop regions. Furthermore, the gene structure is know for three of the dehydrogenase whose 3-D structure is know; alcohol dehydrogenase (Eklund et al., 1976, Dennis et al., 1984, Bränden et al., 1984), glyceraldehyde phosphate dehydrogenase (Moras et al., 1975, Stone et al., 1985) and lactate dehydrogenase (Adams et al., 1970, Li et al., 1985). This domain is divided into two mononucleotide binding domains by loop nr 3, between ß3 and ß4 which changes the strand order. Loop 1 between ß1 and αB and loop 2 between ß2 and αC are on one side of the switch point (Branden, 1980) whereas loops 4 and 5 are on the other side. These four loop regions provide residues involved in NAD binding. The intron positions in loop nr 3 separates these two similar halves and its position is not directly related to the functional residues (Bränden, 1986). Similarly there is an intron position outside strand nr 6 which separates the coenzyme-binding domain from the catalytic domain in these structures.

The positions of the introns in relation to the schematic fold of the nucleotide binding domain are given in Fig.2. It is apparent from this figure that most of the introns in these type B structures are also located either in the functional loop regions or at the beginning of the following helix.

DISCUSSION

The correlation observed here between intron positions and functional residues in the same loops in α/ß proteins is too high to be a coincidence due to a limited sample. We should then ask the question why this correlation exists. One theory can be advanced along the following lines.

We will assume that intron positions reflets ancient evolutionary history along the lines first suggested by Gilbert (1978). Superimposed on this is a certain noise level due to random intron insertions. We will also assume that the genes for ancient proteins were constructed by exon combinations of small structural units, αα, ßß, αßetc as has been suggested by Blake (1985). Using these assumptions we would expect to observe intron positions predominantly in the loop regions at the carboxyl end of the strands if αß units were the building blocks precisely as we observe in these present day enzymes. The introns would be in the loop regions at the other end of the sheet if ßα-units were the building blocks. Units of the type αßα or ßαß would not give a regular α/ß structure.

This mechanism for constructing novel enzymes would introduce variability in the loop regions at the carboxyl edge of the sheet if the splicing mechanism had some frequency of error. The probability of constructing an enzyme with a useful catalytic site around the intron positions at this edge of the sheet wouldbe considerably higher than from random changes around the surface of the molecule. This theory would thus provide a reasonable explanation why we invariably observe the active sites at the carboxyl edge of the sheet and not at the other end which in principle could serve as an active site just as well from a purely structural point of view.

Similar arguments apply to construction at the gene level of enzymes "de novo". Combination of eight α/ß units would produce a barrel type structure. Residues responsible for substrate binding and catalysis could be introduced into the loop regions at one end of the barrel. The lengths and residue types of these loops could easily be changed by introducing restriction sites into positions of the gene that correspond to these loop regions. By suitable changes in these loops one could thus design novel enzymtic activities without affecting the stability of the structural framework.

REFERENCES

Adams, M.J., Ford, G.C., Koekok, R., Lentz, P.J.Jr., Mc Pherson, A. Jr., Rossmann, M.G., Smiley, I.E., Schevitz, R.W. and Wonnocott, A.J. (1970) Structure of lactate dehydrogenase at 2.8 A resolution, Nature 227:1103.
Banner, D.W., Bloomer, H.C., Petsko, G.A., Phillips, D.C., Pogson, C.I. and Wilson, A.I. (1975) Structure of chicken muscle triose phosphate isomerase determined crystallographically at 2.5 A resolution, Nature 255:609-614.

Blake, C.C.F. (1985) Exons and the evolution of proteins, Int. Rev. Cytol. 93:149-185.

Bränden, C.-I. (1980) Relation between structure and function of α/ß-proteins, Quart. Revs. Biophys. 13:317-338.

Bränden, C.-I., Eklund, H., Cambillau, C. and Pryor, A.J. (1984), Correlation of exons with structural domains in alcohol dehydrogenase, EMBO J 3:1307-1310

Bränden, C.-I. (1986), The relation between protein structure in α/ß domains and intron-exon arrangement of the corresponding genes in Nobel Proceedings on Molecular Evolution of life (ed. H. Jörnvall) in press.

Buehner, M., Ford, G., Moras, D., Olsen, K.W. and Rossmann, M. (1974), Three-dimensional structure of D-glyceraldehyde-3-phosphate dehydrogenase, J. Mol. Biol. 90:25-49.

Carrel, H.L., Rubin, B.H., Hurley, T.J. and Glusker, J.P. (1984) X-ray crystal structure of D-xylose isomerase at 4 A resolution, J. Biol. Chem. 259:3230-3236.

Dennis, E.S., Gerlach, W.L., Pryor, A.J., Bennetzen, J.L., Inglis, A., Llewellyn, D., Sachs, M.M., Ferl, J.R. and Peacock, W.J. (1984) Molecular analysis of the alcohol dehydrogenae (ADH 1) gene of maize, Nucl.Acids.Res. 12:3983-4000.

Eklund, H., Nordström, B., Zeppezauer, E., Söderlund, G., Ohlsson, I., Boiwe, T., Söderberg, B.-O., Tapia, O., Branden, C.-I. and Akeson, A. (1976), Three-dimensionnal structure of horse liver alcohol dehydrogenase at 2.4 A resolution. J.Mol.Biol. 102:27-59.

Gilbert, W. (1978) Why genes in pices ?, Nature 271:501.

Gilbert, W. (1985) reported at the second Bio-Technology-Waksman Institute Symposium (see Bialy,H. Genetic engineering in the precambrion), Biotechnology 3:516.

Lebiodea, D.L., Hatada, M.H., Tulinsky, A. and Mavridis, I.M. (1982) Comparison of the folding of 2-keto-3-deoxy-6 phosphogluconate aldolase, triosephosphate isomerase and pyruvate kinase, J.Mol.Biol. 162:445-458.

Li, S., Tiano, H.F., Fukasawa, K.M., Yagi, K., Shimizu, M., Sharief, F.S., Nakashima, Y. and Pan, Y. (1985) Protein structure and gene organization of mouse lactate dehydrogenase-A isozyme, Eur. J. Biochem. 149:215-225.

Lindqvist, Y. and Branden, C.-I. (1985) Structure of gycolate oxidase from spinach, Proc. Natl. Acad. Sci. USA 82:in press.

Lonberg, N. and Gilbert, W. (1985) Intron/exon structure of the chicken pyruvate kinase gene, Cell 40:81-90.

Matsuura, Y., Kusunoki, M., Harada, W., Tanaka, N., Iga, Y., Yasuoka, N., Toda, H., Narita, K. and Kakuda, M. (1980) Molecular structure of Taka-amylase A. Backbone chain folding at 3 A resolution, J.Biochem. (Tokyo) 87:1555-1558.

Richardson, J.S. (1981) The anatomy and taxonomy of protein structure, Adv.Protein Chem. 34:167-339.

Schneider, G., Lindqvist, Y., Anderson, I., Knight, S., Branden, C.-I. and Lorimer, G. (1986) X-ray structural studies of Rubisco from Rhodospirillum rubrum and spinach, Proc.Roy.Soc.Trans. in press.

Stone, E.M., Rothblum, K.N., Alevy, M.C., Kuo, T.M. and Schwartz, R.J. (1985), Complete sequence of the chicken glyceraldehyde-3-phosphate dehydrogenase gene, Proc. Natl.Acad.Sci. USA 82:1628-1632.

Stuart, D.I., Levine, M., Muirhead, H. and Stammers, D.K. (1979), Crystal structure of cat muscle pyruvate kinase at a resolution of 2.6 A., J.Mol.Biol. 134:109-142.

THE THREE-DIMENSIONAL STRUCTURE OF ANTIBODIES

Roberto J. Poljak

Département d'Immunologie, Institut Pasteur
75724 Paris Cedex 15, France

The first part of this article gives a summary of some of the most important results obtained by single-crystal X-ray diffraction techniques in the study of antibody structure. In the second part, recent result obtained in the authors'laboratory are discussed.

THREE-DIMENSIONAL STRUCTURE OF MYELOMA PROTEINS

The first X-ray diffraction studies of immunoglobins were conducted on crystalline materials obtained from human myeloma proteins. These were followed by studies of experimentally induced murine myeloma proteins. The three-dimensional structures of the Fc fragment of rabbit immunoglobulins and of the pFc' fragment of guinea pig immunoglobulins have also been determined. These results have been reviewed (1,2). It is nowadays clear that myeloma immunoglobulins are exactly like antibody molecules in their structure and physicochemical properties. Although their antigen-binding specificity is generally unknown, their homogeneity and abundance make them a very suitable material for the study of antibody structure. The major conclusions obtained by the X-ray diffraction studies of immunoglobulins are summarized below.

1) The Fab and Fc fragments of an IgG (immunoglobulin of class G, mol. wt. 150,000 daltons, the most abundant class in serum) consist of four "homology subunits" of three-dimensional structure, each containing the amino acid sequence (about 110 amino acids of a homology region (see Fig. 1). In Fab these regions are V_L and C_L (variable and constant parts of a light (L) chain), V_H and C_H1 (variable and first constant regions of the heavy (H) chain). In the Fc they are C_H2 and C_H3 (second and third constant regions of the H chain). The association of these subunits in pairs ($V_H + V_L$; $C_H1 + C_L$; $2 C_H2$; $2 C_H3$) gives rise to domains of three-dimensional structure (see Fig. 2) having each a distinct function. Thus the V domain binds antigen, the C_H2 domain binds the first complement factor, $C_H2 + C_H3$ are active in membrane passage, etc.

2) In the homology subunits over 50 % of the amino acid residues fold in ß-sheets. There are two roughly parallel ß-sheets surrounding a tightly packed core of hydrophobic side chains which include an intrachain disulfide bond (see Fig. 2.). The three-dimensional folding

of V and C subunits follows a common pattern, the "immunoglobulin-fold". This folding pattern also occurs in other proteins of the "immunoglobulin superfamily" which perform functions having to do with antigen recognition, such as the T-cell antigen receptor and the major histocompatibility antigens (class I and class II). The same folding pattern is found in other members of the superfamily which do not have a role in antigen recognition, such as the Thy-1 antigen, the poly Ig receptor, etc. The presence of common folding pattern supports the postulate of gene duplication as the mechanism which gave rise to present immunoglobulin genes (3).

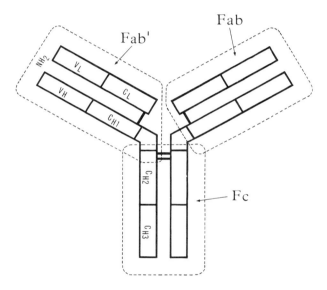

Fig. 1. Diagram of a human immunoglobulin (IgG1) molecule. The light (L) chains (mol. wt. 25,000) are devided into two homology regions, V_L and C_L. The H chains (mol. wt. 50,000) are devided into four homology regions, V_H, C_H1, C_H2 and C_H3. C_H1 and C_H2 are joined by a "hinge" region. Fab, Fab' and Fc fragments (mol. wt. 50,000) are indicated.

3) The core of the contacting region between V_H and V_L is determined by residues which are constant or nearly constant in the aminoacid sequences of different H and L chains. The fact that these residues are constant explains the ability of random L and H chains to form structurally viable pairs, thus increasing and diversifying the number of different antigen combining sites. The most important contacting residues are : Val 37, Gln 39, Leu 45, Tyr 94 and Trp 107 in V_H, and Tyr 35, Gln 37, Ala 42, Pro 43, Tyr 86 and Phe 99 in V_L. A few other residues which interact, belong to the V_H, V_L hypervariable regions and may help explain the preferential association observed in competition experiments between some L chains and some H chains.

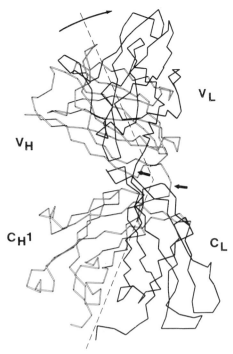

Fig. 2. View of the α-carbon backbone of the Fab fragment of IgG New[6] at 2.0 Å resolution. The L chain (open line), the H chain ($V_H + C_H$, solid line), the local two-fold axes (broken lines) relating V_H to V_L and $C_H 1$ to C_L are shown. The short arrows indicate flexible points of the H and L polypeptide chains. The upper arrow indicates relative motion of the V and C domains.

4) Flexibility is an important property of immunoglobulin molecules. Most of the flexibility resides in the "hinge" region joining $C_H 1$ to $C_H 2$. However, the regions connecting the V and C domains in the Fab part of immunoglobulins are also flexible, giving rise to different conformations as illustrated in Fig. 2. Interpretation of the available immunochemical and physicochemical evidence suggests that the major role of the flexibility of antibodies is to facilitate their primary function, antigen binding, and possibly, the interaction between immunoglobulin G and complement (4).

5) The segments of the polypeptide chain which correspond to the regions of "hypervariability" (5) of the H and L chains occur at one end of the molecule molecule, fully exposed to solvent and in close spatial proximity (6). They define the antigen combining site of antibodies (6-9). The sequence hypervariability of these regions is enhanced by the somatic recombination mechanism that joins the V, D and J gene segments coding for the V regions of H AND L chains (reviewed in references 10-12). The characterization of the antigen combining site has been supported by the determination of the three-dimensional structure of two

hapten-Fab complexes (with vitamin K_1OH (13) and phosphorylcholine (7) as haptens). An important conclusion of these studies (7-13) has been that there is no conformational change in the Fab fragment after specific ligand-binding. This conclusion does not support postulates by which antigen binding provokes conformational changes that are transmitted from the antigen binding site to the molecular sites of antibody effector functions such as complement fixation, which reside in the Fc part. However, these conclusions have been considered, by many workers in the field, as being only suggestive since it was argued that a larger ligand, i.e. a true antigen, would make mor contacts withe combining site and thus produce a conformational change that could not be detected by hapten binding.

X-RAY DIFFRACTION STUDIES OF MONOCLONAL ANTIBODIES OF PREDETERMINED SPECIFICITY

More recently, our laboratory has turned its attention to the study of monoclonal antibodies of predetermined specificity obtained by the technique of cellular hybridization (14). The introduction of this technique has made it possible to obtain large quantities of monoclonal antibodies of predetermined specificity suitable for structural studies including X-ray diffraction analyses. Monoclonal Fab fragments of different specificities have been crystallized and are under investigation. They include anti-azophenylarsonate Fabs carrying a major idiotype (CRI^+) (15) and minor idiotype (CR^-) (16) of the anti-azophenylarsonate antibodies of A/J mice ; an anti-axozolone Fab (17) from BALB/c mice and an anti-lysozyme Fab complexed to its antigen, from monoclonal IgGl BALB/c antibodies (18). Other crystalline Fab have been obtained, as well as Fab-Fab complexes which will be reported elsewhere.

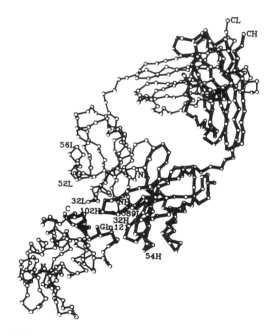

Fig. 3. Diagram of the α-carbon backbone of the Fab-HEL complex at 6 Å resolution (19). The L chain is shown by thinner line, C_H1 and V_H by a thicker line. The contracting area of lysozyme is also shown by a thicker trace

The successful crystallization of an antigen-antibody complex (18) has allowed us to determine its three-dimensional structure at 6 Å resolution (19) and to extend this study to 2.8 Å resolution (currently under way). Hen egg-white lysozyme (HEL) was chosen as the immunizing antigen because 1) it is very immunogenic and its antigenic properties have been investigated in several laboratories (reviewed in reference 20) ; 2) its three-dimensional structure is known (21) ; 3) it is an abundant, easy to purify protein with an isoelectric point well above that of Fab thus facilitating purification of the Fab-HEL complex from the unreacted antigen and Fab molecules ; 4) variant egg-white lysozymes, from other phyllogenetically close avian species, can be used to test the specificity and cross reactivity of the anti-HEL antibodies.

The 6 Å resolution three-dimensional structure of the Fab D1.3-HEL complex has been reported (19). The 6 Å resolution map of the Fab-HEL complex was intrpreted using the α-carbon backbone structures of Fab New (6) and of HEL (21) (see Fig. 3). The unequivocal fit of the HEL backbone to the electron density map indicated that no gross conformational change took place in the antigen moiety of the complex. In the Fab moiety, the axes passing through the middle of V_H-V_L and the middle of C_H1-C_L (see Figs. 2 and 3) are nearly collinear. Thus, the V_H and C_H1 regions are not brought closer together by antegen binding. The interactions between antigen and antibody extend over a loarge area, of about 20 x 25 Å (see Fig. 4). Two polypeptide chain segments of HEL, one around position 20, and another around position 120 to 129 constitute the two most important contacting regions of the antigen. These regions do not correspond to the four proposed antigenic regions of lysozyme (22,23).

Fab D1.3 or the parent IgG antibody inhibit the enzymatic activity of HEL and that of bobwhite quail egg-white lysozyme assayed against the Micrococcus lysodeikticus cell wall substrate. The activity of the lysozymes from egg-whites of California quail, japanese quail, turkey and partridge is not inhibited by Fab D1.3. In agreement with this observation, inhibition tests by enzyme-linked immuno-absorbant assays (ELISA) indicate that only the bobwhite quail lysozyme can inhibit Fab. D1.3 binding to the solid-phase coupled HEL antigen. From this data it is concluded that amino acid sequence alterations at positions 121 and at positions 19-21 interfere with the formation of an antigen-D1.3 antibody complex. The 6 Å resolution, three-dimensional model of the complex obttained by X-ray diffraction methods explains these results since those positions form part of the antigen contacts with the D1.3 antibody. In fact, Gln 121, replaced by His in the sequences of turkey, California quail and partridge lysozyme occupies a central position in the contacting area, surrounded by complementarity-determining regions of the H and L chains.

The results described above are confirmed by the most recent 2.8 Å resolution map of the Fab. D1.3 complex, currently under study (to be reported esewhere).

Acknowledgements

Research on antibody structure and specificity is carried out, in the authors laboratory at the Institut Pasteur, by P.M. Alzari, A.G. Amit, G. Boulot, M.-B. Lascombe-Comarmond, V. Guillon, M. Harper, M. Juy, R.A. Mariuzza, S.E.V. Phillips, C. Rojas A., P. Saludjian, F.A. Saul and P. Tougard. I am grateful to them and to F. Gauthier for their help in the preparation of this manuscript. This research was supported by grants from the C.N.R.S., INSERM and Institut Pasteur.

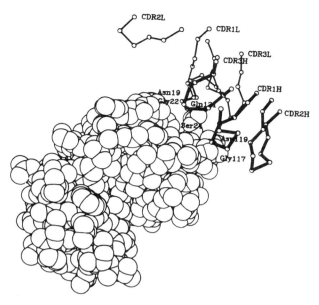

Fig. 4. The Fab D1.3-lysozyme contanc area. The complementarily determining regions of the Fab are shown by their α-carbon backbones. Lysozyme is shown as a space-filling model ; some of its antibody-contacting residues are labelled.

REFERENCES

1. Amzel, L.M. and Poljak, R.J., Three-dimensional structure of immunoglobulins, Ann. Rev. Biochem., 48:961 (1979).
2. Davies, D.R. and Metzger, M., Structural basis of antibody function, Ann. Rev. Immunol. 1:87 (1983).
3. Hill, R.L., Delaney, R., Fellow, R.E and Lebowitz, H.E., The evolutionary origins of the immunoglobulins, Proc. Natl. Acad. Sci. USA, 46:1762 (1966).
4. Iseuman, D.E., Dorrington, K.J. and Painter, R.H., The Structure and function of immunoglobulin domains, J. Immunol., 114:1726 (1975).
5. Kabat, E.A. and Wu, T.T., Attempts to locate complementarity-determining residues in the variable positions of light and heavy chains, Annals N.Y. Acad. Sci., 190:382 (1971).
6. Poljak, R.J., Amzel, L.M., Avey, H.P., Chen, B.L., Phizackerley, R.P. and Saul, F., Three-dimensional structure of the Fab' fragment of a human immunoglobulin at 2.8 Å resolution, Proc. Natl. Acad. Sci. USA, 70:3305 (1973).
7. Segal, D.M., Padlan, E.A., Cohen, G.H., Rudikoff, S., Potter, M. and Davies, D.R., The three-dimensional structure of a phosphorylcholine binding mouse immunoglobulin Fab and the nature of the antigen binding site, Proc. Natl. Acad. Sci. USA, 71:4298 (1974).
8. Schiffer, M., Girling, R.L., Ely, K.R. and Edmundson, A.B. Structure of a λ-type Bence-Jones protein at 3.5 Å resolution, Biochemistry, 12:4620 (1973).
9. Epp, O., Colman, P., Fehlhammer, H., Bode, W., Schiffer, M. and Huber, R., Crystal and molecular structure of a dimer composed of the variable portions of the Bence-Jones Protein Rei, Eur. J. Biochem., 45:513 (1974).

10. Tonegawa, S., Somatic generation of antibody diversity, Nature, 302:575 (1983).
11. Honjo, T., Immunoglobulin genes, Ann. Rev. Immunol., 1:499 (1983).
12. Hood, L., Campbell, J.H. and Elgin, S.C.R., The organization, expression and evolution of antibody genes and other multigene families, Ann. Rev. Genetics, 9:305 (1975).
13. Amzel, L.M., Poljak, R.J., Saul, F., Varga, J.M., and Richards, F.F., The three-dimensional structure of a combining region-ligand complex of immunoglobulin New at 3.5 Å resolution, Proc. Natl. Acad. Sci. USA, 71:1427 (1974).
14. Köhler, G. and Milstein, C., Continuous cultures of fused cells secreting antibody of predefined specificity, Nature, 256:495 (1975).
15. Amit, A.G., Mariuzza, R.A., Saludjian, P., Poljak, R.J., Lamoyi, E. and Nisonoff, A., Preliminary crystallographic study of the Fab fragment of a monoclonal anti-phenylarsonate antibody possessing a major idiotypic specificity, J. Mol. Biol., 169:637 (1983).
16. Amit, A.G., Harper, M., Mariuzza, R.A., Saludjian, P., Poljak, R.J., Lamoyi, E., and Nisonoff, A., Preliminary crystallographic study of the Fab fragment of a monoclonal anti-phenylarsonate antibody, J. Mol. Biol., 165:415 (1983).
17. Mariuzza, R.A., Boulot, G., Guillon, V., Poljak, R.J., Berek, C., Jarvis, J.M. and Milstein, C., Preliminary crystallographic study of the Fab fragments of two monoclonal anti-2-phenyl-oxazolone antibodies, J. Biol. Chem. (in press, 1985).
18. Mariuzza, R.A., Jankovic, D.Lj., Boulot, G., Amit, A.G., Saludjian, P., Le Guern, A., Mazié, J.C. and Poljak, R.J., Preliminary crystallographic study of the complex between the Fab fragment of a monoclonal anti-lysozyme antibody and its antigen, J. Mol. Biol., 170:1055 (1983).
19. Amit, A.G., Mariuzza, R.A., Phillips, S.E.V. and Poljak, R.J., The three-dimensional structure of an antigen-antibody complex at 6 Å resolution, Nature, 133:156 (1985).
20. Benjamin, D.C., Berzofsky, J.A., East, I.J., Gurd, F.R.N., Hannum, C., Leach, S.J., Margoliash, E., Michael, J.G., Miller, A., Prager, E.M., Reichlin, M., Sercarz, E.E., Smith-Gill, S.J., Todd, P.E. and Wilson, A.C., The antigenic structure of proteins: a reappraisal, Ann. Rev. Immunol., 2:67 (1984).
21. Blake, C.C.F., Koening, D.F., Mair, G.A., North, A.C.T., Phillips, D.C. and Sarma, V.R., Structure of hen egg-white lysozyme. A three-dimensional Fourier synthesis at 2 Å resolution, Nature, 206:757 (1965).
22. Atassi, M.Z., Precise determination of the entire antigenic structure of lysozyme: molecular features of protein antigenic structures and potential of "surface-simulation synthesis" - a powerful new concept for protein binding sites, Immunochem. 15:909 (1978).
23. Ibrahimi, I.M., Eder, J., Prager, E.M., Wilson, A.C. and Arnon, R., The effect of a single amino acid substitution on the antigenic specificity of the loop region of lysozyme, Mol. Immunol., 17:37 (1980).

THE 3-DIMENSIONAL STRUCTURES OF INFLUENZA VIRUS NEURAMINIDASE

AND AN ANTINEURAMINIDASE Fab FRAGMENT

P.M. Colman[*], J.N. Varghese[*], W.G. Laver[+] and R.G. Webster[‡]

[*]CSIRO, Division of Protein Chemistry, Parkville 3052
Australia; [+]John Curtin School of Medical Research
Australian National University, Canberra, Australia and
[‡]St. Jude Children's Research Hospital, Memphis
Tennessee

Introduction

Influenza virus attaches to host cells through the binding of its haemagglutinin to sialic acid-containing glycoconjugates. Sialic acid is most commonly found as the terminal sugar residues on an oligosaccharide chain, and it is in this form that it is bound to influenza haemagglutinin. After infection of the cell, progeny virus particles bud out from the plasma membrane where they encounter two immobilising influences. Firstly, cell surface sialic acid will cause these particles to adhere, and, secondly, terminal sialic acid residues placed on the oligosaccharide chains of the newly synthesised viral membrane glycoproteins will lead to self-aggregation of the virus. The role of the neuraminidase, which, like the haemagglutinin, is an integral membrane glycoprotein of influenza, is to trim off the sialic acid (N-acetyl neuraminic acid) from these oligosaccharides to allow elution of progeny virus particles away from infected cells. There are believed to be 50-100 neuraminidase protomers on the viral surface and 10-20 times as much haemagglutinin.

New epidemics of influenza result from the emergence of virus which bears antigenically altered haemagglutinin and neuraminidase molecules. The amino acid sequences between haemagglutinins or neuraminidases of different epidemic-causing strains differ by only 5-15%. Pandemics, on the other hand, result from genetic reassortment, whereby a haemagglutinin or neuraminidase from an animal influenza strain becomes incorporated into a human virus. The sequence differences between different so-called subtypes of haemagglutinin or neuraminidase are typically greater than 50%.

Influenza neuraminidase is a tetramer of four 60,000 dalton polypeptide chains. It is oriented in the viral membrane via an N-terminal hydrophobic anchor sequence. Soluble 'heads' of neuraminidase can be liberated from the viral surface by digestion with pronase. These heads carry the full antigenic and enzymatic capability of the virus, and crystals of heads from two different human (subtype N2) and one avian (N9) neuraminidase have been grown. The heads, amino acids

74-469 in N2 strains, are connected to the transmembrane anchor, amino acids 6-35, via a stalk, residues 36-73. The stalk is believed to render flexible the attachment of the neuraminidase activity to the viral membrane. Deletion mutants of influenza with shortened stalk sequences are observed to elute more slowly off red cells than wild-type suggesting some functional role for the stalk. Here we will concentrate only on the neuraminidase heads and discuss the structure in relation to both antigenicity and enzyme activity. For a review and further background material see Colman and Ward (1985).

Neuraminidase Structure Determination

Two crystal forms were used to determine the structure. A 1967 influenza neuraminidase A/Tokyo/3/67 crystallises in space group I422 with a=139.6Å, c=191.0Å with one subunit per asymmetric unit and the tetrameric enzyme is located on a crystallographic four-fold axis (Colman & Laver, 1981). A 1957 neuraminidase, A/RI/5[+]/57, also of the N2 subtype, crystallises in space group $P4_32_12$ with a=124.1Å, c=181.2Å. Here the tetrameric enzyme is situated on the crystallographic two-fold axis and there are two copies of the subunit per asymmetric unit (Varghese et al., 1983).

Heavy atom derivatives of Tokyo neuraminidase were rarely isomorphous with the native compound. A space group change to a primitive lattice was frequently observed. An uninterpretable image of the protein derived from seven heavy atom compounds was used as a search object in the data of the RI/5[+]/57 enzyme. A rotation function with data between 10Å and 6Å resolution and a search radius of 35Å revealed the angular mapping into the RI/5[+]/57 crystal (Fig. 1). The two largest peaks in

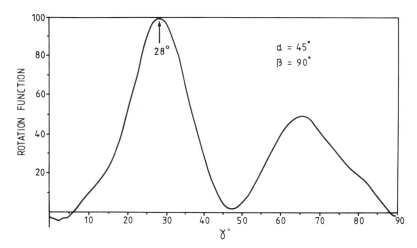

Fig. 1. Cross rotation function of an uninterpretable image of Tokyo/3/67 neuraminidase against crystal data of RI/5[+]/57 neuraminidase. The line $\alpha = 45°$, $\beta = 90°$ is shown. It contains the two largest peaks in the 3-dimensional function.

the three-dimensional rotation function are shown on the line of the function given in Fig. 1, where the correct solution is 2 times the height of the largest false correlation. This angular solution is consistent with the self rotation function study of RI/5$^+$/57 crystal data which showed a four-fold axis parallel to the crystallographic diad with a peak height of 0.40 times the origin peak. Data between 10 and 6Å and vectors of radius less than 35Å were used in the self-rotation function calculation. A translation function, in this case an R-factor calculation, in the two possible space groups of RI/5$^+$/57 neuraminidase fixed the enantiomorph and located the position of the neuraminidase tetramer in the RI/5$^+$/57 crystals (Fig. 2). In the incorrect

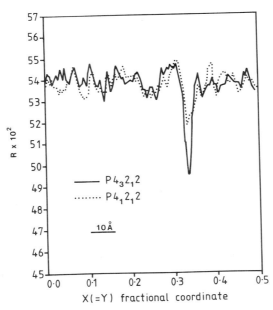

Fig. 2. R-factor analysis of correctly oriented Tokyo/3/67 neuraminidase image as a function of position along the crystallographic diad of RI/5$^+$/57 neuraminidase crystals. Analysis for two enantiomorphs is shown.

space group, $P4_12_12$, one half of the scattering density is correctly located. The minimum R-value in $P4_12_12$ was 0.518 and in $P4_32_12$, 0.496. The correctness of the mapping was confirmed by using the derived phases to solve, via a difference Fourier, the single heavy atom derivative that was collected for RI/5$^+$/57 neuraminidase. SIR phases to 4Å and molecular replacement phases to 3Å were combined to phase the RI/5$^+$/57 data to generate a starting density for 2-fold noncrystallographic symmetry averaging. This procedure converged with an R value of 0.245 after 10 cycles of averaging refinement. Marginal improvement in the image resulted from a single passage of this refined density through the Tokyo crystal and back into RI/5$^+$/57 for further two-fold averaging (Varghese et al., 1983).

Neuraminidase Structure

The subunit folding of influenza neuraminidase is of six four-stranded antiparallel β sheets arranged as if on the blades of a propellor. All sheets have identical topology, (+1,+1,+1) an observation the more remarkable because of the infrequency of this structural unit in other protein structures (Richardson, 1981). An N-terminal arm runs across the bottom of the subunit from the four-fold axis before entering the propellor structure at residue 95 (Fig. 3). The active site,

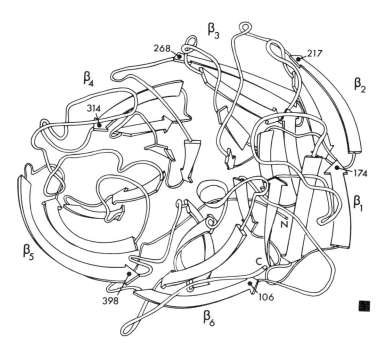

Fig. 3. Schematic polypeptide trace in subunit of neuraminidase. The molecular four-fold is at lower right.

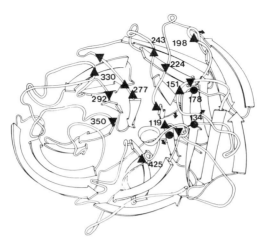

Fig. 4. Conserved residues of catalytic site. Upwards pointing triangles, acidic groups; downwards, basic groups; circles, hydrophobic groups.

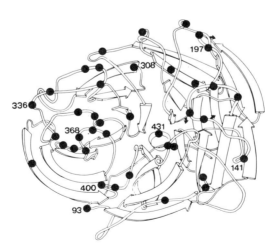

Fig. 5. Location of amino acid sequence changes on surface of subunit between 1957 and 1979 strains of N2 subtype neuraminidase.

identified by soaking sialic acid into the crystals (Colman et al. 1983) is located in a deep pocket on the upper surface of the subunit. To a first approximation it lies directly above the 6 parallel strands of the 6 β-sheets which cluster at the centre of the propellor fold. Unlike the α/β enzymes (see Branden, this volume) the active site is at the N-terminal end of this parallel strand structure. Details of the β-structure of neuraminidase are not available, but it is believed that this central parallel barrel has few elements of regular β-structure; half of the H bonding capacity of each strand is engaged in the anti-parallel sheet structures emanating from the centre.

The active site is tied by 9 conserved acidic groups, 6 conserved basic groups and several conserved hydrophobic and neutral hydrophilic groups. With few exceptions these residues are on loops connecting the β-strands (Fig. 4).

Mapping of amino acid sequence changes found in neuraminidase of naturally-occurring field strains of influenza and in laboratory variants selected with monoclonal antibodies shows a consistent pattern of variation on the upper surface of the subunit (Colman et al., 1983). These altered structures do not segregate into distinct epitopes but rathermore form a continuous surface, encircling the catalytic cavity (Fig. 5). Some of these strain-variable amino acids are immediately adjacent to conserved active site residues. The variable amino acids are typically oriented away from the active centre while the conserved residues are pointing into the cavity. It seems unlikely that antibodies could see residues of the active site at the exclusion of nearby catalytically irrelevant residues, and in this way immune pressure is not brought to bear on the functionally important structures of the enzyme.

Haemagglutinating Neuraminidase

An avian neuraminidase, of the N9 subtype, has been shown to have the capacity to agglutinate red cells (Laver et al., 1984). The neuraminidase has been crystallised in the space group I432 with a=184Å and the gene sequence is known (Air et al., 1985). It is not yet clear whether this neuraminidase acquires a site to which sialic acid containing oligosaccharides bind without cleavage or whether a modification of the active site of the N2 enzyme could be responsible for the agglutinin function. One of the active site residues, asp198, is indeed altered to asn in the N9 sequence. No other changes in the catalytic cavity occur. However, arguing against the modified active site hypothesis is the observation that neuraminidase activity can be abolished with sialic acid analogues without affecting the haemagglutinin activity. High resolution (1.9Å) X-ray diffraction data have been collected and studies with sugars and substrates should clarify the situation.

Antineuraminidase Fab Fragment Structure

Crystallisation of the Fab fragment S10/1 (Webster et al., 1982) from ammonium sulphate has been reported (Colman et al., 1981). The space group is $P3_121$ with a=131.5Å and c=72.2Å. The structure was solved by multiple isomorphous replacement using diamino dinitro platinum, uranyl acetate and potassium hexachloroplatinate as heavy atom derivatives. Date were collected photographically and merging statistics are given in Table 1. Heavy atom sites were located in the usual way by difference Patterson and difference Fourier syntheses and heavy atom parameter refinement resulted in an overall figure of merit

Table 1. Data Collection for S10/1 Antineuraminidase Fab

Compound	No. of Data	No. of Independent Observations	R Merge
Native	17726	8417	9.1
DANP	7621	4061	8.9
UO2A	3894	2666	14.4
UO2B	11379	5977	10.5
PTC6	2833	2021	12.8

DANP: diaminodinitroplatinum (saturated solution)
UO2A: uranyl acetate (10 mM, 5 days)
UO2B: uranyl acetate (10 mM, 18 hours)
PTC6: potassium hexachloroplatinate (10 mM, 18 hours)

$$R_{merge} = \sum_h (I_h - \overline{I}_h) / \sum_h I_h$$

of 0.44 to 3.5Å resolution. The sensitivity of the c-axis dimension to heavy metal reagents is the most likely cause of the poor phasing statistics (Table 2).

The map was interpreted using known Fab structures. The fitting was done in a 6Å resolution electron density map which showed much better contrast between solvent and protein than the 3.5Å image. However, the high resolution image was used to determine possible starting origins for the search procedure. High features of the electron density were selected from a 4.0Å map on the expectation that the disulphide bonds of the immunoglobulin domains might be represented among these peaks. In retrospect, two of the five disulphides are associated with peaks 3 and 4 out of a list of the 15 densest features in the map. In both cases the high density features are associated with Cα positions rather than sulphur atoms indicating that rigid parts of the structure are represented by such features as 4Å resolution. No high density is associated with the two light chain domain disulphides or the interchain disulphide in the 4Å electron density map.

The Cα skeleton of the Kol Fab (Marquart et al., 1980) was then appropriately located at the corresponding position in the 6Å image and a full 3-d orientation search computed. At each high density value the model was positioned with three different origins, representing the centre of the S-S bridge and the two Cα positions either side of it. All four immunoglobulin domain disulphides were searched against every high density feature in this way. $V_L V_H$ were treated as a rigid module as were $C_L C_H$. An example of the sensitivity of the orientational search to the translational placement of the model is demonstrated by the observation that around the correct solution for the C module, no signal >3σ is found for the correct orientation on either the S-S centre or one of the corresponding Cα positions. The alternate Cα position showed a peak of 5σ (cf. max peaks at the alternate origins of <4σ) which was then further explored by finer interval searches in both orientational and translational coordinates. A trace through one angular variable covering the final orientational and translational solution for the $C_L C_H$ module of the Kol protein in the S10/1 Fab 3.5Å image is shown in Fig. 6. The significance of this peak is 9σ measured against all possible orientations for the correct translation. A similar result is found for the V module.

Table 2. Phasing Statistics for S10/1 Antineuraminidase

	Resolution (Å)	18.8	11.6	8.4	6.5	5.4	4.6	4.0	3.5
DANP	RMS f_H	1.19	1.10	0.94	0.85	0.77	0.74	0.73	0.72
	RMS E	1.40	.90	.66	.58	.57	.71	.78	.64
UO2B	RMS f_H	1.28	1.20	1.15	1.13	1.04	.96	.89	.77
	RMS E	1.44	.94	.75	.67	.66	.68	.68	.57
UO2A	RMS f_H	1.93	1.99	1.99	1.92	1.90	1.87	1.84	1.70
	RMS E	1.26	1.35	1.59	1.17	1.25	1.53	1.50	1.24
PTC6	RMS f_H	1.54	1.31	1.28	1.40	1.28	1.27	1.26	1.24
	RMS E	1.56	0.98	1.02	.78	.90	.97	.98	.90

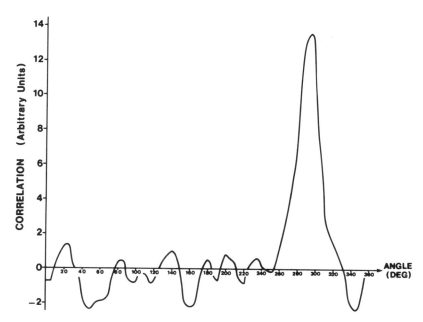

Fig. 6. Trace of correlation between Kol Fab Cα coordinates for C module rotated through one angular variable embracing the final solution for placement of the model into the 6Å electron density of S10/1 Fab.

Table 3. Fit of M603 Fab C^α Coordinates to 6Å Image of S10/1 Antineuraminidase Fab

	Shift 1 (Å)	Eulerian Angles	Shift 2 (Å, trigonal frame of map)
V module −	(56.5, 53.9, −21.8)	(50, 113, 97)	(70.5, 85.8, 9.4)
C module −	(70.5, 54.7, 4.6)	(47, 114, 96)	(79.3, 61.0, 11.2)

Shift 1 places heavy chain atoms $98C^\alpha$ for the V module and $206C^\alpha$ for the C module at the origin of the rotating frame. Shift 2 places the correctly oriented model in the map. Eulerian angle definition is θ_1 around OZ, θ_2 around new OX, θ_3 around new OZ.

Subsequently, a limited search around the known solution using the mouse myeloma M603 Fab coordinates (Segal et al., 1974) resulted in the solution shown in Table 3. The near identity of the orientational part of the solution for the V and C modules testifies to the similarity of the elbow angle in these two proteins. The possibility that heavy and light chains need to be interchanged cannot be excluded. The solution presented has the same sense as the M603 and other Fab fragments with bent elbows, namely with longitudinal contact between V_H and C_H but not between V_L and C_L.

The crystal packing is head to tail with the complementarity determining regions of one molecule associated with the C-terminal end of a 2_1 screw related molecule in the crystal. The amino acid sequence is, at this stage, partly determined (G.M. Air, private communication). Several cycles of constrained crystallographic refinement (Hendrickson & Konnert, 1980) of the model against the 3Å diffraction data have led to an R value of 0.323. The model geometry remains good with RMS deviations from bond length ideality of 0.03Å and bond angle distance ideality of 0.06Å. Further refinement awaits sequence data.

Although the unit cell parameters of this Fab fragment are very similar to those of the human Kol IgG molecule (a=135.6Å, c=82.1Å (6)), the space group enantiomorph is opposite and the quaternary structure of the Fab fragment is grossly different. The relevance of the Fab elbow angle to antibody function remains obscure. The longitudinal contacts, always between CH1 and V_H, are few in number and involve only a small buried surface area. Details of this interface in S10/1 and other bent elbow Fab fragments must await high resolution crystallographic structure refinement.

REFERENCES

Air, G.M., Ritchie, L.R., Laver, W.G., and Colman, P.M., 1985, Gene and protein sequence of an influenza neuraminidase with haemagglutinin activity. Virology, 145:117-122.

Colman, P.M., and Laver, W.G., 1981, The structure of influenza virus neuraminidase heads at 5Å resolution. "Structural Aspects of Recognition and Assembly in Biological Macromolecules", ed. Balaban, M., I.S.S. Rehovot.

Colman, P.M., Gough, K.H., Lilley, G.G., Blagrove, R.J., Webster, R.G., and Laver, W.G., 1981, Crystalline monoclonal Fab fragment with specificity towards an influenza neuraminidase. J. Mol. Biol., 152:609-614.

Colman, P.M., Varghese, J.N., and Laver, W.G., 1983, Structure of the catalytic and antigenic sites in influenza virus neuraminidase. Nature, 303:41-44.

Colman, P.M., and Ward, C.W., 1985, Structure and diversity of influenza virus neuraminidase. Curr. Top. Micro. Immunol., 114:177-255.

Hendrickson, W.A., and Konnert, J.H., 1980, "Computing in Crystallography", eds. Diamond, R.D., Ramaseshan, S. and Ventatesan, K., Indian Academy of Science, Int. Union Crystallogr., Bangalore.

Laver, W.G., Colman, P.M., Webster, R.G., Hinshaw, V.S., and Air, G.M., 1984, Influenza virus neuraminidase with haemagglutinin activity. Virology, 137:314-323.

Marquart, M., Deisenhofer, J., Huber, R., Palm, W., 1980, Crystallographic refinement and atomic models of the intact immunoglobulin Kol and its Fab fragment at 3.0Å and 1.9Å resolution. J. Mol. Biol., 141:369-392.

Richardson, J.S., 1981, The anatomy and taxonomy of protein structure. Adv. Prot. Chem., 34:169-339.
Segal, D.M., Padlan, E.A., Cohen, G.H., Rudikoff, S., Potter, M., and Davies, D.R., 1974, The three-dimensional structure of a phosphorylcholine-binding mouse immunoglobulin Fab fragment and the nature of the antigen binding site. Proc. Nat. Acad. Sci. U.S.A., 71:4298-4302.
Varghese, J.N., Laver, W.G., and Colman, P.M., 1983, Structure of the influenza virus glycoprotein antigen neuraminidase at 2.9Å resolution. Nature, 303:35-40.
Webster, R.G., Hinshaw, V.S., and Laver, W.G., 1982, Selection and analysis of antigenic variants of neuraminidase of N2 influenza viruses with monoclonal antibodies. Virology, 117:93-104.

PERIPLASMIC BINDING PROTEINS: STRUCTURES AND NEW UNDERSTANDING OF PROTEIN-LIGAND INTERACTIONS

F. A. Quiocho, N. K. Vyas, J. S. Sack and M. A. Storey

Department of Biochemistry, Rice University

Houston, Texas 77251

INTRODUCTION

The molecular approach to the study of active transport in bacteria has led to the identification of a group of proteins referred to as "binding proteins", which are initial components of high affinity transport systems for a large variety of carbohydrates, amino acids, and ions. Several of the sugar-binding proteins also act as initial receptors for bacterial chemotaxis. These proteins, which are located in the periplasmic space, exhibit monomeric molecular weights in the range of 25,000 to 45,000 and contain one tight ligand binding site with dissociation constants of about 0.1 μM.

Protein components confined in the cytoplasmic membrane, distinct for either transport or chemotaxis, are additionally required for both processes. It is believed that binding of a substrate to the binding protein causes a conformational change which allows it to be recognized by the the membrane bound protein components. The ensuing interaction initiates active transport or flagella motion.

Our laboratory has been engaged in the crystallographic analysis of the structure and function of six of these binding proteins (1,2,3,4,5). The complete molecular structures of four of these proteins have been determined. The structures of L-arabinose-binding protein has been extensively refined at 1.7 Å resolution to a R-factor of 13.7 % (1) by a restrained least squares method (6). Shortly after refinement of the 2 Å structure of the sulfate-binding protein (2), extensive refinement was further undertaken at 1.7 Å resolution to an R-factor of 14%. The D-galactose-binding protein structure is currently being refined at 1.9 Å resolution. Refinement of the Leu/Ile/Val-binding protein structure at 2.4 Å resolution has reached a R-factor of 17%. A 3.0 Å resolution electron density map of the maltose-binding protein calculated with multiple isomorphous replacement phases is currently being interpreted. The E. coli phosphate-binding protein has recently been crystallized. This paper dwells primarily on the structures of the first four binding proteins. Moreover, as the first three binding protein structures have been solved with their bound ligand, these studies have been a source of new and highly detailed informations on a variety of protein-ligand interactions. We will also describe initial binding studies of leucine to crystals of native Leu/Ile/Val-binding proteins. In this paper, we will elaborate on the function of these proteins.

GENERAL FEATURES OF THE BINDING PROTEIN STRUCTURES

There are remarkable similarities in both the tertiary structure and ligand binding properties of the L-arabinose-, sulfate-, D-galactose- and Leu/Ile/Val-binding proteins. All four structures are ellipsoidal in shape (axial ratios=2:1) and are composed of two similar distinct globular domains, giving an overall dimension of 40 x 35 x 70 Å (see Fig. 1). In all cases, the two domains are connected by three separate peptide segments which are spatially close together. This type of connectivity is very different from those commonly observed in many multi-domain proteins (in which each domain is formed from one continuous polypeptide segment with only peptide connecting any two domains.) In the binding protein structures, the first third of the polypeptide chain constitutes the major part of the N-terminal domain and the second third forms the bulk of the C-terminal. The final third of the chain follows a winding route by traversing both domains. Despite this complex folding pattern, the packing of the secondary structure in both domains are very similar; they have a central β-sheet flanked by at least a pair of α-helices. The α-helices and β-sheet strands approximately alternate along the chain, such that the strands and the N-termini of helices point towards the binding site cleft between the two domains. Since the three connecting segments of the domains in all four structures are not located at the same region of the polypeptide chain, the topologies of these proteins are not exactly identical.

Within each domain are two clusters of hydrophobic residues located at the interface between the sheet and each of the helices. These close-packed residues, which originate from both the sheet and the helices, provide stability to the domain structures protein.

Despite the similarities in the overall structures of the binding proteins, there is minimal sequence homology. Less than 20% of the residues in any given domain are similar to any other.

A two-domain structure is a common structural feature essential to the function of the binding proteins. It provides a cleft between the two domains wherein the ligand binding site is located (Fig. 1). In the structures of the liganded form of the binding proteins, the two domains come together and enclose the substrate. The substrate, even though it is an anion, is bound mainly by hydrogen bonds; there are no salt linkages formed in any of the complexes analyzed thus far. Both domains contribute the hydrogen-bonding residues. There are also numerous van der Waals' contacts between ligand and binding site residues.

PROTEIN-LIGAND INTERACTIONS

L-Arabinose-binding Protein

Fig. 2 shows the details of the hydrogen-bonding between the binding protein and L-arabinose. As a more thorough description of the protein-sugar complex has appeared (1,7), only the salient features of the interactions will be presented here. The most notable feature is the unique ligand site geometry which accommodates either the α or β anomeric form of the L-arabinose substrate. This novel stereospecificity is conferred by the precise alignment of the atom OD2 of Asp 90 which enables it to accept a hydrogen bond from either the α (equatorial) or the β (axial) anomeric hydroxyl. The remaining hydrogen bond and van der Waals' interactions involving both sugar anomers are essentially identical.

Both sugar anomers are held in place by ten hydrogen bonds between all the sugar functional groups and six side chain residues and two

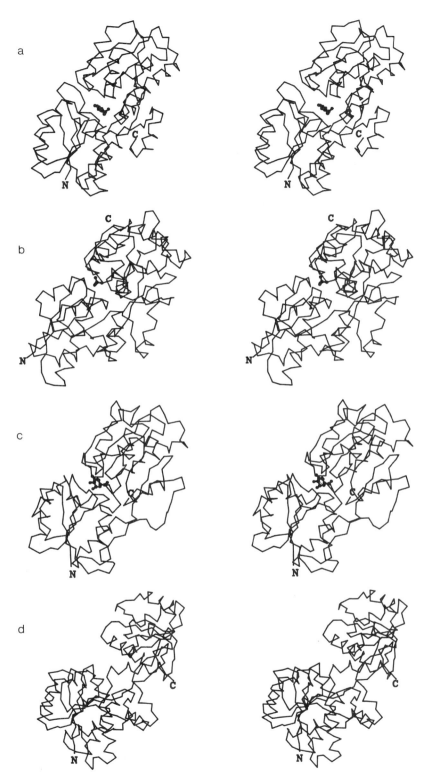

Figure 1. Structures of binding proteins specific for: a, L-arabinose; b, sulfate; c, D-galactose; and d, Leu/Ile/Val.

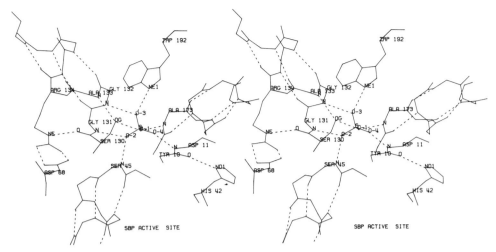

Figure 2. Interactions between L-arabinose-binding protein and α or β anomer of L-arabinose substrate. Hydrogen bonds are indicated by dashed lines. The partial stacking of Trp 16 with the pyranose ring provides some van der Waals' contacts.

isolated water molecules (Fig. 2). These residues are equally divided between the two domains; the N-terminal domain supplies Lys 10, Glu 14, and Asp 90, while the C-terminal domain provides Arg 151, Asn 205 and Asn 232. More hydrogen bonds are formed with the C-terminal domain. The finding that the anomeric hydroxyl is engaged only as a hydrogen bond donor group is consistent with the fact that, because of the anomeric effect, this particular hydroxyl is a stronger hydrogen bond donor and weaker than average hydrogen bond acceptor. The simultaneous participation of the non-anomeric hydroxyls as hydrogen bond donor and acceptor groups can be described simply as:

$$(NH)_n \dashrightarrow OH \dashrightarrow O \qquad [I]$$

where NH and O are hydrogen bond donor and acceptor groups, respectively, OH is a non-anomeric sugar hydroxyl, and n=1 or 2 (see Fig. 2). This "cooperative" hydrogen-bonding leads to stronger than average hydrogen bonds. The geometry of the hydrogen bonds involving the non-anomeric hydroxyl groups are those expected for favorable interactions. For example, the OH-3 is fully coordinated in an arrangement which is essentially tetrahedral, including the sugar C-O bond. On the other hand, the atoms C-4 and O-4 of the sugar, Arg 151 NH2, and Asn 232 OD1 are co-planar, indicating that the hydrogen bonds donated to or by O-4 is shared by both lone pairs of electrons of each acceptor group. The same geometry is formed of the hydrogen bond donated by Lys 10 to O-2.

It is notable that with the exception of Lys 10, all of the residues hydrogen-bonded to the sugar have planar polar side chains with two or more functional groups. Two reasons account for this finding. First, bidentate hydrogen bonds to the the sugar are formed with Arg 151 and Asn 232. Second, these essential planar side chains, together with the sugar molecule, are further engaged in extensive networks of hydrogen bonds with other residues within the binding site (1,7). The formation of the hydrogen bond networks, including Lys 10, leads to full utilization of all functional groups of the essential residues and every one of the potential hydrogen bond donor groups of these residues and, as well, of the L-arabinose.

The binding protein-sugar complex is further stabilized by van der Waals' contacts. There are 47 contacts in the range of 3.2 - 4.0 Å involving the α-anomeric sugar and 46 with the β-anomeric sugar (unpublished data). A cluster of these contacts, which is of interest as it has been shown that the binding of sugar causes a fluorescence change, occurs between a non-polar patch of the L-arabinose (consisting of C-3, C-4, and C-5) and Trp 16 (see Fig. 2).

The structure of ABP is the first demonstration of a protein binding site that can identically accommodate either anomeric form of a sugar substrate. This specificity is fully consistent with the the role of the binding protein in transport: the open-chain, aldehyde form of the L-arabinose is utilized in the biosynthesis of pentose phosphates.

Sulfate-binding Protein

Like in the binding protein-arabinose complex, the sulfate dianion bound in the sulfate-binding protein is linked mainly by hydrogen bonds. As shown in Fig. 3, all the sulfate oxygens are engaged in a total of seven hydrogen bonds with the binding protein (for details, see ref. 2). The O-1 of the sulfate accepts two hydrogen bonds from the peptide NH group of Ala 173 and OG of Ser 130 and the O-2 accepts two hydrogen bonds from Gly 131 NH and Ser 45 NH. The O-3 is the recipient of two hydrogen bonds from Gly 132 NH and Trp 192 NE and O-4 is associated with one hydrogen bond from the amide NH of Asp 11. The peptide CO groups of residues 44, 131, and 172 initiate three α-helices. The N-termini of these helices are close to the sulfate, while the C-termini extend to

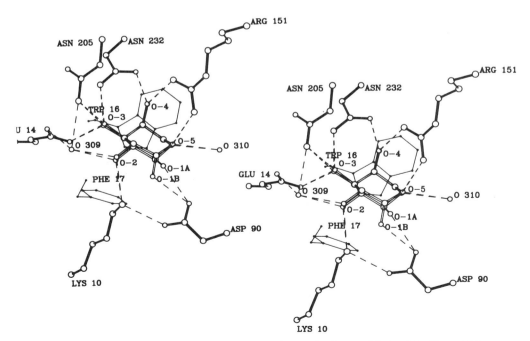

Figure 3. Details of the hydrogen-bonding of the sulfate-binding protein with the sulfate substrate.

the bulk solvent. While only side chains are involved in hydrogen-bonding the sugar in the arabinose receptor, five of the hydrogen bonds to the sulfate are from peptide backbone. The wall of the binding site cleft associated with the N-terminal domain supplies two residues (Asp 11 and Ser 45), while the C-terminal domain provides Ser 130, Gly 131, Gly 132, Ala 173 and Trp 192. There are at least a total of 41 van der Waals' contacts between sulfate and non-hydrogen atoms of the protein between 3.20 and 4.0 Å.

Three unusual features further characterize the complex of the sulfate-binding protein with the dianion. Firstly, there are no positively charged residues, cations, or water molecules within van der Waals' distance to the sulfate, thus precluding the formation of salt links to neutralize the charges of the dianion. Secondly, the sulfate bound in the cleft between the two domains is buried and completely inaccessible to the solvent. Thirdly, the hydrogen bond donor groups to the sulfate are further involved (via peptide bonds) in six separate arrays of hydrogen bonds (see Fig. 3). It is important to note that these arrays contain resonating groups. These arrays commencing from the sulfate oxygens are as follows: i) O-1 to peptide NH of residue 173 to peptide CO of residue 172 which initiates a continuous sequence of alternating peptide units and H-bonds within an α-helix. This left-handed helical line of hydrogen-bonded peptide units has the general form

$$O_i \text{---} NH_{i+4} \qquad [II]$$

where $i=m+3(n-1)$; m is the residue bearing the first peptide O and $n=1,2,3,\cdots$. The helical line in this particular instance has m=172 and n=1 to 3. ii) O-1 to OG of Ser 130 to the peptide NH of residue 133 which is part of a helical line with m=132 and n=1 to 5; iii) O-2 to peptide NH of residue 45 which is part of a helical line with m=44 and n=1 to 3; iv) between O-2 of the sulfate and 131 peptide NH and between 130 peptide CO and Arg 134 NE; v) O-3 to peptide NH of Gly 132 which is part of a helical line with m=131 and n=1 to 5; vi) between O-4 of the sulfate and NH of Asp 11 and between CO of residue 10 and His 42 ND1.

What stabilizes or neutralizes the charges of the sulfate sequestered in a site of low dielectric constant? We have previously proposed that the charges could be stabilized by the fixed dipoles of the hydrogen bonds to the sulfate, augmented through polarization by the strong electronegative field of the sulfate dianion. Furthermore, the six different arrays of hydrogen bonds, which includes resonating groups, radiating from the sulfate could further disperse the charge. The fractional positive charge arising from macrodipoles of the three helices (8,9) close to the sulfate could further provide additional charge stabilization. As the sulfate is not centered on the axis of these helices, the effect of macrodipoles may be only partially achieved.

Leu/Ile/Val-binding Protein

The Leu/Ile/Val-binding protein structure, unlike the other structures, was determined in the absence of any bound ligand (3). In fact, it was necessary to remove bound leucine in order to crystallize the protein in an octahedral form which is suitable for X-ray analysis; otherwise, crystals of the liganded form are needle-shaped. The two domains of the unliganded form of the binding protein are much more separated than in the other structures, thus rendering the binding site cleft wide open and accessible to the bulk solvent (see Fig. 1). This finding is consistent with the proposition that a major conformational change accompanies ligand binding to these proteins. While the two domains of the Leu/Ile/Val-binding are similar to those of the arabinose-, sulfate-, and galactose-binding proteins, their relative orientation is completely different in the absence of bound substrate. In the liganded state, the two domains come close to each other and enclose the bound ligand. The movement of the two domains towards each other about a hinge between the two domains would explain (i) the differences between the unliganded, "open" structure of the Leu/Ile/Val-binding protein and the liganded, "close" structures of the arabinose-, sulfate-, and galactose-binding proteins, and (ii) the sequestering of ligand in the binding site cleft between the two domains in the latter three proteins.

In a further experiment, leucine, isoleucine, or valine was soaked into crystals of the amino acid binding protein (3). Apparently due to crystal packing forces, the presence of any of these branched aliphatic amino acids did not cause the putative conformational change to occur. However, each of the amino acid does bind isomorphously in the cleft but exclusively to the N-terminal domain (e.g., see Fig. 4). It has been suggested that each of these complexes in the crystals could represent an initial or intermediate step to the final, liganded form of the binding protein (3). Consistent with this suggestion is the finding that the dissociation constants of the complexes in the crystals of the Leu/Ile/Val-binding protein are orders of magnitude greater than the complexes in solution (3). Moreover, our observation that one domain (the C-terminal domain) of the liganded, "closed" form of the L-arabinose- and sulfate-binding proteins interacts more strongly with the bound substrate than the other domain indicates preferential binding to one domain in the initial stage of protein-ligand complex formation (unpublished data).

The result of initial 2.8 Å resolution refinement of the binding protein structure with leucine diffused into the crystals indicates that

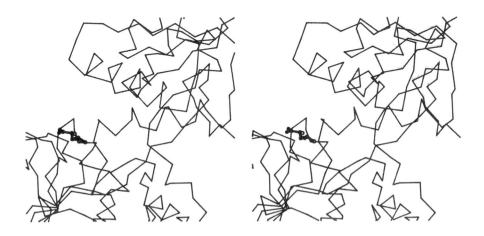

Figure 4. Leucine bound to the "open" form of the Leu/Ile/Val-binding protein. Note that the leucine is confined only to one domain.

the amino acid substrate is bound in a cavity of the N-terminal domain primarily by seven hydrogen bonds: two to the ammonium group and five to the carboxylate group (unpublished data). Four of these hydrogen bonds involve peptide units of the protein and three involve hydroxyl group of Ser and Thr residues. It is notable that neither counter-charged residues nor helix termini are close to the ammonium and carboxylate groups of the bound leucine zwitterion. The groups from the protein hydrogen-bonded to the leucine are in turn engaged in arrays of hydrogen bonds with other residues within the binding site. As in the sulfate-binding protein complex, these arrays are likely to contribute to the charge stabilization of the zwitterion. The side chain of the bound leucine substrate is in a pocket formed primarily by four hydrophobic residues. The size and the non-polar nature of the pocket are fully consistent with the substrate specificity of the binding protein.

DISCUSSION

Periplasmic binding proteins have a wide range of ligand specificity and, have very minimal amino acid sequence homology; nevertheless, there are similarities in the overall structures and locations of the ligand site of the arabinose-, sulfate-, galactose-, and Leu/Ile/Val-binding proteins. Furthermore, it is important to note that the bound substrate, which is often sequestered in the site, is held in place mainly by hydrogen bonds. Of particular significance is the absence of salt linkages in the binding of the sulfate dianion and the leucine zwitterion to their respective binding protein. A further common feature of all binding protein-substrate complexes analyzed thus far is the formation of extensive networks or arrays of hydrogen bonds, radiating from the substrate and extending to several residues within the binding site region or to hydrogen-bonded peptide units within α-helices. These arrays contribute to the stabilization of protein-ligand complexes, especially those involving charged substrates.

As hydrogen bonds are highly directional, they are mainly responsible for conferring stereospecificity on the binding site and ensuring correctness of fit of substrates. An additional feature of hydrogen bonds which is crucial to transport protein components is that these bonds are stable enough to provide significant ligand binding but are of sufficient low strength to allow rapid ligand dissociation.

The complexity and precision in the formation of binding protein-ligand complexes strongly indicate that replacements of essential residues by site-directed mutagenesis offer very little by way of further detailed understanding protein-ligand interactions and, in most cases, of achieving "better" proteins. This observation is particularly pertinent to complexes in which the hydrogen-bonds are formed with peptide groups. On the other hand, binding studies, coupled with crystallographic and theoretical analysis, of the interaction with substrate analogues should prove more straightforward and extremely useful especially in analyzing the energetics of binding. Initial results of these types of studies in our laboratory, which in many ways can be envisioned as "drug design" studies, on L-arabinose binding-protein are very encouraging. For instance, binding studies of L-arabinose-binding protein in solution with D-galactose analogues, each lacking a hydroxyl group have yielded the interesting initial results shown in Table 1. Arabinose-binding protein binds D-galactose with almost equal affinity as L-arabinose (1), and we have shown by structure refinement that the D-galactose is bound to the protein with its pyranose ring in a position identical to that shown in Fig. 2 and its $C-6-CH_2OH$ located in a pocket (unpublished data).

Table 1. Binding Energies of Complexes of L-arabinose-binding Protein with Various Deoxy Derivatives of D-Galactose

Sugar	K_{assoc} (M^{-1})	ΔG (kcal·mole^{-1})	$\Delta\Delta G$ (kcal·mole^{-1})
D-galactose	2.58×10^6	-8.59	—
1-deoxy-D-galactose	1.79×10^3	-4.36	-4.23
2-deoxy-D-galactose	5.81×10^4	-6.39	-2.20
6-deoxy-D-galactose	5.68×10^4	-6.38	-2.21
L-arabinose	1.69×10^7	-9.40	—

Our studies have revealed at least two steps in the protein-ligand complex formation: the substrate binds preferentially to one domain of the "open", unliganded form of the binding protein and this binding triggers a conformational change which results in the closure of the bilobal structure around the substrate shielding it from water. The enclosing conformational change generates the appropriate stereochemistry for the specific interaction of the liganded binding protein with the membrane-bound protein components, in preference to the unliganded form, thus initiating the translocation process or flagella motion.

In addition to conferring substrate specificity, the primary role of periplasmic binding proteins in transport is linked to the ability of these proteins to form tight complexes with their ligands. These roles, together with the finding that binding proteins bind their substrates almost exclusively via hydrogen bonds, suggest a plausible mechanism for the binding protein-dependent active transport (1,3,5,7). This mechanism hinges on the existence of ligand site in the various protein components of the transport system which exhibits different affinities - high in the periplasmic binding protein (uptake side) and low in the site of protein component(s) confined to the plasma membrane (discharge side). A gradient of ligand sites affinities can be easily achieved by decreases in the number of hydrogen bonds that will be formed in each of the sites. Unidirectional transport is achieved by translocation of solute from site to site in synchrony with conformational changes which propagate through the entire system. These changes could alter sites accessibility. The coupling of energy to the binding protein-dependent transport systems requires ATP (8).

ACKNOWLEDGEMENT

We thank B. Kubena for assistance in preparing this manuscript. This work was supported by grants from NIH (GM21371) and the Welch Foundation (C581).

References

1. F.A. Quiocho and N.K. Vyas, *Nature* 310:318 (1984).
2. J.W. Pflugrath and F.A. Quiocho, *Nature* 314:257 (1985).
3. M.A. Saper and F.A. Quiocho, *J. Biol. Chem.* 258:11057 (1983).
4. N.K. Vyas, M.N. Vyas, and F.A. Quiocho, *Proc. Natl. Acad. Sci. USA* 80:1792 (1983).
5. F.A. Quiocho, N.K. Vyas, J.W. Pflugrath, M.A. Saper, M.N. Vyas, and J.S. Sack, *Proc. Int. Symp. Biomol. Struct. Interactions, Suppl. J. Biosci.*, (in the press).
6. J.H. Konnert, and H.A. Hendrickson, *Acta Crystallogr.* A36:244 (1980).
7. F.A. Quiocho, *Annu. Rev. Biochem.*, (in the press).
8. A. Wada, *Adv. Biophys.* 9:1 (1976).
9. W.G.J. Hol, P.T. van Duijnen, and H.J.C. Berendsen, *Nature* 273:443 (1978)
10. E.A. Berger and L.A. Heppel, *J. Biol. Chem.* 249:7747 (1974).

CRYSTAL STRUCTURE OF THAUMATIN I, A SWEET TASTE RECEPTOR BINDING PROTEIN

Abraham de Vos and Sung-Hou Kim

Department of Chemistry, University of California
Berkeley, CA 94720

Among sensory perceptions, taste is one of the least understood. Taste perception is associated with many properties, of which some are similar to, but others quite different from, those in hormone-hormone receptor interaction. These are the following:

(1) <u>Onset concentration and onset time</u>   To elicit sweet sensation, the concentration of the sweet compound has to exceed a threshhold concentration characteristic of each sweet compound. In addition there is a characteristic time lag between the exposure of the tongue to each sweet compound before the brain recognizes sweet taste.

(2) <u>Saturation</u>   As the concentration of the sweet compound increases, sweetness increases until it reaches a saturation point beyond which an increase in the ligand concentration does not register as increased sweetness.

(3) <u>Adaptation</u>   When the sweet receptor is exposed to a given concentration of a ligand, increase in the ligand concentration is required to renew sweet sensation.

(4) <u>Lingering time</u>   After removing the ligand, the sweet sensation lingers on for a period of time depending on the sweet compound.

(5) <u>After sensation</u>   After ligands are removed various different after-tastes characteristic to the ligand are recognized.

(6) <u>Potentiation</u>   Some sweet compounds enhance flavors which are not related to the taste. For example, after the human tongue is exposed to a dilute solution of Thaumatin, the flavor of a small amount of peppermint is tremendously enhanced.

As shown in Table 1, many different compounds can elicit sweet taste in humans. Some examples are sugars, dihydrochalcones, and the synthetic sweeteners saccharin and clyclamate. Several D-amino acids taste sweet, as do some dipeptide esters. Until fairly recently no sweet-tasting macromolecules had been found, but in the early 70's, the sweet-tasting proteins, Thaumatin and Monelin, were isolated and characterized [1-3]. Thaumatin is about 200,000 times sweeter than sucrose on a molar basis, and several thousand times sweeter on a weight

Table 1: Some sweet compounds and their sweetness

compound name	sweetness intensity (x sucrose, weight basis)
sucrose	1
D-glucose	0.7
D-fructose	1.1
sorbitol	0.5
stevioside	300
neohesperidin dihydrochalcone	2000
naringin dihydrochalcone	300
perylaldehyde anti-oxime	2000
dulcin	250
Na-saccharin	200-700
Na-cyclamate	30-80
acesulfam-K	130
aspartame	100-200
glycine	1.5
D-tryptophan	35
D-phenylalanine	7.3
D-tyrosine	5.5
monellin	2000
thaumatin	4000

Table 2: Refinement statistics for the heavy atom derivates

resolution (Å)	12.6	9.2	7.2	5.9	5.0	4.4	3.9	3.5	total
# reflections	58	106	184	262	363	476	626	774	2849
$KAu(CN)_2$									
R-Cullis	.54	.50	.52	.52	.61	.68	.73	.80	.63
rms F_H/residual	1.55	1.81	1.75	2.09	1.72	1.48	1.18	1.04	1.57
$K_2Pd(CN)_4$									
R-Cullis	.41	.54	.56	.51	.55	.65	.61	.71	.59
rms F_H/residual	2.17	1.60	1.75	2.03	1.58	1.25	1.26	1.23	1.62
K_2PtI_6									
R-Cullis	.51	.37	.44	.55	.51	.57	.57	.65	.54
rms F_H/residual	1.90	2.00	2.32	2.24	1.92	1.90	1.50	1.52	1.89
$K_2Pt(SCN)_6$									
R-Cullis	.51	.51	.64	.62	.71	.70	.75	.78	.66
rms F_H/residual	1.37	1.39	1.71	1.91	1.55	1.58	1.43	1.23	1.64
$K_2Hg(SCN)_4$									
R-Cullis	.61	.58	.49	.47	.64	.63	.67	.66	.59
rms F_H/residual	1.09	1.13	1.55	1.99	1.98	2.02	1.77	1.64	1.53
K_2HgI_4									
R-Cullis	.53	.45	.46	.49	.47	.62	.68	.70	.54
rms F_H/residual	1.99	2.79	3.17	3.27	3.08	1.77	1.26	1.38	2.33
figure of merit	.78	.83	.85	.85	.82	.77	.72	.69	.76

Table 3: Solvent flattening and phase extension

	cycle 1	cycle 4	cycle 8
m	0.64	0.75	0.77
R_{mapinv}	0.398	0.269	0.248
$\Delta\alpha°$	38	5.1	1.3
ρ_{sol}	24.6	24.9	23.9
ρ_{max}	263	267	256
$f(prot)$[1]	0.32	0.28	0.28
$f(solv)$[1]	0.19	0.08	0.07

[1] $f(prot)$ is the fraction of protein points above $2\rho_{sol}$
$f(solv)$ is the fraction of solvent points above $2\rho_{sol}$

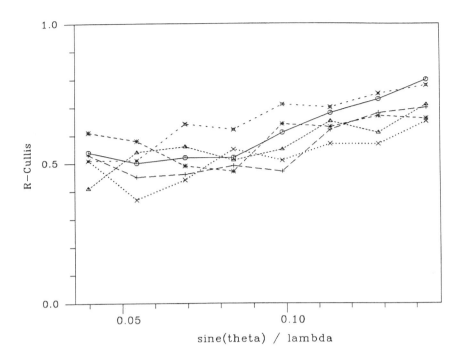
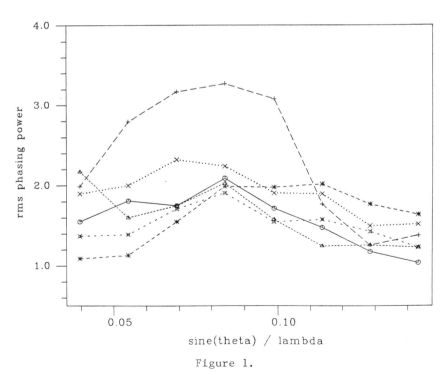

Figure 1.

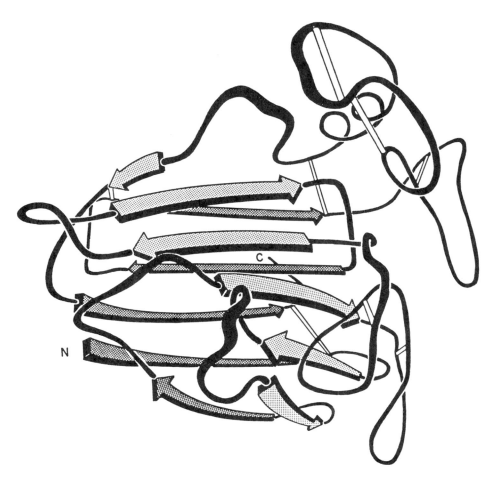

Figure 2.

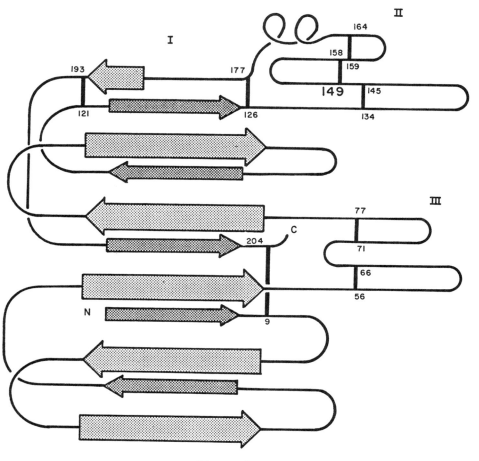

Figure 3.

basis. In fact, it is the sweetest compound known to mankind at the present time. Thaumatins have been isolated from the fruit of a west African rain forest shrub, Thaumatococcus daniellii benth, which has been used for centuries by the inhabitants of the region to sweeten food such as bread and palm wine. There are two major sweet proteins in the fruit, Thaumatins I and II, with an almost identical molecular weight of 22,000. There are 207 amino acid residues, and 5 residues are different between Thumatins I and II [4, 5]. No histidine residues but 8 disulfide bonds are present in the protein. Thaumatin, with many basic groups, has a pI of 12. The three-dimensional conformation of the molecules is important to the sweetness, since denaturation is accompanied by loss of sweetness.

Thaumatin I has been crystallized in space group $P2_12_12_1$ with cell parameters of 74.42 x 53.38 x 52.25 Å [6]. There is one protein molecule per asymmetric unit and 50% of the crystal volume is protein. X-ray diffraction data were collected at 4°C on a Nicolet 4 circle automatic diffractometer modified with an extended detector arm with helium filled incident and diffracted beam collimators. Nickel filtered CuKα X-ray was used for data collection with an omega step scan technique.

The initial electron density map at a resolution of 3.5 Å was calculated based on 6 heavy atom derivatives (listed in Table 2). The R factors between the native and these derivatives range from 12% to 19% on structure factor amplitude. The refinement statistics for these heavy atom derivatives are listed in Table 2 and shown in Figure 1. The electron density map based on MIR method showed most of the structural features but there were a few confusing areas. These areas, compounded by the 8 disulfide bonds, made it impossible to trace the backbone uniquely from this map. Subsequently the phases were improved and extended to 3.1 Å by flattening the solvent regions outside the molecular boundary using a program written by B.-C. Wang; this resolved most of the ambiguities. Statistics of the improvement are shown in Table 3, where m, R and $\Delta\alpha°$ stand for figure of merit, R factor between observed and calculated (inverse transform of map) F's, and phase changes, respectively.

A schematic drawing of the structure of Thaumatin I is shown in Figure 2. The overall dimension of the molecule is approximately 43 x 40 x 30 x $Å^3$. The molecule can be roughly divided into three domains designated as I, II and III in the schematic topological diagram shown in Figure 3. Domain I is a large domain consisting of beta structure. Domains II and III are small and contain many loops and disulfide bonds plus several short beta ribbons. A small alpha helix is present in domain II. The large domain I consists of 11 beta strands forming two beta sheets. All beta strands are antiparallel to their neighbors except the N terminal and C terminal strands, which are parallel to each other. An unusual feature of the beta barrel is that the strands of one sheet are almost parallel to, and on top of, the strands of the other sheet. It is also interesting to note that the beta strand between Cys 121 and Cys 126 and the strand immediately preceding it together with the connnecting loop form one of the most hydrophilic parts of the molecule.

Domains II and III, consisting of beta ribbons and small loops stabilized by disulfide bonds, have some similarities to regions found in snake venom toxins and wheat germ agglutinin [7]. Other small disulfide-rich proteins include cytotoxins and ragwood pollen allergens. Since these proteins all bind to membrane-bound receptors, it is tempting to speculate that domain II and/or domain III of Thaumatin is important for binding to the membrane-bound sweet receptors.

Monellin, the second intensely sweet protein, is also isolated from an African berry. Although these two proteins have very little sequence homology, antibodies raised against one cross-react to the other suggesting some similarities among the antigenic determinants of both proteins. Furthermore, these proteins prebound with antibodies no longer taste sweet. The crystal structure determination of Monellin is in progress.

REFERENCES

1. J. A. Morris and R. H. Cagan, Purification of Monellin, the sweet principle of Dioscoreophyllum cumminsii, Biochem.Biophys.Acta 261, 114 (1972).
2. H. van der Wel, Isolation and characterization of the sweet principle from Dioscoreophyllum cumminsii (Stapf) Diels, FEBS Lett. 21, 88 (1972).
3. H. van der Wel and K. Loeve, Isolation and characterization of Thaumatin I and II, the sweet-tasting proteins from Thaumatococcus daniellii Benth, Eur.J.Biochem. 31, 221 (1972).
4. R. B. Iyengar, P. Smits, F. van der Ouderaa, H. van der Wel, J. B. van Brouwershaven, P. Ravestein, G. Richers and P. D. van Wassenaar, The complete amino acid sequence of the sweet protein Thaumatin I, Eur.J.Biochem. 96, 193 (1979).
5. L. Edens, I. Bom, A. M. Ledeboer, J. Maat, M. Y. Toonen, C. Visser and C. T. Verrips, Synthesis and processing of the plant protein Thaumatin in yeast, Cell 37, 629 (1984).
6. A. M. de Vos, M. H. Hatada, H. van der Wel, H. Krabbendam, A. F. Peerdeman and S.-H. Kim, Three-dimensional structure of Thaumatin I, an intensely sweet protein, Proc.Nat.Acad.Sci.USA 82, 1406 (1985).
7. J. Drenth, B. W. Low, J. S. Richardson and C. S. Wright, The toxin-agglutin fold, J.Biol.Chem. 255, 2652 (1980).

STRUCTURES OF PYRUVOYL-DEPENDENT HISTIDINE DECARBOXYLASE AND MUTANT-3 PROHISTIDINE DECARBOXYLASE FROM LACTOBACILLUS 30A

M.L.Hackert, K.Clinger, S.R.Ernst, E.H.Parks and E.E.Snell

Department of Chemistry and Clayton Fndt. Biochem. Inst
University of Texas, Austin, TX 78712

INTRODUCTION - BIOCHEMISTRY

In man, the decarboxylation of histidine to form histamine is most often associated with the clinical treatment of colds, allergies, and ulcers. However, mammalian histidine decarboxylase activity is low, and attempts to isolate and characterize it have been frustrated by instability (Boeker and Snell, 1972; Snell, 1977; Tran and Snyder, 1981). On the other hand, much histidine decarboxylase activity is found in mammalian intestines due to the presence of lactic acid bacteria similar to Lactobacillus 30a. This organism produces large amounts of histidine decarboxylase, which has been studied extensively. Although the biological function of the histamine produced is not yet understood, histamine is known to be a powerful regulator of the vascular and digestive systems. Histamine of bacterial origin may cause a selective advantage to the microorganisms by altering the physiology of the host (Recsei and Snell, 1984).

Enzymatic amino acid decarboxylation proceeds by the formation of a Schiff's base with a bound carbonyl group. This adduct is cleaved to yield carbon dioxide and the corresponding amine (Boeker and Snell, 1972). Most amino acid decarboxylases contain pyridoxal-5'-phosphate as cofactor but in recent years a growing number of bacterial enzymes have been identified as containing a bound keto acid residue at the active site (Snell, 1977; Recsei and Snell, 1984). These include S-adenosylmethionine decarboxylase (Markham, et al., 1982) and phosphatidylserine decarboxylase (Dowhan, et al., 1974) from Escherichia coli and histidine decarboxylase (HDC) (Recsei and Snell, 1984) from Lactobacillus 30a. Many non-pyridoxal phosphate decarboxylases contain a covalently bound pyruvoyl moiety as the keto acid prosthetic group.

The Lactobacillus 30a histidine decarboxylase was reported in 1953 to retain activity when the cells were grown at low levels of vitamin B_6 (Rodwell, 1953), suggesting that this amino acid decarboxylase was not dependent on pyridoxal phosphate. Homogeneous preparations lacked vitamin B_6, but yet were inhibited by such carbonyl reagents as phenylhydrazine (Rosenthaler, et al., 1965) to yield the labelled phenylhydrazone of N-pyruvoylphenylalanine (Riley and Snell, 1968). Recsei and Snell isolated the substrate-enzyme complexes formed upon binding of histidine and histamine to histidine decarboxylase by borohydride reduction and acid hydrolysis. The secondary amines produced confirmed that pyruvate was the prosthetic group in histidine decarboxylase (Recsei and Snell, 1970).

Furthermore, this evidence indicated that the enzyme catalyzes the decarboxylation via a Schiff's base mechanism analogous to that of pyridoxal phosphate containing enzymes (Figure 1).

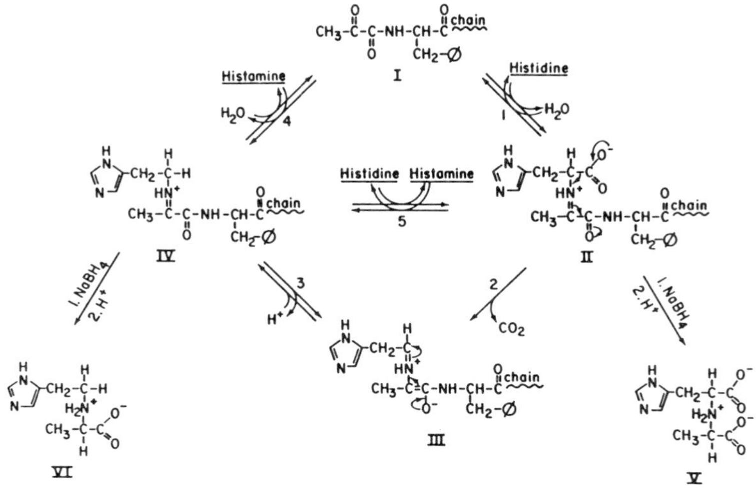

Figure 1. Proposed reaction mechanism for pyruvoyl-dependent histidine decarboxylase (after Recsei and Snell, 1984).

In order to determine the origin of the pyruvoyl group, possible precursors were labelled and introduced into the cell medium. Labelled pyruvate was not incorporated, but the use of ^{14}C-serine resulted in the same specific activity for the pyruvoyl residues as for the serine residues of the enzyme (Riley and Snell, 1970). A search was made to determine whether the pyruvate was inserted into a precursor protein or if the pyruvate arose from the processing of a serine residue in an intact protein chain. However, mutants were found which synthesized inactive proteins which were cross-reactive to anti-histidine decarboxylase antibodies (Recsei and Snell, 1972; Recsei and Snell, 1973). One mutant protein (Lactobacillus 30a mutant-3) was found in good yield which exhibited less than 10% of the activity of wild-type enzyme. However, upon isolation and crystallization at pH 7.5, the mutant enzyme became essentially as active as wild-type decarboxylase (Recsei and Snell, 1973). However, if stored at pH 4.8, this mutant enzyme did not activate over a period of years. The wild-type prohistidine decarboxylase has also been isolated from Lactobacillus 30a (Recsei and Snell, 1982) which activates three times faster at pH 7.0 than the mutant-3 prohistidine decarboxylase.

The wild-type histidine decarboxylase (HDC) and mutant-3 proenzyme (pHDC) proteins were found to be quite similar, but yet had important differences. Wild-type HDC and pHDC have identical molecular weights as determined by sedimentation equilibrium and electrophoretic migration on polyacrylamide gels (PAG) under non-denaturing conditions (Recsei and Snell, 1973). However, the proteins differ on electrophoresis on PAG under denaturing conditions. Wild-type HDC has two bands, while pHDC has only one band of higher molecular weight. Also, the pHDC does not form a phenylhydrazone with phenylhydrazine or yield pyruvate upon acid hydrolysis, indicating that the pyruvate has not yet been formed. Upon incubation at pH 7.5, the mutant enzyme not only activates, but its electrophoretic pattern in the presence of sodium dodecyl sulfate (SDS) indicates two subunits very similar to those of wild-type enzyme (Recsei and Snell, 1973). Thus the lone subunit (π) of the proenzyme is cleaved to form the two subunits (α- and β-subunits) of mature HDC (Figure 2).

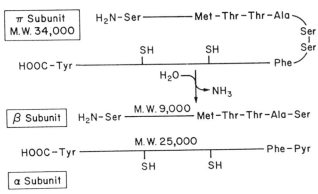

Figure 2. Cleavage of prohistidne decarboxylase π chains to form the α and β subunits of active HDC (after Snell, 1977).

Upon activation of the proenzyme, one mole of ammonia was released and one pyruvoyl group was formed per mole of peptide chain (Recsei and Snell, 1973; Snell, Recsei, and Misono, 1976). This activation is first order and directly proportional to proenzyme concentration (Recsei and Snell, 1973). Proenzyme immobilized on a Sepharose column and washed extensively with 9 M urea to remove contaiminating proteins is still capable of activation (Snell, Recsei, and Misono, 1976). The immobilized enzyme activates readily at pH 7.6 and 37°C in the presence of potassium ion or other monovalent cation. From this, it may be deduced that the conversion is non-enzymatic or intramolecular. It is also reported that high concentrations of alkali metal ions or ammonium ion speed the activation process (Recsei and Snell, 1981) in order of effectiveness as follows:

$$K^+ > NH_4^+ > Rb^+ > Na^+ \approx Cs^+ > Li^+$$

Studies of the pH effect on activation imply that two groups of pK 6.8 and 8.2 are involved in the activation process (Recsei and Snell, 1981).

The mechanism of activation of prohistidine decarboxylase has been studied by the use of ^{18}O labelling experiments (Recsei, Huynh, and Snell, 1983). Activation of mutant prohistidine decarboxylase in the presence of $H_2^{18}O$ did not label the carboxyl oxygen of serine 81. Therefore proenzyme chain cleavage is not a hydrolytic reaction as in most known enzymatically catalyzed proteolytic reactions. When ^{18}O carboxyl labelled serine was added to the growth media of Lactobacillus 30a mutant-3, ^{18}O was present in the carboxyl of ser-81 as well as other serine residues, but was absent in all other residues. In similar experiments using ^{18}O hydroxyl labelled serine it was found that the serines of active enzyme all contained ^{18}O in their hydroxyl group upon hydrolysis of the protein, but serine 81 also contained labelled oxygen in its carboxyl group (Recsei, Huynh, and Snell, 1983). Thus proenzyme activation occurs by an unusual serinolysis combined with an α,β-elimination at ser-82 to form the pyruvoyl group of the α chain.

HDC contains two cysteine residues in each α chain, but there are no disulfide bonds in the protein (Lane and Snell, 1976). One of the sulfhydryls can be titrated directly with 5,5'-dithiobis (2-nitrobenzoic acid) (DTNB), at a pH of 6.5 or above. This reaction is blocked by the presence of the non-competitive inhibitor histidine methyl ester (Lane, Manning, and Snell, 1976) and slowed by the competitive inhibitor histamine (Recsei and Snell, 1982). The amount of DTNB bound is paralleled by a loss of activity in the wild-type enzyme and thus this sulfhydryl is probably located near the active site.

STRUCTURE OF wild-type HDC

Ultracentrifugation experiments indicate that histidine decarboxylase is a hexamer in its most stable form, existing in the subunit composition $(\alpha\beta)_6$ and $(\pi)_6$ for the mature and precursor enzymes respectively (Hackert, et al., 1981). Sedimentation equilibrium results implied both hexamers are of $M_r \approx 208,000$, with the π-subunit $M_r \approx 37,000$, the α-subunit $M_r \approx 28,000$, and the β-subunit $M_r \approx 9,000$.

Histidine decarboxylase and prohistidine decarboxylase tend to dissociate into smaller particles at a pH above 4.8 and at low ionic strength (Hackert, et al., 1981). The mutant proenzyme dissociates to some degree even at pH 4.8 (Recsei and Snell, 1973; Recsei and Snell, 1982). The smaller protein particles demonstrate the existence of stable $(\alpha\beta)_3$ and $(\pi)_3$ species which are similar to particles reported for Micrococcus histidine decarboxylase, which is also a pyruvate containing enzyme (Prozorovski and Jornvall, 1974; Gonchar, et al., 1977).

The amino acid sequence of Lactobacillus HDC has been determined (Vaaler, et al., 1982; Huynh, et al., 1984). Proenzyme is cleaved between serines 81 and 82 with ser-82 converted into the pyruvic acid residue and ser-81 becomes the carboxyl terminal residue of the β chain. This places the pyruvoyl group at the amino terminal end of the α chain. The 81 residues of the β chain correspond to a chemical molecular weight of 8,840. The α chain was reported to contain 225 amino acid residues, but subsequent DNA sequence studies (Robertus, 1985) have shown there to be 228 amino acid residues, corresponding to a molecular weight of $\approx 25,260$. The molecular weight for the native $(\alpha\beta)_6$ molecule is therefore calculated to be $\approx 204,600$, which is in good agreement with the experimental result of 208,000 obtained by ultracentrifugation.

There is evidence that the mutant proenzyme π chain has only two amino acid differences from the wild-type enzyme, an alanine instead of ser-51, and an aspartic acid residue in the place of gly-58 (Vaaler, et al., 1982). The intact proenzyme π chain contains 310 amino acid residues yielding a chemical molecular weight of $\approx 34,100$.

The x-ray structure of histidine decarboxylase from Lactobacillus 30a has been determined by the method of multiple isomorphous replacement to 3.0 Å resolution (Parks, et al., 1985). The wild-type enzyme crystallizes in space group $I4_122$ with with a = 221.7Å and c = 107.1Å with one $(\alpha\beta)_3$ half molecule per asymmetric unit. The low resolution (5.5Å) x-ray data were collected using a four-circle autodiffractometer. A pHMB derivative was prepared by titrating the Class I sulfhydryls in solution before crystallization. These heavy-atom positions were unraveled by interpreting the results of a direct phase determination (MULTAN) within the constraints of the non-crystallographic, molecular symmetry (Parks, et al., 1983). Other heavy atom derivatives were prepared by soaking crystals in solutions containing the heavy atoms. The higher resolution data (3.0Å resolution) from native and two heavy atom derivative crystals were collected using the multiwire area detector facility at UCSD (Xuong, et al., 1978). Each data set included $\approx 250,000$ reflections measured from each crystal in less than two days.

A 3.0Å resolution electron density map was calculated using best phases obtained from the isomorphous and anomalous contributions from the Pt and diHg derivatives, m = 0.79 (Parks, et al., 1985). Averaged maps, calculated using Bricogne's double-sort technique (Bricogne, 1976) were rewritten with sections perpendicular to the molecular 3-fold axis. The signal-to-noise ratio was higher in the averaged map and it was possible

to complete the chain tracing unambiguously. The amino acid residues for one
(αβ) unit were fitted to the electron density map on an Evans and
Sutherland Multi-Picture Display System interfaced to a VAX 11/750 using
the FRODO software of T. Alwyn Jones.

The structure of the αβ unit contains six α-helices and fifteen
β strands (Figure 3). There are two long α-helices, one from the β
chain on one side and another from the α chain on the other side of the
αβ unit. Sandwiched between these two helices are several strands of
β-sheet, in two prominent anti-parallel sheets (Parks, et al., 1985).

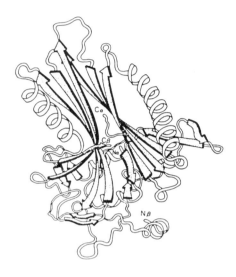

Figure 3. Schematic drawing of one αβ unit of HDC. The molecular
3-fold axis is vertical and near the viewer in this representation.

The $(\alpha\beta)_3$ half molecule is roughly spherical with a diameter of
approximately 65Å. There is a large central cavity around the
three-fold molecular axis which extends from the outermost portions of the
molecule inward about 30Å towards the molecular two-fold axis. This
central cavity is formed by the three-fold repeat of the large β subunit
helix and two edge strands of the β-sandwich structure. The active site
pyruvoyl groups, at the amino termini of the α chains, are located in
pockets around the bottom of the large central cavity (Figure 4). These
pyruvoyl groups are separated from their 3-fold related mates by about 23
Å. The pyruvoyl carbonyl groups are not found in a Schiff base linkage
with the ε-amino group of a lysyl residue as is the case with most
pyridoxal phosphate dependent enzymes.

There are three fairly well defined regions of interaction between the
(αβ) units in forming the $(\alpha\beta)_3$ particle: a lock-washer arrangement
of β-chain loops close to the center of the $(\alpha\beta)_6$ molecule; a region
near the carboxy termini of the β-chains in the center of the $(\alpha\beta)_3$
particles; and a region on the surface of the $(\alpha\beta)_3$ particle where the
two large helixes from the neighboring (αβ) units overlap (Figure 4).
These interactions involve both main-chain and side-chain contacts. None
of the interactions appears to involve all three (αβ) units at once. Two
$(\alpha\beta)_3$ particles associate to form a dumbbell shaped $(\alpha\beta)_6$ molecule
of an overall length of ≈120 Å. Each (αβ) unit interacts with only
one of the (αβ) units in the two-fold related $(\alpha\beta)_3$ particle.

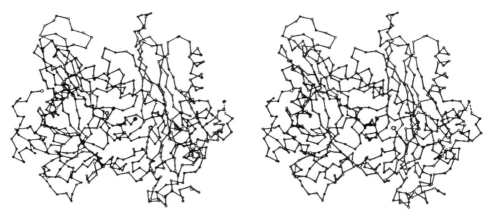

Figure 4. Stereoview of an alpha carbon drawing of an $(\alpha\beta)_2$ unit from HDC. The molecular 3-fold is vertical and the active site pyruvoyl group is marked as a large, open circle in this view.

STRUCTURES OF proHDC

Four crystal forms of prohistine decarboxylase from mutant-3 were grown, none of which were isomorphous with crystals of wild-type HDC. Two forms were selcted for further study: space group $C222_1$; a=101, b=110, c=210Å with a $(\pi)_3$ particle/a.u. and space group P321; a=100, c=164Å with two π subunits/a.u. Native data were collected to 3.5Å and 2.5Å resolution, respectively, for the orthorhombic and trigonal crystal forms using the multiwire area detector facility at UCSD. Both structures have been solved by the molecular replacement technique using the wild-type HDC coordinates as a starting model. The orientations derived from the self-on-self rotation functions were confirmed with the aid of cross-rotation functions between the three crystal forms. The translation function as described by Crowther and Blow (1967) was used to correctly position the molecules. These translation solutions were confirmed by R-factor searches and improved by rigid-body refinement.

It was apparent from these results that the structures of mutant-3 pHDC were very similar to that of wild-type HDC. To lessen the bias in the resulting electron density maps, the phases employed for the trigonal system were derived from a model which did not contain amino acid residues 80 through 83 near the site of cleavage during self-activation. The resulting (F_o-F_c) map, computed initially at 3.4Å resolution, showed continuous density across this region indicating that the π chain was still intact within the crystal. Model building using FRODO and refinement to extend these results to 2.5Å are in progress. Assigning hydrogen bonds at the present stage of model building would be premature.

ACTIVE SITE DESCRIPTION

Kinetic studies have shown that the maximum velocity of the enzyme is essentially constant between the pH values of 2.9 to 7.6, indicating that ionizable groups with pK values in this pH range are not directly involved in the catalytic process (Chang and Snell, 1968). Residues required in the catalytic process are an electron withdrawing group (E:) to increase the electrophilicity of the pyruvoyl moiety, and a proton donor (BH). The K_m for histidine is almost constant between pH 4.8 and 6.0 and increases outside of this range. The increase in K_m at pH values less than 4.8 indicates that a single negatively charged carboxyl group is involved in binding the substrate.

To localize the substrate binding site an active site derivative was prepared by soaking already existing crystals of the native enzyme in a solution of histamine. 3.0Å resolution data were collected and a difference electron density map computed. Figure 5 shows an active site region of the enzyme superimposed on the corresponding portion of the difference electron density map. There is continuous density between the alpha carbon of histamine and the carbonyl carbon of the pyruvoyl residue.

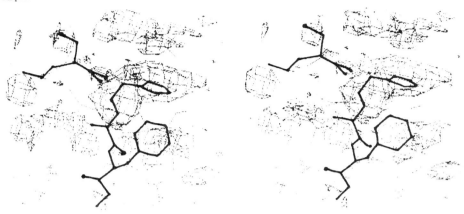

Figure 5. Stereo view of the histamine product complex superimposed on its difference electron density map.

A schematic representation of the active site is given in Figure 6. The most likely candidates for the negatively charged carboxyl group involved in binding the substrate can be seen to be Glu 197, Ser 81, Glu 66, or Asp 63. In the catalytic process, the best candidate for the electron with-drawing group is the amide N-H of Phe 195 and/or Tyr 262 while the proton donor is most likely Lys 155. These residues are conserved in all pyruvoyl dependent HDCs thus far examined (Huynh and Snell, 1985 a,b).

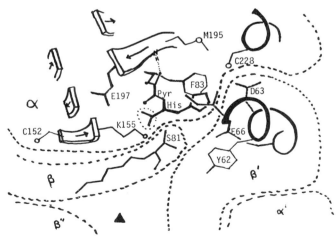

Figure 6. Schematic representation of one active site of HDC. Subunit boundaries are indicated by heavy dashed lines and the carboxyl group of the substrate circled by a light dashed line.

Studies on the self-activation process had indicated involvement of a proton donor of $pK_a \approx 8.2$ (Recsei and Snell, 1981), it was anticipated that cys 152 or cys 228 would serve in that capacity. However, both residues are too distant from the site of activation (>7Å) to take part in the activation process. Furthermore, sequence evidence from the cysteinyl peptides isolated by disulfide chromatography indicates that cys 228 is not conserved in the pyruvoyl dependent HDC from <u>Clostridium perfringens</u>, which is presumably activated in much the same manner as HDC from <u>Lactobacillus</u> (Huynh and Snell, 1985a). Residues which are closer to the activation site which could perhaps act as a proton donor are lys 155, tyr 262, and tyr 62 from a neighboring π subunit. These amino acid residues would require altered pK_a values as the isolated individual side chain pK_as are not in this range.

In like manner, histidinyl residues 142 or 233 were expected to act as the proton acceptor in the activation reaction due to the $pK_a \approx 6.8$. These residues, though, are not close to the activation site. Proton acceptors which are at the site of activation and could possibly have a role in activation are glu 197, tyr 262, and two residues of the neighboring π chain, tyr 62 and glu 66. Again, altered pK_a values would be required to conform to the observed kinetic data. Such modified pK_a values in proteins are not uncommon.

There is also a possibility that the ser 81 carbonyl group and a proenzyme nucleophilic amino acid side chain could form an acyl-enzyme intermediate (e.g., a thioester, ester, or anhydirde) (Recsei and Snell, 1983). Such an intermediate may be preferred as it may promote α,β-elimination since the carboxylate anion is an excellent leaving group. Residues which may perform this function are cys 152, glu 197, cys 228, plus glu 66 from neighboring π chain.

Kinetic experiments indicate a half-order dependence on monovalent cation concentration (Recsei and Snell, 1981). This evidence indicates that a bound cation interacts with pHDC in the rate limiting step of activation. The order of effectiveness has a good correlation with crystal ionic volume, but not the hydrated volume. This implies that a non-hydrated ion is bound near the activation site during activation (Recsei and Snell, 1981). The cation may form a complex with the carbonyl oxygen atom of ser 81 which would facilitate nucleophilic attack at the carbonyl roup or serve as a general charge neutralizer for some of the many acidic side chains in the vicinity of the activation site.

CONCLUSION

The structures of pyruvoyl-dependent histidine decarboxylase and prohistidine decarboxylase have been determined. The structure of HDC is unusual in that the catalytic sites are located around the bottom of a large central cavity present in each half of the dumbbell-shaped molecule. A histamine product complex has been analyzed to locate the imidazole binding site and thus permit description of the active site environment. Data to 2.5Å resolution for both HDC and proHDC are available and phase extension and refinement to this resolution are in progress to enable a more complete analysis and comparison of these structures. These results, together with related biochemical and site-directed mutagenesis studies underway, should permit us to better understand the mechanisms of self-activation and action of this enzyme.

ACKNOWLEDGEMENTS

We gratefully acknowledge Drs. N.-h Xuong and R. Hamlin and their colleagues at the regional area detector facility at UCSD for assistance in data collection. This project is supported by NIH grant (GM30105).

REFERENCES

Boeker, E. A. and Snell, E. E. (1972) in *The Enzymes*, 3rd ed., Vol. 6 (P.D. Boyer, ed.), pp. 217-253. Academic Press, New York.
Bricogne, G. (1976) *Acta Crystallogr.* A32:832-847.
Crowther, R. A. and Blow, D. M. (1967) *Acta Crystallogr.* 23:544-548.
Chang, G. W. and Snell, E. E. (1968) *Biochemistry* 7:2012-2020.
Dowhan, W., Wickner, W. T. and Kennedy, E. P. (1974) *J. Biol. Chem.* 249:3079-3084.
Gonchar, N. A., Katsnelson, A. A., L'vov, Y. M., Semina, L. A. and Feigin, L. A. (1977) *Biofizika* 22:801-805.
Hackert, M. L., Meador, W. E., Oliver, R. M., Salmon, J. B., Recsei, P. A. and Snell, E. E. (1981) *J. Biol. Chem.* 256:687-690.
Huynh, Q. K., Recsei, P. A., Vaaler, G. L. and Snell, E. E. (1984) *J. Biol. Chem.* 259:2833-2839.
Huynh, Q. K. and Snell, E. E. (1985a) *J. Biol. Chem.* 260:2798-2803.
Huynh, Q. K. and Snell, E. E. (1985b) *J. Biol. Chem.* 260:2794-2797.
Lane, R. S. and Snell, E. E. (1976) *Biochemistry* 15:4175-4179.
Lane, R. S., Manning, J. M. and Snell, E. E. (1976) *Biochemistry* 15:4180-4185.
Markham, G. D., Tabor, C. W. and Tabor, H. (1982) *J. Biol. Chem.* 257:12063-12068.
Parks, E. H., Ernst, S. R., Hamlin, R., Xuong, Ng. H., and Hackert, M. L. (1985) *J. Mol. Biol.* 182: 455-465.
Parks, E. H., Clinger, K. and Hackert, M. L. (1983) *Acta Crystallogr.* B39:490-494.
Prozorovskii, V. and Jornvall, H. (1974) *Eur. J. Biochem.* 42:405-409.
Recsei, P. A. and Snell, E. E. (1970) *Biochemistry* 9:1492-1497.
Recsei, P. A. and Snell, E. E. (1972) *J. Bacteriol.* 112:624-626.
Recsei, P. A. and Snell, E. E. (1973) *Biochemistry* 12:365-371.
Recsei, P. A. and Snell, E. E. (1981) in *Metabolic Interconversion of Enzymes 1980* (Holzer, H., ed.) pp. 335-344. Springer-Verlag, Berlin.
Recsei, P. A. and Snell, E. E. (1982) *J. Biol. Chem.* 257:7196-7202.
Recsei, P. A., Huynh, Q. K. and Snell, E. E. (1983) *Proc. Natl. Acad. Sci. USA* 80:973-977.
Recsei, P. A. and Snell, E. E. (1984) *Ann. Rev. Biochem.* 53:357-387.
Riley, W. D. and Snell, E. E. (1968) *Biochemistry* 7:3520-3528.
Riley, W. D. and Snell, E. E. (1970) *Biochemistry* 9:1485-1491.
Robertus, J. D. (private communication).
Rodwell, A. W. (1953) *J. Gen. Microbiol.* 8:233-237.
Rosenthaler, J., Guirard, B. M., Chang, G. W. and Snell, E. E. (1965) *Proc. Natl. Acad. Sci. USA* 54:152-158.
Snell, E. E., Recsei, P. A. and Misono, H. (1976) in *Metabolic Interconversion of Enzymes 1975* (Shatiel, S., ed.) pp. 213-219. Springer-Verlag, New York.
Snell, E. E. (1977) *Trends Biochem. Sci.* 2:131-135.
Tran, V. T. and Snyder, S. H. (1981) *J. Biol. Chem.* 256:680-686.
Vaaler, G. L., Recsei, P. A., Fox, J. L. and Snell, E. E. (1982) *J. Biol. Chem.* 257:12770-12774.
Xuong, Hg-H., Freer, S., Hamlin, R., Nielson, C. and Vernon, W. (1978) *Acta Crystallogr.* A34:289-296.

STATE OF X-RAY STRUCTURE DETERMINATION OF ASCORBATE OXIDASE FROM GREEN ZUCCHINI SQUASH

A. Messerschmidt[a], M. Bolognesi[b], A. Finazzi-Agrò[c], and R. Ladenstein[a]

a Max Planck Institut für Biochemie
   D-8033 Martinsried, Federal Republic of Germany

b Department of Genetics and Microbiology
   University of Pavia
   I-27100 Pavia, Italy

c Institute of Biochemistry
   University of Rome "La Sapienza"
   I-00185 Roma, Italy

The copper enzyme ascorbate oxidase (E.C. 1.10.3.3) belongs to the group of "blue oxidases". According to recent investigations of Avigliano et al.(1983) the enzyme has a molecular weight of 145,000 with a copper content of $8/146,000 \, M_r$. The molecule consists of two non-covalently linked subunits of slightly different molecular weight (75,000 and 72,000 respectively). The spectroscopic properties of ascorbate oxidase indicate that the enzyme contains all three different types of copper atoms (Fee, 1975; Malmstroem, 1982).

Two different crystalline modifications of ascorbate oxidase have been used for the crystallographic investigations. The protein from green zucchini squash was purified according to Marchesini & Kronek (1979) for crystallization of modification I, and according to Avigliano et al.(1972) for crystallization of modification II.

Modification I was crystallized at 4°C by the vapour diffusion method (Ladenstein et al., 1979) using 1.9M sodium-potassium phosphate buffer (pH 7.0) as reservoir solution. The crystals, as in modification II, had a deep blue color and belong to the orthorhombic space group $P2_12_12_1$ with $\underline{a} = 190.7$ Å, $\underline{b} = 125.2$ Å, $\underline{c} = 112.3$ Å. Assuming 2 molecules/asymmetric unit a packing density of $V_M = 2.4$ Å^3/ dalton is obtained, a value fitting well into the range found for other protein crystals (Matthews, 1968). The crystals diffract to 2.5 Å resolution.

The crystals of modification II were obtained by dialysis of protein solutions (15 mg/ml) against 12% (v/v) 2-methyl-2,4-pentane-diol in 0.05 M phosphate buffer at 4°C and pH 5.4-5.8 (Bolognesi et al., 1983). The crystals are orthorhombic, space group $P2_12_12$ with $\underline{a} = 106.7$ Å, $\underline{b} = 105.1$ Å, $\underline{c} = 113.5$ Å. The asymmetric unit accommodates one molecule of the enzyme with a packing density parameter of $V_M = 2.27$ Å^3/dalton. The crystals diffract to 2.5 Å resolution.

TABLE 1

Modification I	Modification II
native compound	native compound
2 mercury-derivatives	1mM K_2OsO_4- derivative
obtained by treatment	1mM K_2ReCl_6-derivative
with HgCl	1mM K_3IrCl_6-derivative
(crystallisation and data collection in collaboration with A. Marchesini, Torino)	

The complete X-ray diffraction data sets listed in Table I were collected on Enraf-Nonius rotation cameras in a cold room at 4°C, using graphite monochromatized CuK_α radiation produced by a Philips 1.5 kW sealed off fine focus tube.

The films were evaluated using the FILME program system (Schwager et al., 1975). One data set of modification II consists of about 40,000 independent reflections. Data reduction, calculation of Patterson search functions, difference Patterson maps, difference Fourier maps and vector verification were performed with the PROTEIN program system (Steigemann, 1974).

In order to get information about the arrangement of the subunits in both modifications real space Patterson search techniques were applied. Self rotation functions for both crystal forms were calculated using polar angles ψ, ϕ, and k. A cross rotation function was produced to verify the results of the self rotation functions. In this case Eulerian angles θ_1, θ_2, θ_3 according to Rossmann & Blow (1962) were used. From the results

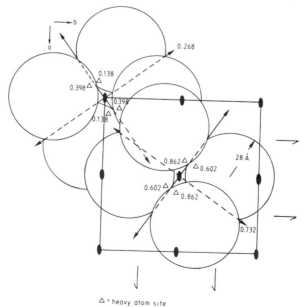

Fig.1: Arrangement of the subunits in the unit cell of modification II

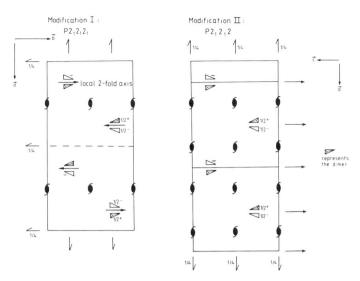

Fig.2: Comparison of the arrangement of the subunits in both modifications.

the quaternary structure of both modifications could be proposed as shown in Figures 1 and 2. The situation in modification II is very clear. The subunits form a tetramer of 222-symmetry. One 2-fold axis is coincident with the crystallographic 2fold axis parallel to c. The other two lie in the a-b-plane, forming an angle of 54.3° and -35.7° with respect to the b--axis. In modification I tetramers are present as well. It can be expected that both modifications exhibit a similar packing. This is achieved if the unit cells are arranged in the way shown in Figure 2. The intersection Point of the three 2-fold axes should have a z-value of about 0.75 in modification II, and a y-value of about 0.25 in modification I, implying that the intersection point should lie on the 2_1 axis parallel to a of modification I. The only difference in packing between modification I and II is the continuation in the lower half of the unit cells drawn. In modification I the tetramer is rotated about the 2_1 axis parallel to a, and then translated along this axis by an amount of a/2 (modification I) or approximately a (modification II). In modification II the tetramer is simply translated by an amount of a. The three non-crystallographic 2-fold axes of the tetramer in modification I do not have exactly the same orientation as they should have after applying the coordinate transformation depicted in Figure 2. The orientation deviates by about 10°. This results from a careful interpretation of the self rotation function of modification I and the cross rotation function between modification I and II.

Isomorphous difference Patterson maps of 5 heavy-atoms derivatives were calculated (Tab.1). Only those of the Rhenium and Iridium derivatives could be interpreted. The other derivatives obtained appear to lack isomorphism. From the model of the quaternary structure 2 heavy atom sites could be expected; they show up quite clearly in the Rhenium derivative map. All the necessary peaks in the Harker sections as well as the peaks corresponding to cross vectors between symmetry-related sites are present and represent the strongest peaks. The Iridium derivative unfortunately shows the same heavy-atoms sites and is of lower quality. An electron density map calculated with single isomorphous replacement phases of the Rhenium derivative wasn't good enough to show the molecular boundaries. From the heavy-atoms sites of the Rhenium derivative the z-value of the intersection point of the tetramer 2-fold axes can be deduced. Its value is z = 0.268 and matches very well the predicted value which would be necessary to get the similar packing shown in Figure 2. The existence of tetramers also in modification II (pH 5.4-5.8) is in excellent agreement with the recent results of Avigliano et al. (1983) which show that the tetramers are present in solution also at lower pH values.

Search for further derivatives is underway.

REFERENCES

Avigliano, L., Gerosa, P., Rotilio, G., Finazzi-Agrò, A., Calabrese, L. and Mondovi, B. (1972)
    Ital. J. Biochem. 21, 248-255.
Avigliano, L., and Mondovi, B. (1983).
    Mol. Cell. Biochem. 56, 107-112.
Bolognesi, M., Gatti, G., Coda, A., Avigliano, L., Marcozzi, G., and Finazzi-Agrò, A. (1983).
    J. Mol. Biol. 169, 351-352.
Fee, J.A.,(1975) In Structure and Bonding, vol. 23.
    pp. 1-60. Springer Verlag. Berlin und Heidelberg.
Ladenstein, R., Marchesini, A. and Palmieri, S. (1979)
    FEBS Letters 107, 407-408.
Malmstroem, B.G. (1982). Annu. Rev. Biochem. 51, 21-59.
Marchesini, A. and Kroneck, P.M.H. (1979)
    Europ. J. Biochem. 101, 65-76.
Matthews, B.W. (1968). J. Mol. Biol. 33, 491-497.
Rossman, M.G. and Blow, D.M. (1962)
    Acta Cryst. 15, 24-31.
Schwager, P., Bartels, K. and Jones, I.A. (1975).
    J. Appl. Cryst. 8, 275-280.
Steigemann, W. (1974) Doctoral Thesis. Technical University. Munich.

X-RAY STRUCTURE OF THE LIGHT-HARVESTING BILIPROTEIN C-PHYCOCYANIN FROM M. laminosus

Wolfram Bode, Tilman Schirmer, Robert Huber
Walter Sidler* and Herbert Zuber*

Max-Planck-Institut fur Biochimie
D 8033 Martinsried, F.R.G.
*Institut fur Molekularbiologie und Biophysik
ETH Zurich, CH-8093 Zurich

Blue-green algae (cyanobacteriae) possess specialized antenna organelles, the phycobilisomes, to collect light of a broad spectrum and to transfer its energy very efficiently towards the chlorophyll containing photosynthetic reaction centers. Their major constituents are phycobiliproteins, intensely coloured protein-pigment complexes, composed of polypeptide chains covalently linked with open-chain tetrapyrroles (bilins).

We (Schirmer et al., 1985) have determined the spatial structure of one component, C-phycobilisomes from the cyanobacterium Mastigocladus laminosus. This oligomeric protein consists of α- and β-chains containing 162 and 172 amino acids (Sidler et al., 1981). Conical hexagonal crystals of space group $P6_3$ could be grown from concentrated C-phycocyanin solutions at 0.15M ammonium sulfate, pH8.5, using the vapour diffusion method. Their cell constants are a=b=154.6 Å. c=40.5Å. With a modified Ficoll density method (Bode and Schirmer, 1985) the apparent crystal density was determined to 1.11 g/CM^3, corresponding to a protein mass of 36kDa per asymmetric unit, i.e. one ($\alpha\beta$)-monomer. With native crystals X-ray reflections to 2.2Å resolution were collected on films. Self-rotation correlation of the native Patterson map indicated the presence of a local twofold axis within the ($\alpha\beta$)-monomer. Six different heavy metal derivatives (mercury and/or platinum compounds) were included in phase refinement and contributed to phasing. The final overall figure-of-merit was 0.59 for data to 3 Å resolution. In the first electron density map protein and solvent regions could be distinguished, and many secondary structure features, especially Å-helices, were recognized. This map could be further improved by solvent flattening. The polypeptide chain could be traced using the known amino acid sequences of the α- and the β-chain (Frank et al., 1978). The ($\alpha\beta$)-monomers are arranged around the crystallographic triads. The resulting trimers are disc shaped and have a diameter of about 110 Å and a thickness of about 30 Å. The discs exhibit a central hole with a diameter of about 35 Å. Two trimers are related by the crystallographic two-fold screw axis running parallel to c through the center of the cell.

Both subunits, α and β, exhibit the same tertiary structure (Fig.1) and are related by a local two-fold rotation axis. Their molecular conformation is characterized by the predominance of α-helices and the absence of β-sheet elements. Each subunit is composed of 8 helices assigned XΣ,Y,A,B,E,F,G and H, and irregular loops connecting them. Helices A and E as well as G and H are running almost anti-parallel to one another. Helices A to H build up the globular part or each subunit. Helix-pair X/Y of each subunit protrudes and makes extensive contacts with helices A and F of the other subunit of the monomer. This mutual interaction is of basic importance for the α-β-association.

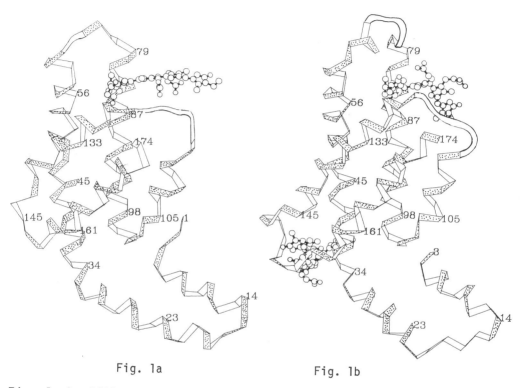

Fig. 1a,b. Ribbontype α-carbon plot of (a) α-C-PC and (b) β-C-PC.

Chromophores α84 and β84 are topologically equivalent (Fig.1a,b). Their tioether bonds and their first three pyrrole rings are defined by electron density. Chromophore β155 can be traced completely. All chromophores show an extended conformation, but are considerably twisted around the methene bridges, so that the pyrrole rings are not coplanar. Some of the pyrrole rings and of the propionic side chains interact specially with polar (charged) residues in the surrounding. Free bilinpigments (Kratky, 1983) as well as bilins in denatured C-PC (Scheer and Kufer, 1977) have a strong tendency to adopt cyclic-helical conformations and have their absorption maxima in the near UV. Thus the intense visible absorption of native biliproteins can be attributed to the elongated conformation stabilized by the protein. In accordance with a deletion of 10 residues in the α-chain, the backbone around Å155 has no counterpart in the α-chain.

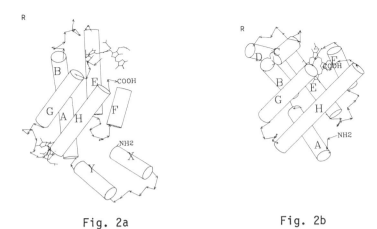

Fig. 2a   Fig. 2b

Fig. 2. Arrangement of helices and chromophores in C-PC ß-chain (a) and in myoglobin (b). Both figures were produced using a progam of Lesk and Hardman, 1982.

The globular part (helices A to H) of each C-PC subunit structure resembles closely the globin fold. In Fig. 2 the structures of the C-PC ß-chain and of myoglobin are displayed in the same orientation using the helix nomenclature for globins. Helices A to H of myoglobin possess closely related counterparts in C-PC. A major difference is the angle formed by helices A/E and G/H in C-PC (50°) compared with myoglobin (90°). Within the globin family these angles vary, however, also to 30° Lesk and Chothia, 1980. The attachment site of the tetrapyrroles $\alpha 84$ and $\beta \text{Å}'$ is topologically equivalent to the attachment of the haem via the distal histidine. Helices X and Y represent, however, unique features of C-PC. The sequence homology of C-PC and myoglobin was found to be 3.7 standard deviations above random alignment (much lower than values obtained from comparisons within the various globins) using the Mutation Data Matrix Dayhoff et al., 1979 as a measure for the similarity of the amino acids.

In crystals of C-PC from <u>Agmenellum quadruplicatum</u> (the structure analysis of which is underway in our laboratory in cooperation with M.L. Hackert, University of Texas, Austin) trimers are associated face to face forming hexamers. These hexamers as well as their crystal stack are very probably alike the discs and the rods observed in native phycobilisomes. Whereas the intra-hexamer contact is very extensive, there are only a few, weak inter-hexamer contacts. The increased order of some segments in the hexamer structure compared with the trimeric structure in <u>M. Laminosus</u> C-PC might be a consequence of the tight trimer-trimer association. The results of the X-ray analyses of C-PC are consistent with the phycobilisome models suggested by Moerschel et al., 1977 and others on the basis of electron micrographs after negative staining. In these models the rods are built from hexameric $(\alpha\beta)_6$-double-discs of height 60 Å, consisting of two $(\alpha\beta)_3$-trimers of height 30 Å stacked upon one another. The linker polypeptides Tandau de Marsac and Cohen-Bazire, 1977, additional components in phycobilisomes which are closely associated with phycobiliproteins, might extend through the central channel of the C-PC-trimers and hexamers.

From the distances and angles between the chromophores the dipole coupling between the various chromophores in phyucobilisome rods could be estimated. Accordingly, the light energy should be transferred towards the innermost β84-chromophores, which seem to be adaquately arranged to channel the energy towards the reaction center.

REFERENCES

1. T. Schirmer, W. Bode, R. Huber, W. Sidler and H. Zuber. In : Optical Properties and Structure of Tetrapyrroles, Eds. G. Blauer and H. Sund, pp. 445-449, de Gruyter, Berlin (1985).

2. W. Sidler, J. Gysi, E. Isker and H. Zuber, Hoppe-Seyler's Z. Physiol. Chem., 362:611-628 (1981).

3. W. Bode, T. Schirmer, Biol. Chem. Hoppe-Seyler, 366:287-295 (1985).

4. G. Frank, W. Sidler, H. Widmer and H. Zuber, Hoppe-Seyler's Z. Physiol. Chem., 359:1491-1507 (1978).

5. C. Kratky, C. Jorde, H. Falk and K. Thirring, Tetrahedron, 39:1859-1863 (1983).

6. H. Scheer and W. Kufer, Z. Naturforsch. 32c:513-519 (1977).

7. M.O. Dayhoff. In : Atlas of Protein Sequence and Structure, Ed. Dayhoff, M.O., 5:353-358, Biomedical Research Foundation, Washington (1978).

8. E. Moerschel, K.P. Koller, W. Wehrmeyer and H. Schneider, Cytobiologie, 16:118-129 (1977).

9. N. Tandeau de Marsac and G. Cohen-Bazire, Proc. Nat. Acad. Sci. USA, 74:1635-1639 (1977).

10. T. Schirmer, W. Bode, R. Huber, W. and Zuber, H. J.Mol.Biol., 184:257-277 (1985).

THE CRYSTAL STRUCTURE OF THE PHOTOSYNTHETIC REACTION CENTER FROM <u>RHODOPSEUDOMONAS VIRIDIS</u>

J. Deisenhofer and H. Michel

Max-Planck-Institut fuer Biochemie
D-8033 Martinsried, FRG

The photosynthetic reaction center (RC) from the purple photosynthetic bacterium <u>Rhodopseudomonas viridis</u> consists of the four protein subunits H ("hea<u>vy</u>"), M (med<u>ium</u>"), L ("light"), and cytochrome. It contains as prosthetic groups four bacteriochlorophyll-b (BChl-b), two bacteriopheophytin-b (BPh-b), one non-heme iron, two quinones, and four heme groups. The total molecular weight of the RC complex is about 140000. In vivo, the RC is located in the inner membrane of the bacterium and, after absorption of light energy, performs the primary charge separation of the photosynthetic process. It is know from spectroscopic studies that an electron is transferred from a primary donor (a "special pair" of BChls) to a primary acceptor (a quinone) <u>via</u> a BChl and a BPh.

The X-ray analysis at 3 Å resolution of well ordered crystals (1) of the RC from <u>Rps. viridis</u> was started with collection of about 55000 independent re<u>flections f</u>rom native crystals. The phase problem was solved using the method of multiple isomorphous replacement with heavy atom compounds. Five different heavy atom derivatives allowed the determination of phases to 3 Å resolution. The electron density distribution calculated from these phases, and from the measured structure factor amplitudes was improved by solvent flattening. Interpretation of this map led to the construction of atomic models of the prosthetic groups, and of the protein subunits in the RC complex. In the following we briefly summarize results of the X-ray work which have been described in more detail elsewhere (2), (3), (4).

Fig.1 shows the four BChl-b molecules, the two BPh-b molecules, one quinone (menaquinone), the non-heme iron, and the four heme groups located in the electron density map. In Fig. 2 these molecules are shown together with the polypeptide chains of the RC's protein subunits L, M, H, and cytochrome. the heme groups are covalently linked to the cytochrome subunit. The remaining chromophores are associated with the subunits L and M in the central part of the RC. In this region the structure shows a high amount of local twofold symmetry which relates large parts of the polypeptide chains of L and M, and also the pyrrole ring systems of corresponding BChl-bs and BPh-bs. the non-heme iron sits on this L-M local diad.

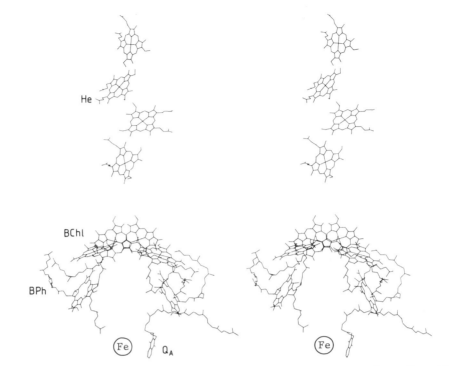

Figure 1: Stereo pair showing the arrangement of the chromophores in the RC.

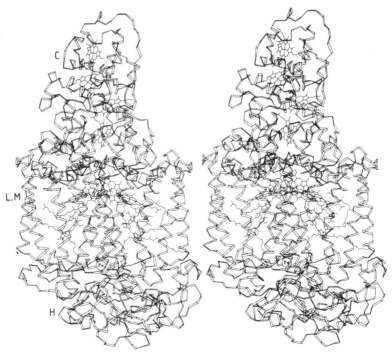

Figure 2: Stereo pair showing the structure of the protein subunits (represented as ribbons), together with the chromophores in the RC.

Close to the L-M local diad two BChl-bs are in intimate contact, and form the "special pair", the primary electron donor. Their pyrrole rings I are stacked with a distance of about 3 Å; the acetyl groups at these rings interact with the Mg^{2+} ions of the other BChl-bs. On either side of the L-M local diad an "accessory" BChl-b is in contact with the special pair. The BPh-b are located next to these accessory BChl-bs. So, we find in the RC two strings of chromophores, which both originate at the special pair, and appear to be suitable as electron pathways. The local twofold symmetry ends at the level of the quinones : only the tightly bound quinone (Q_A) was found in the electron density map near the BPh-b positioned on the L-side of the L-M complex. Most of the second quinone (Q_B) was lost during preparation of the RCs (Sinning and Michel, unpublished). Binding studies in RC crystals with o-phenanthroline, terbutryn, and ubiquinone-1 revealed a common binding site for these three compounds which we believe to be the Q_B binding site. This Q_B site is related to the Q_A site by the L-M local diad. It remains an intriguing question why the electron pathway in the RC leads almost exclusively from the special pair first to Q_A, and then to Q_B.

The four heme groups of the cytochrome form an almost linear string leading from the tip of the subunit towards the special pair. Two pairs of heme group are related by another local diad; about one third of the polypeptide chain also obeys this internal local symmetry of the cytochrome subunit. Outstanding structural feaures of each of the subunits L and M are five helices running in the general direction of the L-M local diad with lengths between 24 and 30 residues. From the hydrophobic character of these helical segments we conclude that they span the bacterial membrane. On both sides of the helix region of L and M the polypeptide segments connecting the transmembrane helices, and the terminal segments form flat surfaces perpendicular to the L-M local diad. The cytochrome is bound to the surface close to the special pair; the carboxy-terminal globular domain of the H-subunit is in contact with the surface that is near the non-heme iron and the quinone. Near its amino-terminus the H-subunit forms another transmembrane helix; the remainder of this subunit is the only part of the RC with a significant amount of beta-sheet.

Crystallographic refinement of the RC model at 2.9 Å resolution is in progress.

REFERENCES

1. Michel, H., J. Mol. Biol. 158:567 (1982)
2. Deisenhofer, J., Epp, O., Miki, K., Huber, R. and Michel, H., J. Mol. Biol. 180:385 (1984)
3. Deisenhofer, J., Michel, H. and Huber, R., Trends Biochem. Sci. 10:243 (1985)
4. Deisenhofer, J., Epp, O., Miki, K., Huber, R. and Michel, H., Nature 318/618 (1985).

STRUCTURE AND FUNCTION OF SOME ELECTRON TRANSFER-PROTEINS AND-COMPLEXES MONOHEME C-TYPE-AND MULTIHEME C_3-CYTOCHROMES

Richard Haser, Michel Frey and Francoise Payan

CRMC-2CNRS, Campus de Luminy, case 913
13288 Marseille cedex 09, France

INTRODUCTION

Electron transfer processes are essential mechanisms in many biochemical pathways like photosynthesis and respiration in general. Among the components of the different electron flow systems, a number of monoheme c-type cytochromes are well characterized, both in terms of sequence and of high-resolution three-dimensional structures.

Since the first crystal structure of a cytochrome c, from horse heart (DICKERSON et al., 1971), the structures of nine (three from eukaryotic sources and six from bacterial) c-type cytochromes have been determined to atomic resolution, including the recent investigation of *Azotobacter* cytochrome c_5 (CARTER et al.,1985). All these structures share what is known as "the mitochondrial-type cytochrome c fold", which accounts for the highly conserved amino acid residues and which suggests evolutionary relationships between these proteins. Indeed on the basis of sequence and structure comparisons, together with what has been learnt about bacterial metabolism, it was possible to trace the evolution of various prokaryotic families and to propose that the cytochrome c ancestor molecule was one of the small bacterial cytochromes involved in photosynthesis (DICKERSON, 1980).

The c-type cytochromes have been the subject of extensive reviews elsewhere (MEYER and KAMEN, 1982; MATHEWS, 1985). Therefore we will focus here on some recent results concerning-the models of electron transfer complexes, -some mechanistic proposals related to observed heme-heme interactions, -the structural features characteristic of the multiheme c_3 family and its apparent relations with other cytochromes. Moreover, crystallographic, spectroscopic and electrochemical studies of the four-heme cytochrome c_3 show that this electron carrier offers itself as an excellent probe to investigate the mechanisms of intra-and inter-protein electron transfer.

MODELS OF ELECTRON TRANSFER COMPLEXES WITH HEMOPROTEINS

Although many electron-transferring proteins, representative of different structural classes, have been determined at high resolution, there is, up to now, no crystal structure available for an electron transfer complex, as between two hemoproteins, or between a hemoprotein and an other redox macromolecule (ferredoxin, flavodoxin, rubredoxin...). Clearly such a structure would be of importance for a better understanding of the intermolecular interface which favors the electron exchange within the complex.

The first approach to get a coherent picture of an electron transfer complex between two hemoproteins, on the basis of their three-dimensional structures, concerned the interaction of cytochrome b_5 with cytochrome c (SALEMME, 1976). This approach uses computer graphics simulation to optimize the electrostatic interactions and the complementary fit of the molecular surfaces. The resulting inter-molecular interface is formed around the exposed heme edges, through specific interactions between negatively charged residues on cytochrome b_5, i.e. Glu 48, Glu 44, Asp 60, the most exposed propionate group and, respectively lysyl residues 13, 27, 72, 79 on cytochrome c. This hydrogen bonding pattern leads to a nearly coplanar arrangement of the two porphyrin rings, with their edges separated by about 8 Å (Fig. 1).

Fig. 1 The hypothetical model between cytochrome b_5 and cytochrome c (SALEMME, 1976). Residues involved in charge-charge interactions are shown. Reproduced from MATHEWS (1985).

A variety of experimental studies have provided support for this model, for example : kinetic sudies of lysine-substituted derivatives of cytochrome c (NG et al., 1977) and recent NMR studies of the complex in solution (ELEY and MOORE, 1983). However sequence alignments of b_5-like proteins (flavocytochrome b_2 and sulfite oxidase) suggest that, if Glu 44 and the heme propionate group are key positions for coupling these proteins with cytochrome c, the electrostatic interactions may not play a large role in stabilizing the complexes (GUIARD and LEDERER, 1979).

Within the last few years, and using a similar approach, several other putative electron transfer complexes have been proposed : cytochrome c peroxidase.cytochrome c (POULOS and KRAUT, 1980; SHERIDAN et al., 1982); cytochrome b_5 with the α and β subunits of methemoglobin (POULOS and MAUK, 1983); cytochrome c.flavodoxin (SIMONDSEN et al., 1982).

In all cases it turns out that complex formation involves charge and surface complementarities and a nearly parallel alignment of the prosthetic groups. In the above complexes, it should be noted that some of the lysyl residues of the cytochrome c binding area (residues 13, 72, 25, 27) are indeed among the crucial amino acids defining the interaction domain of cytochrome c with its oxidase and reductase (KOPPENOL and MARGOLIASH, 1982).

In the cytochrome c peroxidase.cytochrome c complex the heme edges are about 16 Å apart and nearly coplanar. Subsequent chemical modification and physical studies have provided much support for the proposed model. Of particular interest, cross-linking between the two proteins has led to a covalent one-to-one association which displays the features of a virtual electron transfer complex (WALDEMEYER and BOSSHARD, 1985). Moreover differential chemical modification of carboxyl groups of cytochrome c peroxidase have demonstrated that some of these groups are of great importance for the recognition of the cytochrome c active site domain (BECHTOLD and BOSSHARD, 1985). Recently also, covalent cross-linking of cytochrome c and flavodoxin has been performed and it is suggested that the cross-linked complex is similar to the kinetically significant entity (DICKERSON et al., 1985). Clearly these impressive results increase the hope to isolate and characterize stable analogues of precursor complexes to electron transfer, suitable for crystallization experiments and for further three-dimensional structure determinations.

We summarize in Table 1. some geometrical parameters concerning the prosthetic group arrangements in different complexes of electron-transferring proteins.

Table 1. GEOMETRICAL PARAMETERS CHARACTERISTIC OF ELECTRON TRANSFER COMPLEXES INVOLVING HEMOPROTEINS

	$b_5.c$ (a)	$HEM.b_5$ (b)	$FLA.c$ (c)	$CCP.c$ (d)	$c_3.c_3$ (e)
INTERMOLECULAR					
FE-FE DISTANCE (Å)	16.	16.		24.6	18.9 (2-4) 19.2 (1-4) 21.7 (2-3)
EDGE-TO-EDGE DISTANCE (HEME TO HEME OR TO FLAVIN)	8.4	7.-8.	5.	16.5	14. (1-4)
ANGLE (°) BETWEEN HEME PLANES (OR BETWEEN HEME AND FLAVIN PLANES)	15.	coplanar	30.	coplanar	35. (1-4)
INTRAMOLECULAR					
FE-FE DISTANCE D(Å)					10.9<D<17.3 (1-3) (1-4)
EDGE-TO-EDGE DISTANCE E(Å)					4.<E<14. (1-3) (1-4)

(a) SALEMME, 1976. Complex between cytochrome b_5 and cytochrome c
(b) POULOS and MAUK, 1983. Interaction of cytochrome b_5 with the α or β subunit of Methemoglobine (HEM) shows that in both cases the orientation of the heme groups and the distance between their edges are essentially identical.
(c) SIMONDSEN et al., 1982. FLA = Flavodoxine
(d) POULOS and KRAUT, 1980. CCP = Cytochrome c peroxidase
(a) (b) (c) and (d) are computer-graphics generated models.
(e) PIERROT et al., 1982. Data from the crystal structure of the four heme cytochrome c_3 from *Desulfovibrio desulfuricans* Norway. In brackets are indicated the corresponding heme numbers (see also Figs. 2, 3)

Before discussing the possible functional implications of the inter-and intra-molecular interactions, we shall first describe the major structural features of the multiheme cytochrome c_3.

THE ELECTRON TRANSPORT CHAIN IN DESULFOVIBRIO.CYTOCHROME C_3

Cytochrome C_3 is a four heme periplasmic protein found in all anaerobic sulfate-reducing bacteria belonging to the genus *Desulfovibrio* (for a recent review on the bioenergetics of *Desulfovibrio*, ODOM and PECK, 1984). It acts as the natural electron donor and acceptor for hydrogenase, an iron-sulfur cluster containing enzyme responsible for the oxidation of hydrogen. This reaction is coupled to the generation of a proton gradient that is utilized for the synthesis of ATP by means of a reversible ATPase.

Other redox components, belonging to the above electron-transport system are : ferredoxin, flavodoxin, rubredoxin, a monoheme cytochrome (C_{553}), and at least three new types of low-molecular-weight redox proteins unique to *Desulfovibrio*.

It has been demonstrated that the reduction of ferredoxin, flavodoxin and rubredoxin by hydrogenase plus H_2 requires cytochrome C_3 (BELL et al., 1978). More recently, ESR (CAMMACK et al., 1985) and preliminary NMR studies (GUERLESQUIN et al., 1985a) have shown the formation of a one-to-one complex between cytochrome C_3 and ferredoxin I, both components having been isolated from the same species (*Desulfovibrio desulfuricans*, strain Norway, hereafter D.d.N). In this context our overall goal is to elucidate the spatial structures of several electron-transferring proteins from the same organism, the first in this case having concerned cytochrome c_3. Preliminary crystallographic data of one of the redox partners for this cytochrome, ferredoxin, have been reported (GUERLESQUIN et al., 1983).

CYTOCHROME C_3

The amino acid sequences of the protein from seven species have been determined, including the *Desulfuromonas acetoxidans* cytochrome (c_7) containing three hemes. The sequence of cytochrome c_3 D.d.N is shown hereafter (Fig. 2). For clarity and further structure comparisons we have adopted the heme numbering previously proposed (MEYER and KAMEN, 1982). Therefore the ancient (french or japanese) heme labels should be abandoned...

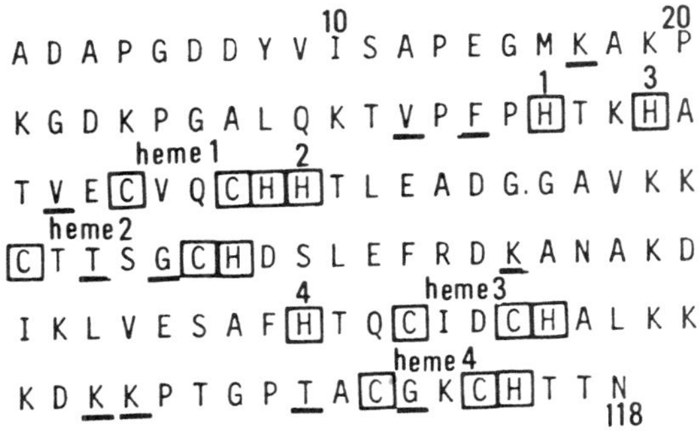

```
                        10                    20
        A D A P G D D Y V I S A P E G M K A K P
                              1           3
        K G D K P G A L Q K T V P F P[H]T K[H]A
             heme 1      2
        T V E[C]V Q[C H H]T L E A D G.G A V K K
          heme 2
        [C]T T S G[C H]D S L E F R D K A N A K D
                        4         heme 3
        I K L V E S A F[H]T Q[C]I D[C H]A L K K
                         heme 4
        K D K K P T G P I A[C]G K[C H]T T N
                                           118
```

Fig. 2 : Amino acid sequence of cytochrome c_3 D.d.N (BRUSCHI, 1981). The heme attachment sites and the iron ligands (His in all cases) are indicated. The residues we consider as highly conserved on the basis of sequence alignment and of the three dimensional structure comparisons are underlined. Throughout the paper we will use the D.d.N. numbering for the other cytochromes c_3.

The crystal structures of cytochrome c_3 D.d.N (Mr = 15066) (HASER et al., 1979; PIERROT et al., 1982) and from *Desulfovibrio vulgaris* Miyasaki (Mr = 13995) (hereafter D.v.M; HIGUCHI et al., 1979; HIGUCHI et al., 1984) have been determined, respectively at 2.5 Å and 1.8 Å resolution. The structure of cytochrome c_3 is illustrated on figure 3 which shows the α-carbon backbone together with the heme core active center.

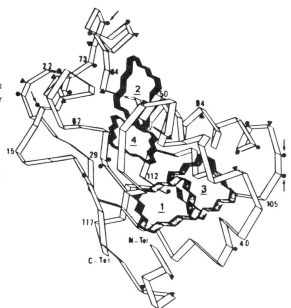

Fig. 3 : Schematic diagram of cytochrome c_3 D.d.N. Cα positions of all charged amino acids are labeled :
• basic side chain;
▼ acidic; and → highly conserved lysine residues. Iron to iron distances are Fe_1-Fe_2 12.8 Å; Fe_2-Fe_3 16.8 Å; Fe_1-Fe_3 10.9 Å; Fe_4-Fe_2 16.3 Å; Fe_1-Fe_4 17.3 Å; Fe_3-Fe_4 12.7 Å.

As in cytochrome c, each heme group is bound to the polypeptide chain through thioether linkage to two cysteine side chains. However, unlike in cytochrome c, and like in cytochrome b_5 the iron ligands are both histidyl residues. The overall folding may best be described as forming two domains connected by a single peptide (residues 71-73). Each domain accomodates one heme and the interdomain groove encompasses the two other redox centers (hemes 2 and 3).

The conformation of the polypeptide chains of the two c_3's (D.d.N and D.v.M) differ to a significant extent in a few regions, an observation which appears consistent with the poor degree of sequence homology among these two proteins (about 28% identity). However despite these differences, the geometries of the heme clusters are remarkably conserved : comparison with the D.v.M.c3 structure shows that the relative positions and orientation of the porphyrins are very similar and that the relative iron-iron distances differ only by 0.4 Å on the average.

An other interesting common feature of the two stuctures concerns the solvent accessibility of the different heme groups which, on average, is much higher (~ 140 Å^2) than that of heme in c-type cytochromes (~ 42 Å^2). Indeed the accessible surface areas of the four heme groups have been calculated and the values, in the order of the heme numbering, are 136, 168, 136 and 127 Å^2 for the structure of D.v.M.c$_3$ (HIGUCHI et al., 1984). In both models heme 2, which is one of the interdomain redox centers, appears to be the more exposed one to the external medium. This particularity might confer a distinct role to heme 2, for example in the recognition process of other redox partners. The different heme environments have been described in detail and in the next section we will focus on some structural features which may be relevant to the intra-and inter-molecular heme-heme electron transfer.

HEME-HEME INTERACTIONS

The hypothetical models of various hemoprotein complexes which have been briefly described above, fall into two categories according to the distances between the active redox sites (Table 1) :

- those for which the redox groups are in close proximity (<10 Å), like in the b$_5$.c complex, the hemoglobin.b$_5$, the flavodoxin.cytochrome c.
- those which are characterized by widely separated hemes like in cytochrome c peroxidase.cytochrome c (CCP.c).

As we will see cytochrome c$_3$ with its four heme sites may be considered as belonging to both categories.

For complexes of the first type, it appears to be generally accepted that direct electron transfer operates between the two partners via a classical outer-sphere mechanism. Such a mechanism requires, at least during the electron transfer event, some orbital overlap of the exposed edges of the two reacting hemes. Optimal overlap could be facilitated by protein conformational flexibility in the intermolecular region. The outer-sphere electron transfer process has been demonstrated for cytochrome c, in reactions with numerous nonphysiological redox agents (FERGUSON et al., 1979 and references therein). If direct orbital coupling is precluded, the electron propagation between the porphyrin moieties may be ensured by what is known as a short-range tunneling mechanism (HOPFIELD, 1974).

In the case of the CCP.c complex, the distance between the heme moieties (~ 16 Å) does not enable direct electron transfer between them. It has been suggested that bridging groups belonging to the intervening protein loops could contribute to the formation of an extended πorbital system which couples the two hemes (POULOS and KRAUT, 1980). The proposed supramolecular conduction orbital is the result of overlapping π-orbitals contributed by the porphyrins and by the parallel conjugated rings of Phe 82 (from cytochrome c) and His 181 (from CCP). In addition, an intricate hydrogen bonding pattern and a number of ionic interactions are observed in the intermolecular region of the model. All these interactions are probably important for mediating the transfer of an electron from cytochrome c to cytochrome c peroxidase (POULOS, 1982). Indeed, the crucial importance of His 181 for the heme-heme communication has been recently demonstrated by chemical modification experiments (BOSSHARD et al., 1984). On the other hand, site directed mutagenesis of cytochrome c has shown that the invariant Phe 82 (which has been changed into Ser, Tyr or Gly) is not an absolute requirement for electron transfer in the CCP.c system, although its substitution produces a decrease of the reduction potential by as much as 50 mV (PIELAK et al., 1985). Besides this experimental evidence of an aromatic side chain beeing invoved in the control of the polarity of the heme environment, there is also strong evidence for a major influence of at least one heme propionate on the electrochemical behavior of cytochrome b_5 (REID et al., 1984). Clearly the electron transfer reactions between hemoproteins are higly depending on the stuctural features which control the reduction potentials.

With cytochrome c_3 we will see hereafter that we have a system which is designed to display both intra-and inter-molecular heme-heme interactions. A number of physicochemical techniques such as ESR (DER VARTANIAN, 1973; CAMMACK et al., 1985), NMR (SANTOS et al., 1984; GUERLESQUIN et al., 1985; KIMURA et al., 1985), Mossbauer spectrometry (UTUNO et al., 1980), spectral and kinetic studies (YAGI, 1984), have been used to study the electron exchange processes in cytochrome c_3 from different species. In fact the question of intra-or/and inter-molecular electron transfer continues to give rise to much controversy, probably because of the weakness of the above techniques to discriminate between the two mechanisms. A strong indication that both heme-heme interactions do occur is the high electrical conductivity of cytochrome c_3 (KIMURA et al., 1979).

On the basis of the two three-dimensional structures available for cytochrome c_3 we first survey some possibilities of intra-molecular electron exchange. (A more detailed description will be published elsewhere). To clarify the discussion we use the following linear representation of the interconnected heme groups in the isolated molecule (Fig. 4).

									84	101		
His−Fe−His	−	His−Fe−His.	Phe.	His−Fe−	His	...	His−Fe−His					
n° 67 (2)	49	48 (1)	36	34	39	(3)	96	89	(4)	115		
d(Å) Fe−Fe	12.8		10.9				12.7			C_3 DdN		
	12.2		11.0				12.0			C_3 DvM		

Fig. 4 : Linear representation of the interconnected heme groups. The helix (84-101) in D.d.N.c_3 corresponds closely to the two helices 82-89 and 91-97 found in D.v.M.c_3.

For example, in both structures the arrangement of hemes 1 and 3 is nearly identical and the strictly invariant (even in cytochrome c_7) Phe 34 is found in the same location, and almost parallel to His 39 as shown in Fig. 5.

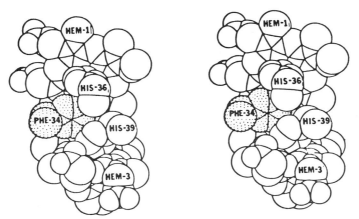

Fig. 5 : Stereo space-filling model of the hemes 1 and 3 and the intervening side chains. Reproduced by permission from HIGUCHI et al., (1984).

This close proximity gives rise to an extensive π-orbital overlap. In addition, the iron ligands, Phe 34, and some of the heme propionates are involved in a hydrogen bonding network, similar in both structures and probably important for maintaining the proper orientations of the different groups and for the control of the reduction potentials. Results on the reduction of cytochrome c_3 by hydrogenase suggest also that the above hemes are the most strongly interacting ones (YAGI, 1984). An other intramolecular interaction is "seen" between hemes 2 and 1. The highly conserved connecting segment His 48 His 49 brings histidyl ring 48 close to the porphyrin ring of heme 2 (~ 5.5 A°) and the closet heme edge-to-edge distance is 7 Å. Here again the structural features observed favor an outer sphere electron exchange mechanism between the two hemes.

In D.d.N. cytochrome c_3 the porphyrin rings of hemes 3 and 4 are nearly parallel to the α-helix running from residue 84 to residue 101 (Fig. 6). Moreover the helix provides for each heme an iron ligand (His 89 and His 96).

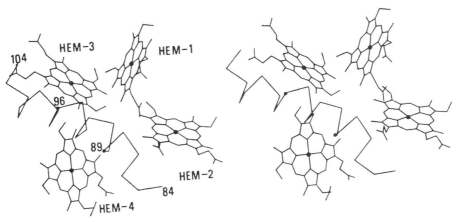

Fig. 6 Stereo diagram of the hemes 3 and 4 (in D.d.N.c_3) packing against the α-helix (residues 84-101). The black dots indicate the iron atoms and α-carbons of the histidyl iron ligands.

This configuration, despite a short inter-iron distance (12.7 Å), does not lead to a π-orbital overlap between the heme groups. However we would like to suggest that a significant interaction could be mediated by the strong internal field generated by the α-helix dipole and through the iron ligands which are provided by the helix. By analogy with what is observed for many enzymes, this "active site helix" might influence the reactivities of the nearby heme moieties and therefore could modulate the electron transfer rates (for a recent review on the role of the α-helix dipole in protein function and structure, see HOL, 1985). Indeed, on the basis of electrostatic calculations, it has been proposed that within the mitochondrial proton pump, electrons and protons may travel along the axis of α-helices (OVCHINNINKOV and UKRAINSKII, 1979).

It is important to note here that in D.v.M.c_3 the polypeptide chain corresponding to the D.d.N.c_3 helix is organised in two short helices (82-89 and 91-97). We suggest that this difference might well be responsible for the differences in heme redox potentials observed for the two cytochromes. In D.v.M.c_3 the individual midpoint potentials are -240 mV, -297 mV, -315 mV and -357 mV (KIMURA et al., 1985), whereas for D.d.N.c_3 the following values are reported -150mV, -270mV, -325mV and -355mV (GAYDA et al., 1985). The significant increase in redox potential for one of the hemes may be related to a weaker electric field in the vicinity of the porphyrin ring and produced by the interrupted helix after residue 89 (D.d.N. numbering) in D.v.M. c_3. All the above proposals for the intra-molecular interactions may be put together to produce a plausible picture of the electron exchange within cytochrome c_3 D.d.N. Here we will refer again to figures 3 and 4. As heme 2 has the highest exposure to the external medium and likely the lowest redox potential value (for a general discussion on oxidation-reduction potentials in various cytochromes see MATHEWS, 1985), the electrons would enter cytochrome c_3 preferentially via the porphyrin ring of heme 2 and further would reduce the other heme groups, following the sequence shown in figure 4. This scheme, although still speculative, is consistent with the suggested two sites (about 30 Å apart) on the surface of cytochrome c_3 for binding other redox partners (HASER et al.,1981; PIERROT et al., 1982). The proposed interaction domains contain Lys 17 and Lys 75 in the vicinity of heme 2 and the highly conserved Lys 103, Lys 104 near heme 3 (Fig. 3). It should be noted that the latter lysine residues are found in the C-terminal region of the α-helix which contains the remarkable strech of sequence Lys 99 Lys 100 Lys 101 Asp 102 Lys 103 Lys 104.

Of particular interest are also Lys 100 and Lys 101 which interact with the heme 2 propionates of another cytochrome c_3 molecule. These complementary charge interactions are similar to those observed in the hypothetical hemoprotein complexes described above. Moreover the structural consequence of these interactions is a close approach between heme groups at the molecular interface and most striking, the tendency to bring heme groups in nearly parallel alignment, a common feature to all computer-generated hemoprotein models studied far. In this repect, the cytochrome c_3 structure provides the first compelling experimental evidence for intermolecular interactions leading to nearly parallel heme groups, an arrangement most probably responsible for the high electrical conductivity observed for this electron-transferring protein.

However inspection of the intervening medium between the closest hemes does indicate that direct electron transfer via the assistance of a π-orbital cloud is not possible, nor are present specific groups, which like in the CCP.c model, could mediate the electron exchange. It might be that the sulfur atoms of the exposed heme thioether linkages (as observed in both c_3 structures) play a role in the intermolecular electron propagation.

STRUCTURE RELATIONSHIPS BETWEEN CYTOCHROME C_3 AND OTHER CYTOCHROMES

Inspection of various hemoprotein complexes leads us to underline a striking similarity between the heme intermolecular cluster of cytochrome c_3 (D.d.N.) and the heme arrangement in the c-type cytochrome of the reaction center complex of the purple bacteria *R. viridis*, the structure of this complex having been determined recently at the atomic level (DEISENHOFER et al., 1984; 1985). In figure 7 we compare both arrangements.

The similarity holds for the relative heme orientations and for the inter-iron distances. Indeed a detailed comparison based on refined coordinates should be performed in order to see how different (or close) these spatial arrangements are and which other structural properties are eventually in common.

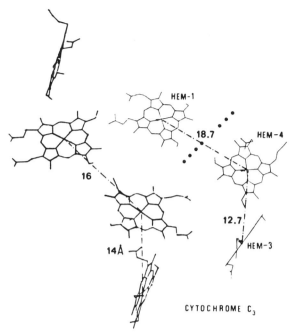

Fig. 7 : Comparison between the heme arrangements in the *R. viridis* reaction center complex (on the left) and in cytochrome c_3 (D.d.N.). The dotted line indicates the intermolecular interface.

Nevertheless it is interesting to note that similar devices have been tailored to allow efficient electron transfer in such different organisms as strictly anaerobic bacteria and photosynthetic bacteria.

ACKNOWLEDGMENTS

This work has been supported by grants from CNRS and from MRT (Ministère de la Recherche et de la Technologie).

REFERENCES

Bechtold, R. and Bosshard, H.R. (1985) J. Biol. Chem. 260, 5191-5200.
Bell, G.R., Lee, J.P., Peck, H.D., Jr. and Le Gall, J. (1978) Biochimie, 60, 315-320.
Bosshard, H.R., Bänziger, J., Hasler, T. and Poulos, T.L. (1984) J. Biol; Chem. 259, 5683-5690.
Bruschi, M. (1981) Biochim. Biophys. Acta, 671, 219-226.
Cammack, R., Fauque, G., Moura, J.J.G. and Le Gall, J. (1984) Biochim. Biophys. Acta, 784, 68-74.
Carter, D.C., Melis, K.A., O'Donnell, S.E., Burgess, B.K., Furey, W.F., Wang, B-C. and Stout, C.D. (1985) J. Mol. Biol. 184, 279-295.
Deisenhofer, J., Epp, O., Miki, K., Huber, R. and Michel, H. (1984) J. Mol. Biol. 180, 385-398.
Deisenhofer, J., Michel, H. and Huber, R. (1985) Trends Biochem. Sci. 10, 243-248.
Der Vartanian, D.V. (1973) J. Magn. Reson. 10, 170-178.
Dickerson, J.L., Kornuc, J.J. and Rees, D.C. (1985) J. Biol. Chem. 260, 5175-5178.
Dickerson, R.E., Takano, T., Eisenberg, D., Kallai, O.B., Samson, L., Cooper, A. and Margoliash, E. (1971) J. Biol. Chem. 246, 1511-1535.

Dickerson, R.E. (1980) Sci. Am. 242, 99-110.
Eley, C.G.S. and Moore, G.R. (1983) Biochem. J. 215, 11-21.
Ferguson-Miller, S., Brautigan, D.L. and Margoliash, E. (1979) In The Porphyrins, Vol. VII B (Ed. D. Dolphin) pp. 149-240, Academic Press, N.Y.
Gayda, J.P., Bertrand, P., More, C., Guerlesquin, F. and Bruschi, M. (1985) Biochim. Biophys. Acta, 289, 262-267.
Guerlesquin, F., Noailly, M. and Bruschi, M. (1985a) Biochem. Biophys. Res. Comm. 130, 1102-1108.
Guerlesquin, F., Bruschi, M. and Wüthrich, K. (1985b) Biochim. Biophys. Acta, 830, 296-303.
Guerlesquin, F., Bruschi, M., Astier, J.P. and Frey, M. (1983) J. Mol. Biol. 168, 203-205.
Guiard, B. and Lederer, F. (1979) J. Mol. Biol. 135, 639-650.
Haser, R., Pierrot, M., Frey, M., Payan, F., Astier, J.P., Bruschi, M. and Le Gall, J. (1979) Nature, 282, 806-810.
Haser, R., Pierrot, M., Frey, M., Payan, F. and Astier, J.P. (1981) In Stuctural Aspects of Recognition and Assembly in Biological Macromolecules (Balaban ed.) pp. 213-223. International Science Services, Philadelphia.
Higuchi, Y., Kusunoki, M., Yasuoka, N., Kakudo, M. and Yagi, T. (1981) J. Biochem., Tokyo, 90, 1715-1723.
Higuchi, Y., Kusunoki, M., Matsuura, Y., Yasuoka, N. and Kakudo, M. (1984) J. Mol. Biol., 172, 109-139.
Hol, W.G.J. (1985) Prog. Biophys. molec. Biol., 45, 149-195.
Hopfield, J.J. (1974) Proc. natn. Acad. Sci. U.S.A. 71, 3640-3644.
Kimura, K., Nakahara, Y., Yagi, T. and Inokuchi, H. (1979), J. Chem. Phys. 70, 3317-3323.
Kimura, K., Nakajima, S., Niki, K. and Inokuchi, H. (1985) Bull. Chem. Soc. Jpn. 58, 1010-1012.
Koppenol, W.H. and Margoliash, E. (1982) J. Biol. Chem. 257, 4426-4437.
Mathews, F.S. (1985) Prog. Biophys. Molec. Biol. 45, 1-56.
Meyer, T.E. and Kamen, M.D. (1982) Advances in Protein Chemistry, 35, 105-212.
Moura, J.J.G., Santos, H., Moura, I., Le Gall, J., Moore, G.R., Willams, R.J.P. and Xavier, A.V. (1982) Eur. J. Biochem. 127, 151-153.
NG, S. Smith, M.B., Smith, H.T. and Millett, F. (1977) Biochemistry 16, 4975-4978.
Odom, J.M. and Peck, H.D. (1984) Ann. Rev. Microbiol. 38, 551-592.
Ovchinnikov, A.A. and Ukrainskii, I.I. (1979) Doklady Acad. Sci. USSR 244, 751-754.
Pielak, G.J., Mauk, A.G. and Smith, M. (1985) Nature 313, 152-154.
Pierrot, M., Haser, R., Frey, M., Payan, F. and Astier, J.P. (1982) J. Biol. Chem. 257, 14341-14348.
Poulos, T.L. and Kraut, J. (1980) J. Biol. Chem. 255, 10322-10330.
Poulos, T.L. and Mauk, A.G. (1983) J. Biol. Chem. 258, 7369-7373.
Poulos, T.L. (1982) in Molecular Structure and Biological Activity (Griffin, J.F. and Duax, W.L., eds.) Vol. 2, pp 217-221.
Reid, L.S., Mauk, R.M. and Mauk, A.G. (1984), J. Am. Chem. Soc. 106, 2182-2185.
Salemme, F.R. (1976) J. Mol. Biol. 102, 563-568.
Sheridan, R.P., Levy, R.M. and Salemme, F.R. (1982) Proc. natn. Acad. Sci. U.S.A. 79, 4545-4549.
Simondsen, R.P., Weber, P.C., Salemme, F.R. and Tollen, G. (1982) Biochemistry 21, 6366-6375.
Waldemeyer, B. and Bosshard, H.R. (1985) J. Biol. Chem. 260, 5184-5188.
Yagi, T. (1984) Biochim. Biophys. Acta, 767, 288-294.

AUTHOR INDEX

Alber, T., 251
Anderson, J.E., 319
Arn, R., 337
Arnold, E., 263

Baker, E.N., 251
Baker, T., 179
Baldwin, J.M., 101
Bell, J.A., 251
Berriman, J., 345
Blow, D.M., 241
Blundell, T.L., 153
Bode, W., 417
Bolognesi, M., 413
Branden, C.I., 359
Brick, P., 241
Brown, K.A., 241
Brown, R.S., 345
Burnett, R.M., 309

Ceska, T.A., 101
Chow, M., 281
Clinger, K., 403
Colman, P.M., 373
Cook, S.P., 251
Cooper, S., 153

Daopin, S., 251
de Vos, A., 395
Deisenhofer, J., 421
Dickerson, R.E., 209
Dodson, E., 179
Dodson, G., 179

Elliott, R., 167
Erickson, J.W., 263
Ernst, S.R., 403

Fersht, A.R., 241
Filman, D.J., 281
Finazzi-Agro, A., 413
Fourme, R., 27
Frankenberger, E.A., 263
Frey, M., 425

Garavito, R.M., 3
Getzoff, E.D., 131

Giegé, R., 15
Glaeser, R.M., 101
Glover, I.D., 153
Goodfellow, J.M., 167
Goodsell, D., 209
Gray, T.M., 251
Griffith, J.P., 263

Hackert, M.L., 403
Harrington, M., 293
Harrison, S.G., 319
Haser, R., 425
Hecht, H.J., 263
Heinemann, U., 337
Helliwell, J.R., 45
Henderson, R., 101
Hendrickson, W.A., 81
Hodgkin, D., 179
Hogle, J.M., 281
Hol, W.G.J., 223
Hosur, M.V., 293
Howell, P.L., 167
Hubbard, R., 179
Huber, R., 417
Husain, J., 153

Jacrot, B., 117
Jansonius, J.N., 229
Jenkins, J.A., 3
Johnson, J.E., 263
Johnson, J.E., 293
Jones, T.A., 125

Kahn, R., 27
Kamer, G., 263
Kim, S.H., 395
Kopka, M.L., 209

Ladenstein, R., 413
Laver, W.G., 373
Lehn, J.M., 193
Leonard, K., 345
Levitt, M., 197
Luo, M., 263

Maslowska, M., 337
Matthews, B.W., 251

Messerschmidt, A., 413
Michel, H., 421
Mosser, A.G., 263

Oefner, C., 327
Olson, A.J., 131

Pahler, A., 337
Parks, E.H., 403
Payan, F., 425
Pitts, J.E. 153
Pjura, P., 209
Podjarny, A.D., 63
Poljak, R.J., 365

Quiocho, F.A., 385

Rabinovich, D., 353
Ranghino, G., 193
Roberts, M.M., 309
Romano, S., 193
Rossmann, M.G., 263
Rueckert, R.R., 263

Sack, J.S., 385
Saenger, W., 337
Schirmer, T., 417
Schmidt, T., 293
Sharon, R., 197
Sherry, B., 263
Sidler, W., 417

Snell, E.E., 403
Stauffacher, C.V., 293
Sternberg, M.J.E., 141
Storey, M.A., 385
Suck, D., 327

Tainer, J.A., 131
Threharne, A.C., 153
Tikle, I.J., 153

Usha, R., 293

Van Heel, M. 89
Van Oostrum, J., 309
Varghese, J.N., 373
Vriend, G., 263
Vyas, N.K., 385

Weaver, L.H., 251
Webster, R.G., 373
Wilson, K., 251
Winkler, F.K., 345
Winter, G., 241
Wipf, G., 193
Wood, S.P. 153
Wozniak, J.A., 251

Yoon, C., 209

Zuber, H., 417

SUBJECT INDEX

Accuracy in X-ray measurements, 47
Adenovirus,
  architecture of, 309
  outer capsid organization of, 311
Amino acid sequence,
  Protein structure prediction from, 141
Anisotropic thermal motion
  of pancreatic polypeptide and deamino-oxytoxin, 1
Anomalous scattering, 40, 81
Anti-tumor drugs, 209
Antibodies, 365
Antigenic sites [of poliovirus], 272, 289
Antigenic sites in polio and FMDV, 272
Antineuraminidase Fab fragment, 373, 378
Aqueous solutions,
  Simulations, 172
Architectural principles [viruses], 313
Ascorbate oxidase, 413
Aspartate aminotransferase, 234
Assembly [of picornaviruses], 274, 288, 289
Atomicity, 66

Bacteriophage,
  Repressor/Operator complex of a, 319
Bacteriorhodopsin, 101
Base recognition sites in RNases A and T1, 337
Beam optics, 49
Bias minimisation, 72
Biliprotein C-Phycocyanin, 417
Binding proteins, 385

C-Phycocyanin from M. laminosus, 417
Capsid of adenovirus, 311
Capsid protein subunits in poliovirus, 284

Catalysis,
  covalent, 234
Cleavage of RNA, 337
Colloidal effects, 9
Computer graphics, 125, 131, 132
Conformation and biology, 163
Cooling cell, 50
Covalent catalysis, 234
Cowpea mosaic virus, 293
Crystal growth, 17, 21
Crystal hydrate simulations, 171
Crystallization
  and data collection, 1-59
  of integral membrane proteins, 3
  of proteins and nucleic acids, 15
Cytochrome C3, 428, 429
  Structure relationships with other cytochromes, 435
Cytochromes, 425

Data collection and crystallization, 1-59
Data collection, 33, 36, 41, 45
Data compression, 92
Deamino-Oxytocin, 153
Density modification methods, 63
Density Modification, 63
  Algorithms, 64
  Implementation, 65
Desulfovibrio.Cytochrom C3, 428
Detectors, 35, 50, 51, 54
Detergent, 5, 7
Diffuse scattering, 39
Distamycin and netropsin, 209
DM, see Density Modification
DNA fragments, 36
  Complexes of ECO RV endonuclease with, 345
DNA helices,
  Errors in, 353
DNA minor groove-binding anti-tumor drugs, 209
DNA,
  Interaction of DNase with, 327
DNase I, 327, 329

441

Drug design – Site directed
    mutagenesis, 207-260
Drug design, 222
Dynamics and water, 151-205
Dynamics of proteins in solution,
    197

ECO RV endonuclease, 345
Electron crystallography, 89, 101
Electron transfer –proteins and
    –complexes, 425
Electron transport chain, 428
Electrostatic forces, 137
Endonucleases, 345
Energy calculations
    in tertiary structure
        prediction, 142
Enzyme mechanism
    and X-ray crystallography, 229
Evolution of proteins, 258
Evolution of viruses, 288

FRODO, 128
Functional loop regions, 359
Functional residues and intron
    positions, 362

GRAMPS, 134
GRANY, 134
Graphics and structure analysis,
    123-149

Haemagglutinating neuraminidase,
    378
Heme-heme interactions, 431
Hemoproteins,
    Electron transfer complexes
        with, 425
Hexon,
    Structure of, 309
Histidine decarboxylase, 403
Hoechst 33258, 214
Hoechst/DNA complex, 218
    Structure analysis of, 214
Homology
    in tertiary structure
        prediction, 143
Hydrates, 168
Hydrogen bonds
    and protein stability, 258
Hydrolysis mechanisms of RNA, 339

Immunogenic sites on HRV14, 271
Influenza virus neuraminidase,
    373
Inhibition, 225, 226
Insulin [2Zn] crystals,
    Crystal structure, 180
    Water structure in, 179
Interfaces [of biomolecules],
    Water at, 167
Intron positions, 359, 362
Isolexins, 219

L-Arabinose-binding protein, 386
Leu/Ile/Val-binding protein, 390
Lexitropsins, 213
Light-harvesting biliprotein, 417
Local environment
    and protein stability, 257
Local structure,
    Prediction of, 141
Loop regions, [functional], 359
Lysosyme, 251, 252

MAD, see Multiwavelength
    Anomalous dispersion
Map continuity [macromolecular
    stereochemistry], 68
Maximum value [of electron
    density], 66
Membrane proteins, 3
Micelle, 5, 9
Microheterogeneities, 16
MIR, see Multiple Isomorphous
    Replacement
Molecular dynamics, 39
Molecular modeling,
    Computer graphics in, 131
    Tools for, 132
Monochromatic techniques, 37, 48,
    83
Monoclonal antibodies, 368
Monomer fluidity, 5
Monte Carlo, 193
Multiple Isomorphous Replacement,
    63
Multiwavelength Anomalous
    dispersion, 85
Multiwavelength diffraction, 41
Mutagenesis,
    site-directed, 241, 254
Mutants,
    selected and directed, 251
Myeloma proteins, 365

Netropsin and distamycin, 209
Neuraminidase of influenza virus,
    373,
Neutron diffraction, 117
Non-crystallographic symmetry, 68
Nucleic Acids – Protein, 317-355
Nucleic acids,
    microheterogeneities in, 16

Oligonucleotide crystals, 169
Omit map technique, 72
Optical systems for data
    collection, 33
Oxytocin, 154

Pancreatic polypeptide, 153, 159
Papain, 231
Parallax, 54
Periplasmic binding proteins, 385
Phase problem , 40, 61-121

Phases [in diffraction experiments], 70, 71
Photosynthetic reaction center, 421
Picornavirus, 263
Poliovirus,
 capside protein subunits, 284
 structure, 281
Polychromatic techniques, 37, 48, 83
Polypeptide hormones, 153
Positivity, 65
Precipitating agents, 18
Precollision orientation, 137
Prediction of protein structure, 141
Pressure cell, 50
Prohistidine decarboxylase, 403
PROLOG, 146
Protein – Nucleic Acids, 317–355
Protein crystals, 169
Protein dynamics in solution, 197
Protein evolution, 258
Protein hydrogen bonding, 184
Protein structure prediction, 141
Protein–DNA specificity, 321
Protein–Water contacts,
 Geometry of, 185
Protein/Water interaction, 202
Proteins, 357–438
 microheterogeneities in, 16
Proteolytic enzymes, 230
Purification procedures, 17
Pyruvoyl-dependent histidine decarboxylase, 403

Receptor binding site, 272
Repressor/Operator complex, 319
Restriction endonucleases, 345
Rhinovirus 14,
 Immunogenic sites on, 271
 Structure of, 263
Ribonucleases A and T1, 337
RNA cleavage, 337
RNA hydrolysis mechanisms, 339

Sample holder, 50
Scattering and measurement efficiency, 46
Secondary structure,
 Prediction of, 141
 Stability of, 258
SFs, see Structure Factors
Short wavelengths,
 use of, 37

Site directed mutagenesis – Drug design, 207–260
Site-directed mutagenesis, 241, 254
Small crystals, 37
Solvent flatness, 67
Solvent,
 effect of protein on, 202
 effect on protein, 199
Stability of proteins, 257
Storage rings, 27, 30
Structure factors, 64
 calculated and observed, 69
Sulfate-binding protein, 389
Sweet taste receptor binding protein, 395
Synchrotron radiation, 27, 30

T4 phage lysosyme,
 Mutants of, 251
Temperature factors, 39
Temperature-sensitive lysosyme mutants, 252
Templates, 144
Tertiary structure prediction
 by α/β docking, 145
 by energy calculations, 142
 by homology, 143
 by templates, 144
Thaumatin I, 395
Thermal stability [of proteins], 257
Thermolysin, 232
Time dependent X-ray diffraction studies, 37
Topology [of proteins], 146
Trypsin inhibitor, 197

Undulators, 30, 33

Virion structure, 287
Virus evolution, 288
Viruses, 261–316

Water and dynamics, 151–205
Water around a macrocyclic receptor, 193
Water at biomolecule interfaces, 167
Water molecules,
 Identifying procedures, 181
Water structure, 185
 in 2Zn insulin crystals, 179
Water–Water interactions, 187
Water...Biomolecule interactions, 171
Wiggler, 30, 57